P9-DNC-252

Dietary Fibre:
Chemical and Biological Aspects

Special Publication No. 83

Dietary Fibre: Chemical and Biological Aspects

Edited by
D. A. T. Southgate, K. Waldron, I. T. Johnson and G. R. Fenwick
AFRC Institute of Food Research, Norwich

ROYAL SOCIETY OF CHEMISTRY

The Proceedings of Fibre 90: Chemical and Biological Aspects of Dietary Fibre organised by the Food Chemistry Group of the Royal Society of Chemistry, held 17–20 April 1990 in Norwich, England.

British Library Cataloguing in Publication Data
Fibre '90. Conference : Norwich, England
Dietary fibre.
1. Man. Diet. Role of Fibre. Physiological aspects
I. Title II. Southgate, D. A. T. III. Series
612.3

ISBN 0-85186-667-0

Published by the Royal Society of Chemistry, Thomas Graham House, The Science Park, Cambridge, CB4 4WF

Printed in Great Britain by Whitstable Litho Printers Ltd., Whitstable, Kent

Preface

Whilst Hippokrates (460-377 BC) recommended the eating of wholemeal bread for its 'salutary effects upon the bowels' it is only in recent years that the potential health implications of dietary fibre have begun to be fully investigated. During this period, thousands of research papers, reports and articles have been published and many meetings organised.

The decision of the Food Chemistry Group of the Royal Society of Chemistry to hold a meeting on the chemical and biological aspects of dietary fibre (FIBRE 90) was a recognition both of the important role of chemistry in the study of fibre, and of the need for coherent, interdisciplinary research in this area. For these reasons, the meeting had the objective of bringing together workers in the chemical and biological fields, food technologists, representatives of industry and the regulatory authorities.

The advances of the last two decades have provided increasing insights into the chemical complexity of dietary fibre and have led to hypotheses being proposed to explain the resultant physiological effects of fibre-containing foods and products. The reasons why these hypotheses have been difficult to evaluate include the complexity and variability of the dietary fibre mixture, the difficulties of relating structure and physicochemical properties to physiological effects and, especially, the need to develop a proper conceptual framework to relate the composition of the dietary intake to indicators of health and disease.

This area, offering challenges to the scientist, opportunities to the food technologist and ingredient supplier, and benefits to the consumer, will continue to develop in the present decade. There will be a need, however, to facilitate the transfer of basic research data into the applied area, and to ensure that the public is kept fully informed of the real significance of fibre for health and wellbeing.

The editors are grateful to the Food Chemistry Group, to staff at the AFRC Institute of Food Research and Norwich City College who assisted in the planning and hosting of FIBRE 90. Particular thanks are due to the patience and support of Ms Catherine Lyall of the Royal Society of Chemistry. Above all the editors would like to thank all those who attended FIBRE 90 - this volume is dedicated to them.

D.A.T. Southgate
K.W. Waldron
I.T. Johnson
G.R. Fenwick

June 1990

The following are thanked for their generous sponsorship of FIBRE 90:

British Bakeries Ltd

Dionex Ltd

Kelloggs Ltd

Rank Hovis Ltd

Unilever Plc

Weetabix Ltd

The Royal Society of Chemistry

The Royal Society of Chemistry, Food Chemistry Group

Foreword

The number of deaths in the UK directly attributable to food poisoning was over 200 and during the past year, the fear of food poisoning in the UK has almost overtaken sex and football in its extent of media exposure. Salmonella, listeria and poisonous shellfish are a few of the examples that have made the headlines.

This represents, however, only one part of a wider picture. Over the past few decades there have been profound changes in lifestyle and diet of individuals, especially in the Western World. This has come about through new technology and increased affluence that enables consumers to change their eating habits to meet individual tastes and choice.

It is therefore something of a paradox that relatively little publicity is devoted to the increasing numbers of people who now die each year from cancer and circulatory disease which can be directly attributed to diet and lifestyle. In England and Wales alone, the numbers amount to 250,000 each year! Only Ireland, Scotland, Finland, Czechoslovakia and Hungary have higher figures. In Japan the figure is lower by a factor of 10. Why is this? What are the biological mechanisms? How precisely does food control our health? Our understanding of these links is primitive in the extreme. This conference, which follows the highly successful meeting on Bioavailability in 1988, provides a platform for contributing to that understanding and making a real contribution to health, quality of life and the associated role of diet.

Professor P. Richmond, Royal Society of Chemistry, Food Chemistry Group.

Contents

Dietary Fibre, Health and the Consumer

CONCEPTS OF DIETARY FIBRE

K.W. Heaton

University Department of Medicine
Bristol Royal Infirmary
Bristol BS2 8HW

Hellendoorn called dietary fibre an abstraction.[1] Perhaps he went too far but it is certainly better to think of fibre as a concept than as a substance. Indeed it is a series of concepts, each with its own validity and its own limitations. The term dietary fibre is simply a convenient shorthand expression which covers a wide variety of conceptions.

Trowell traced the history of the fibre hypothesis and of fibre definitions in great detail.[2] His own 1972 definition of dietary fibre as "that portion of food which is derived from the cellular walls of plants which is digested very poorly by human beings" played an important part in stimulating scientific thinking on the subject. The term dietary fibre had been coined in 1953 and the idea of plant cell walls being indigestible was far from new, as witnessed by the term unavailable carbohydrate, but the conjunction of a reasonably appropriate name and a definition that seemed clearcut, physiological and relevant was irresistible at a time when speculation was running wild. The term and the definition have faced no serious competition till 1990. But in 1990 the British Nutrition Foundation's authoritative report will advocate the term dietary fibre be dropped in scientific discourse. Instead it will urge us to refer to the material in question in precise terms like wheat bran, ispaghula husk, citrus pectin, guar gum, carboxymethyl cellulose etc.

This is fine in terms of describing experiments but it is highly limiting in terms of hypotheses linking diseases to eating habits. Yet it was the publication of such hypotheses by Cleave[3] and Burkitt[4] in 1969 which set off the modern wave of interest and research into dietary fibre. The striking characteristic of these hypotheses was their enormous scope and potential. Perhaps they were overambitious, and it is easy now to smile at them as naive, but at least they widened people's horizons. The new substance-orientated approach seems more scientific but if it involves us putting on blinkers, we could end up getting lost. Horses with blinkers don't see road signs!

Cleave's concept - "what God hath joined together let no man put asunder"

Cleave was one of the first people to advocate raw wheat bran as a laxative but it is quite wrong to talk of his ideas as the bran hypothesis. What he advocated was not that we add fibre to our food but that we stop taking bran and other fibrous elements away from our food.[3,5] In other words he indicted man-made fibre-depleted foods, chiefly sugar and white flour, as causing disease. He called these foods refined carbohydrates but what he meant was foods from which a fibre-rich fraction had been artificially split off. Such foods, like broken marriages, cause all kinds of trouble. Firstly, they rob the large intestine of the natural stimulant to its activity and so cause constipation and all the ills which go with constipation. Secondly, by being too easy to eat and digest they encourage overnutrition or, as we might term it, excess energy intake and so cause obesity and all the ills which go with obesity. Thirdly, when refined foods are eaten, sugars enter the bloodstream abnormally fast (hence his term saccharine disease, meaning sugar-related) and this leads to hyperglycaemia, diabetes and coronary heart disease. Many of Cleave's ideas were crude but they were supremely logical and fitted with the broad patterns of disease and of eating habits in the world as far as they were known at that time. Actually the epidemiology of chronic diseases was then in its infancy and nutritional epidemiology had not been invented. Both received a great stimulus from the Cleave and Burkitt hypotheses. Too often Cleave is not given credit but it was he who galvanised Burkitt, Trowell, Kritchevsky and Avery Jones, and many less famous people.

Burkitt's concept - "nature's laxative"[4,6,7]

Burkitt extended and popularised Cleave's ideas but emphasised the protective value of fibre as opposed to the harmfulness of fibre-depleted foods. This more "positive" approach had great appeal to the public and provided scientists with apparently easy hypotheses to test, but it also distorted some people's perceptions. Bran came to be seen as a panacea and the malnutritious aspects of sugar and white flour were overlooked.

Burkitt produced a coherent and still attractive hypothesis linking large bowel cancer to small, concentrated, slowly moving colonic contents.[8] He also produced plausible arguments blaming straining at stool for hiatus hernia, haemorrhoids and varicose veins, though these ideas are still unproven and not widely held. Appendicitis "had to" be due to caecal stasis and blockage of the appendiceal lumen by inspissated contents;[9] but proof is still lacking and the epidemiology fits better with a "hygiene hypothesis"[10].

The central feature of Burkitt's message was the danger of a small faecal output, so characteristic of western societies. The essence of his concept of fibre was nature's laxative. Here he seemed to be on safe ground. No-one disputes that fibre bulks the stools. People have even published regression equations relating stool output to fibre intake, e.g. 24 h stool weight = 5.2 x dietary fibre intake + 48.[11]

Collecting and weighing stools may be simple in theory but it has seldom been done as an epidemiological exercise. As a result there are very few data relating the stool weights of populations to their risk of disease, indeed no prospective data whatsoever. So, Burkitt's hypothesis remains largely untested.

It has, however, emerged that within a British population variation in fibre intake accounts for only a small proportion of the variation in stool weight,[12] that within an American population, personality factors account for as much variation in stool weight as fibre intake,[13] and that some constipated people do not respond to fibre.[14] So it is not surprising that some iconoclastic people feel constipation is largely independent of fibre intake.[15] This may be true <u>within</u> a population eating generally small amounts of fibre but the fact remains that groups, like vegetarians, who eat more than the average amount of fibre, have bulkier faecal outputs[16] and it must be the case that, were Britons to double their fibre intake, the average British stool would be heavier and softer. Whether this would translate into fewer Britons complaining of constipation is less certain. Some of the elements of constipation are symptoms of irritable bowel syndrome (abdominal discomfort relieved by defecation, unproductive calls to stool and feelings of incomplete evacuation) and many cases of IBS are stress-related, not diet-related.[17]

So, the concept of fibre as the answer to constipation and its associated ills is true but not the whole truth.

Trowell's concept - "the colon's portion"

In 1972 a definition was what fibre needed to be scientifically respectable but it soon became apparent that Trowell's definition of indigestible polysaccharides and lignin[18] was too restrictive. It had long been known that starchy foods gave some people diarrhoea, colic and excess flatus.[19] In 1981 it was reported that even normal people malabsorb some of the starch in white bread and macaroni[20] and, on direct measurement, this unabsorbed fraction was found to amount to 10% on average, ranging from 2 to 20%.[21] Since, even in the West which has largely switched from starchy staples to fatty and sugary foods, average starch intakes are ten to fifteen times higher than average fibre intakes, it was immediately apparent that of the polysaccharides entering the colon half or more must be starch. At the same time it was becoming accepted that the faecal bulking effect of fibre (and other laxatives) was in large part due to the fermentation of undigested polysaccharide in the colon with consequent multiplication of colonic bacteria and an increase in the bacterial component of faeces.[22,23] Since starch too is fermented by colonic bacteria it is likely that undigested starch has similar effects on the colon as fibre. The scanty data available suggest this is the case[24] but much more research is needed.

So indigestibility is no longer a secure foundation for a definition of fibre. The Trowell definition has had its day.

Various factors reduce starch digestibility - coarse particle size with seeds like cereals and pulses,[25-27] roasting and toasting, cooling after cooking (with potatoes)[28] and of course eating raw or unripe foods.[29,30] All of these donate fibre-like properties to foods and need to be put into the equation if one is trying to relate polysaccharide intake to human physiology and disease. It is a daunting task. In the meantime, knowledge of these complexities should deter us from expecting simple or close relationships between fibre intake and disease incidence in different populations.

The structural concept - "prison walls"

Look at a photomicrograph of any plant food in its raw unprocessed state and what you see is a cellular structure. The nourishment is inside the cells. It is imprisoned within the cell walls. Plant-eating animals except herbivores have to break down these walls, that is disrupt the fibre, in order to obtain nourishment. The tools they have to do this are their teeth and the antrum of the stomach (the muscular distal third) which, respectively, chop up and churn up the food until it is reduced to particles of 1-2 mm or less. The wall-breaking is clearly incomplete but presumably this does not matter since teeth and antrum are all that is available to do the job - except in the case of man who has learned to cook roots and grind seeds, that is, to mill cereal grains. But milling was a crude process until at least the 17th century.

So man, like all other plant-eating animals has, until modern times, eaten his food with a largely intact cellular structure which is only partly destroyed by the time the food is discharged still in particulate form into the small intestine.

There is abundant evidence that the intact cellular structure of foods delays or impedes digestion and absorption. This has been shown in terms of reduced glycaemia and insulinaemia with apples,[31] oranges,[32] raisins,[33] lentils,[34] rice,[26] wheat,[25] maize,[25] and rye.[35] It has also been shown in terms of reduced fat absorption with peanuts.[36] There seems little reason to doubt this is a general law, though there are apparently odd exceptions, like grapes[32].

If slow or incomplete absorption and muted insulinaemia are the norm then the more rapid absorption and higher insulin levels which result from eating fibre-depleted <u>and</u> fibre-disrupted products must be abnormal.

Tearing down walls between nations is obviously a good thing. Tearing down the walls nature has put in our food may be less desirable. Insulin has wide-ranging effects on metabolism and hyperinsulinaemia is increasingly being linked with major western diseases - hypertension, atherosclerosis, gallstones and renal stones.[37]

Conclusion

Dietary fibre is not a substance or even a collection of substances. It is not a concept but a spectrum of concepts. In planning an experiment it is usually necessary to focus on one very restricted manifestation of fibre but in interpreting one's observations it is important to take off one's blinkers and look at fibre in as broad a way as possible. Undigested starch and intact cellular architecture may turn out to be more important than dietary fibre as normally defined. A diet which is naturally rich in starch and intact cell walls is likely to have other benefits - it will be rich in micronutrients and potassium and low in fat, sugar and sodium. Such a diet combines virtually all the internationally accepted guidelines to healthy eating.

REFERENCES

1. E.W. Hellendoorn. Am. J. Clin. Nutr., 1981, 34, 1437.
2. H. Trowell. 'Dietary Fibre, Fibre-Depleted Foods and Disease', Eds. H. Trowell, D. Burkitt and K. Heaton, Academic Press, London, 1985, Chapter 1, p.1.
3. T.L. Cleave, G.D. Campbell and N.S. Painter. 'Diabetes, Coronary Thrombosis and the Saccharine Disease', Wright, Bristol, 2nd edn., 1969.
4. D.P. Burkitt. Lancet, 1969, 2, 1229.
5. T.L. Cleave. 'The Saccharine Disease', John Wright, Bristol.1974
6. D.P. Burkitt. Br. Med. J., 1973, 1, 274.
7. D.P. Burkitt and H.C. Trowell, Eds. 'Refined Carbohydrate Foods and Disease. Some Implications of Dietary Fibre', Academic Press, London, 1975.
8. D.P. Burkitt. Cancer, 1971, 28, 3.
9. D.P. Burkitt. Br. J. Surg., 1971, 58, 695.
10. D.J.P. Barker. Br. Med. J., 1985, 290, 1125.
11. J.L. Kelsay and W.M. Clark. Am. J. Clin. Nutr., 1984, 40, 1357.
12. M.A. Eastwood, W.G. Brydon, J.D. Baird, R.A. Elton, S. Helliwell, J.H. Smith and J.L. Pritchard. Am. J. Clin. Nutr., 1984, 40, 628.
13. D.M. Tucker, H.H. Sandstead, G.M. Logan, L.M. Klevay, J. Mahalko, L.K. Johnson, L. Inman and G.E. Inglett. Gastroenterology, 1981, 81, 879.
14. G.K. Turnbull, J.E. Lennard-Jones and C.I. Bartram. Lancet, 1986, 1, 767.
15. S.A. Müller-Lissner. Br. Med. J., 1988, 296, 615.
16. G.J. Davies, M. Crowder, B. Reid and J.W.T. Dickerson. Gut, 1986, 27, 164.

17. K.W. Heaton. 'Recent Advances in gastroenterology 7' Ed. R.E. Pounder, Churchill Livingstone, 1988 Chapter 13, p.291.
18. H. Trowell. Lancet, 1974, 1, 503.
19. T.C. Hunt. 'Price's Textbook of the Practice of Medicine', Ed. D. Hunter, Oxford University Press, London, 9th edn., 1956, p.629.
20. I.H. Anderson, A.S. Levine and M.D. Levitt. New Eng. J. Med., 1981, 304, 891.
21. A.M. Stephen, A.C. Haddad and S.F. Phillips. Gastroenterology, 1983, 85, 589.
22. A.M. Stephen and J.H. Cummings. J. Med. Microbiol., 1980, 13, 45.
23. A.M. Stephen and J.H. Cummings. Gut, 1979, 20, A457.
24. P.S. Shetty and A.V. Kurpad. Am. J. Clin. Nutr., 1986, 43, 210.
25. K.W. Heaton, S.N. Marcus, P.M. Emmett and C.H. Bolton. Am. J. Clin. Nutr., 1988, 47, 675.
26. K. O'Dea, P. Snow and P. Nestel. Am. J. Clin. Nutr., 1981, 34, 1991.
27. P. Würsch, S. DelVedovo and B. Koellreutter. Am. J. Clin. Nutr., 1986, 43, 25.
28. H.N. Englyst and J.H. Cummings. Am. J. Clin. Nutr., 1987, 45, 423.
29. H.N. Englyst and J.H. Cummings. Am. J. Clin. Nutr., 1986, 44, 42.
30. T.M.S. Wolever et al. Nutr. Res., 1986, 34, 349.
31. G.B. Haber, K.W. Heaton, D. Murphy and L. Burroughs. Lancet, 1977, 2, 679.
32. R.P. Bolton, K.W. Heaton and L.F. Burroughs. Am. J. Clin. Nutr., 1981, 34, 211.
33. G.J. Oettle, P.M. Emmett and K.W. Heaton. Am. J. Clin. Nutr., 1987, 45, 86.
34. D.J.A. Jenkins, M.J. Thorne, K. Camelon, A. Jenkins, A.V. Rao, R.H. Taylor, L.U. Thompson, J. Kalmusky, R. Reichert and T. Francis. Am. J. Clin. Nutr., 1982, 36, 1093.
35. D.J.A. Jenkins, T.M.S. Wolever, A.L. Jenkins, L. Giordano, S. Giudici, L.U. Thompson, J. Kalmusky, R.G. Josse and G.S. Wong. Am. J. Clin. Nutr., 1986, 43, 516.
36. A.S. Levine and S.E. Silvis. New Engl. J. Med., 1980, 303, 917.
37. L.J.D. O'Donnell, P.M. Emmett and K.W. Heaton. Br. Med. J., 1989, 298, 1616.

DIETARY FIBRE AND HEALTH

D. A. T. Southgate

AFRC Institute of Food Research
Colney Lane
Norwich NR4 7UA

1 INTRODUCTION

Many of the details of the physiological effects of dietary fibre that are relevant to a discussion of the relationships between the consumption of dietary fibre and health have been discussed in earlier papers and I would like to draw them together and try to present an overview of the current evidence.

The topic of dietary fibre has raised a large number of questions since it was first suggested that the intake of dietary fibre was related to the aetiology of a range of chronic disease states. In any discussion of the topic I think that it is useful to start from the original hypothesis that developed from the writings of Burkitt and Trowell in the early 1970's[1].

This hypothesis contained two separate propositions; First, that diets that were rich in plant foods containing plant cell wall material were protective against a number of diseases whose prevalence was higher in Western developed communities. The diseases included obesity, diabetes, coronary heart disease and some cancers. The second proposition was that diets poor in these foods were causative for some conditions particularly those that were related to the development of excessive abdomenal pressures: diverticular disease, haemorrhoids, and appendicitis, and that for other diseases provided the conditions where other aetiological factors were active[2].

The hypothesis therefore related to the effects of types of diets and the differences in the dietary fibre contents of these diets were identified as the key factor;

primarily because plausible subsiduary hypotheses could be developed regarding possible modes of action.

Following this line dietary fibre research focussed on the study of the specific component; dietary fibre and many important physiological effects of dietary fibre (which I am using in the sense of plant cell wall material) have been identified as a result of this research.

The focus on dietary fibre per se, while valuable in itself, has diverted attention away from the central "dietary" hypothesis. This is because the protective diets which formed the basis for the observations that led to the original hypothesis differed in many other attributes in addition to the amounts of plant cell walls they contained[3].

The observed physiological effects of a dietary fibre-rich diet are an integration of the effects of the diet itself and any specific effects due to the dietary fibre component. It is this integration that I would like to explore in discussing the relation between dietary fibre and health.

This approach raises some fundamental nutritional concepts that apply to all aspects of the relationships between diet and health and the aetiology of disease. This is not an obscurantist view that sees the diet as a whole as a mystical concept but merely a reflection of the fact that the nutritional/physiological effects of any diet are an integration of the effects of many variables. The composition of the diet varies in a multidimensional domain; defined by a large number of food dimensions and a range of dimensions that are defined by patterns of consumption[4].

This is of particular importance in any discussion of the differences between dietary fibre-rich diets and dietary fibre-poor diets. The concentrations of plant cell wall material in many plant foods are low and therefore the dietary fibre is invariably associated, in the foods, with many other components; starch in cereal foods and potato, and a myriad of other components found throughout the plant kingdom; pigments, vitamins, inorganic constituents and a range of biologically active components such as glucosinolates, glycoalkaloids and flavonoids[5].

In epidemiological studies of populations there are

therefore many confounding variables that are not strictly independent of dietary fibre since they are present in the foods which provide the dietary fibre. Epidemiological studies in themselves, can only form part of the evaluation of the dietary fibre hypothesis[6].

Limitations of Epidemiological Studies

There are many limitations of epidemiological studies of the relation between diet and disease that are of a general nature. The most important of these is the uncertainty concerning the validity of measurements of food intake, especially habitual food intake over long periods of time by free-living populations, where it is probable that a nutritional "uncertainty" principle applies since it is impossible to measure habitual food intake accurately.

The second relates to the translation of food intake data into nutrients especially by calculation from food composition tables or nutritional databases where the natural variability of foods places a limit on the predictive accuracy of the calculated intakes[7]. This has been particularly difficult in the case of dietary fibre where the complexity of dietary fibre, the controversy over its definition and the consequential debate over the analytical measurement of dietary fibre have served to delay critical studies of intake in populations with differing patterns of disease incidence and prevalence[8,9]. Thus 20 years after the original hypothesis was proposed we are only now in a position to compute intakes of the non-starch polysaccharides (a good index of plant cell wall material that characterises the composition of the components)[10] in a few countries and total dietary fibre[11] intakes in a few more. The studies that have been carried out have been limited in number and in their capacity to relate the types of dietary fibre consumed to the incidence of the diseases in question.

At the present time the epidemiological evidence for coronary heart disease and colorectal cancer shows negative association between intake and incidence, that is, one that is consistent with a protective effect[12]. The crucial test at the epidemiological level requires the demonstration of a high incidence of disease associated with a high level of intake; such evidence would provide refutation of the hypothesis. One must therefore seek evidence for mechanisms of action and evidence from intervention.

The Physiological Effects of Diets Rich in Dietary Fibre

The study of these effects and relating them to the aetiological mechanisms of specific disease states is a central part of the evidence for a relation between diet and disease. Many of the diseases linked to dietary fibre are strictly of unknown aetiology and the dietary fibre studies may, in themselves, be of great value in establishing the aetiological processes because dietary fibre provides a powerful experimental tool for modifying conditions in the small and large intestines and for modifying the metabolism of carbohydrates and lipids. The experimental studies with isolated polysaccharides as "models" of the components of plant cell walls and plant cell wall preparations have given new insights into the determinants of digestion in the small intestine and to the physiology of the colon, as the other papers in the symposium have shown. But in many cases, for sound experimental reasons, these studies are, de facto, studies of the pharmacology of polysaccharides; because isolation of the polysaccharide involves separating the polysaccharide from the food matrix and from the physical structures of the plant cell wall and there is substantial evidence that this alters the chemical and physical properties of the polysaccharides[13].

Modes of Action

There is substantial evidence that diets that are rich in dietary fibre exert effects throughout the intestinal tract and that some of these effects are specifically due to the dietary fibre present in the diets and others to the diet, the mixture of foods consumed.

Effects on Ingestion

These effects are related to the diet because the foods making up the diet possess specific and characteristic textural and physical properties that modify acceptability and patterns of ingestion. Fibre-rich diets are characteristically more bulky, that is their density is lower, and they may have lower energy densities, kcal per gram, because of the lower fat intake and the higher water content of many plant foods. The ingestion of isocaloric intakes of dietary fibre-rich and dietary fibre-poor diets will therefore require very different amounts of foods in total; this is especially true for the traditional diets eaten in the third world. Experimentally, using semi-synthetic diets and isolated fibre sources these dietary features may not be observed.

Dietary fibre rich diets containing relatively unprocessed foods also require considerable mastication before they can be swallowed attenuating the rate at which they can be eaten.

It is not established with certainty that these properties act to reduce voluntary food consumption; and experimentally the design of studies to test these effects are extremely difficult because of the difficulty of disguising the diets that differ in these properties.

Gastric Effects

The bulky diets, usually rich in starchy foods, hydrate with the gastric secretions and produce increased gastric bulking. The predominately insoluble plant cell wall materials are emptied more slowly from the stomach. Whether these effects are of physiological significance remains to be established with diets but the studies with some viscous polysaccharides suggests that a satiating effect may arise from these bulking effects.

Small Intestinal Effects

Studies with isolated polysaccharides show that some, particuarly the soluble, viscous polysaccharides, have effects on the rates of absorption of water-soluble nutrients such as glucose and amino-acids and in reducing fat absorption[14]. The effects of diets may thus be expected to depend on the proportion of the dietary fibre that can be solubilised in the small intestine and that has the capacity to modify the physical properties of the intestinal contents. Assessing the importance of this feature of the diets in epidemiological studies depends on a proper characterisation of the dietary fibre present in the diet. Most studies of the dietary fibre consumption in different population groups have not characterised the diets sufficiently to establish whether or not the diet contained sufficient of the particular fractions to elicit these types of effects. Some studies with a limited range of foods show that the integrity of the plant cell wall structures is important in determining the rates of sugar absorption[15], an effect that may be due to the steric effects of the cell wall structure in delaying difusion to the absorbing mucosal surfaces. Some starch digestion is also slowed or inhibited by the cell wall structures which have remained intact in the foods, thus legume starches are more slowly digested because of the thicker, more resistant, walls they possess in comparison with potato or cereals[16].

Other fractions of the starch may also be resistant to enzymatic hydrolysis in the small intestine because of physical changes or characteristics, especially those resulting from heat processing; retrograde amylose appears to be an important part of this resistant starch, although other forms of starch appear to escape digestion in the small intestine[17].

As small intestinal digestion proceeds the dietary fibre materials become more important quantitatively in the contents and binding to functional groups exposed during digestion or adsorption to the surfaces of the cell wall materials can occur. In vitro binding of many divalent cations and bile salts has been observed but these effects have been difficult to establish in vivo with high-fibre diets; this apparent anomaly may be due to the fact that the preparation of the fibre sources for the in vitro studies exposed functional groups not exposed during in vivo digestion or that the binding effects were dissipated by bacterial degradation in the large intestine. The studies with ileostomists suggest that the latter explanation is more probably correct[18].

Thus it is evident that the small intestinal effects will depend on the composition of the dietary fibre; the types of polysaccharides present and their physical properties. These will be determined by the mixtures of foods consumed. In addition there are effects of the integrity of the plant cell wall structures present which alter rates of solution and diffusion and therefore release of nutrients. The consumption of a high fibre diet will therefore produce effects over and above those seen in studies of isolated dietary fibre components.

Large Bowel Effects

Dietary fibre forms a major part of the organic matter entering the large bowel although it is accompanied by intestinal secretions including proteins and mucus, probably some amino acids and lipids. The traces of other indigestible substances present in the diet, salts, pigments, mineral oil and such like, will also be present[19]. These materials are mixed with the contents of the caecum and provide the microflora with substrates. A high dietary fibre intake will thus present the bacteria with carbohydrate substrate and may shift metabolism to a condition where the supply of nitrogen will be a limiting factor for bacterial growth. The fermentation will produce short chain fatty acids, carbon dioxide, methane and hydrogen and produce a substantial modification of the physico-chemical environment. Undigested material will contribute bulk, which together with the bacterial mass arising from the fermentation of the carbohydrate substrates, is responsible for the increased faecal matter characteristic of an increased dietary fibre intake[20].

The extent of fermentation is dependent on the composition of the dietary fibre ingested especially the extent of lignification and cutinisation or suberinisation. The particle size is also an important determinant because the bacterial attack is a surface effect so the surface area of the dietary fibre particles is a critical feature.

The physical properties and fermentability of dietary fibre also have some direct effects on mucosal structure and function where some isolated polysaccharides increase cell turnover and the maturation of mucosal enzyme systems.

Relation between Modes of Action and the Aetiology of Disease

The physiological effects of dietary fibre and some specific polysaccharides are generally consistent with the postulated protective effects although it is not possible to construct a generalised hypothesis that applies to dietary fibre as a whole. All the observed effects are dependent on the composition and physical properties of the polysaccharides present and most are modified by the physical organisation of the polysaccharides into cell wall structures[16]. It is not possible to predict with certainty the relation between structure, properties and physiological effects, although it is possible to make some tentative statements, for example, the effects on the rate of absorption of glucose and on serum cholesterol levels seem to require soluble components that are capable of forming viscous solutions. Faecal bulking effects appear to be most strongly associated with dietary fibre sources that are insoluble, poorly fermentable and with good water-binding capacity. Although virtually all sources produce some increases in stool weight the increases with highly fermentable soluble forms are usually small.

Direct evidence of prevention of diseases such as cancer produced by chemical carcinogens is equivocal, and some studies show enhanced tumour formation in the presence of some polysaccharides. In these types of study the tumour rates are very high and it is possible that any protective effect can be overcome by very active initiators, alternatively the studies using isolated sources may not contain protective substances associated with the dietary fibre in the plant.

High Dietary Fibre Diets : The Constraints Imposed by the Composition of Foods

The foods forming diets that are rich in dietary fibre exert a number of effects that are relevant to the discussion of the relation between intake and health or the incidence of disease.

Firstly, the foods chosen determine the types of polysaccharides present in the diet. Most plant foods contain a range of polysaccharides. In general the pattern of polysaccharides present is characteristic of the major groups of plant foods and reflect the types of tissues and especially the maturity of the cell walls present[16]. Cereal foods are usually consumed in a mature state so that the levels of pectic substances are low and the soluble components are principally arabinoxylans in wheat, rye and barley with substantial contributions of beta glucans in oats. The insoluble components contain arabinoxylans, xyloglucans and cellulose. In fruits and vegetables the cell walls are usually less mature and the soluble components are pectic substances and arabinogalactans and the insoluble components include arabinogalactans, galacto- and gluco- mannans, xyloglucans and cellulose. The diets therefore contain a complex mixture of soluble and insoluble polysaccharides and therefore the effects of the diet will depend on the proportions of the different foods present and not on the absolute amounts of total dietary fibre present.

The concentrations of dietary fibre in the various food sources vary considerably and the ratio of dietary fibre to energy has implications for the amounts of different foods present in high fibre diets[4]. Thus an intake of say 30g per day requires a dietary fibre: energy ratio of 1.2 g DF per 100 kcal in a intake of 2500 kcal and this cannot be achieved with refined cereal foods alone, the presence of vegetables providing between 13 and 3 percent of the energy is essential and this diet would only meet the target if virtually no fibre-free foods were eaten. Thus a high fibre diet for practical reasons should include some high extraction cereal foods and a substantial contribution from other plant foods, fruits and vegetables. The implications of these quantitative constraints are that a high fibre intake requires a different choice of foods and that the inclusion of a substantial proportion of the energy intake from vegetable sources permits a greater choice of foods that are low in fibre or fibre free. This also implies that a high-fibre intake is invariably associated with intakes of a wide range of other components of plants and of these the anti-oxidant vitamins, vitamin C and the carotenoids and other bioactive components in plants may be the principal reason why these diets are protective against cancer. The epidemiological evidence for many cancers strongly suggest a protective effect of vegetable consumption[21].

The Protective Effects of Dietary Fibre

At the present time it is not possible to say that the protective effects of a high intake of dietary fibre have been established unequivocally, the observed physiological effects are generally consistent with the protective hypothesis and there is no real evidence that refutes it. In addition to specific effects of the polysaccharide components there are additional effects due to their consumption in plant cell wall structures.

The diets that provide increased intakes of dietary fibre are different in many ways from diets that are low in dietary fibre and it is clear that dietary fibre may be a "marker" that identifies diets that contain other protective properties[22].

The ultimate test of the dietary fibre hypothesis involves taking up the challenge of relating the composition of the diet, as a whole, to the aetiology of the disease processes and understanding how the various effects are integrated; only from these studies can the evidence be drawn to provide sound guidance on the choice of diets to minimise the risks of chronic disease.

REFERENCES

1. D.P. Burkitt and H.C. Trowell. (editors) Refined carbohydrate foods and disease. Some implications of dietary fibre Academic Press. New York. 1975.
2. D.A.T. Southgate. In. G.V. Vahouny and D. Kritchevsky (editors) Dietary Fibre in Health and Disease Plenum Press. New York. 1982. 1.
3. D.A.T. Southgate. Reports of Internal Association of Cereal Chemists 1980. 10. 79.
4. D.A.T. Southgate. In. J. Dobbing (editor) A Balanced Diet. Springer Verlag. London. 1988. 117.

5. P.R. Cheeke (editor) Toxicants of Plant Origin. CRC Press, Boca Raton. 1989.
6. D.A.T. Southgate and J.M. Penson. In G.C. Birch and J.J. Parker (editors) Dietary Fibre. Applied Science Publishers, London, 1983. 1.
7. A.A. Paul and D.A.T. Southgate. In. M.E. Cameron and W.J. Van Staven (editors) Manual of Methodologies for Food Studies. Oxford Medical Publications, Oxford. 1988 121.
8. D.A.T. Southgate and H N Englyst In. H. Trowell, D. Burkitt and K. Heaton (editors) Dietary Fibre, fibre-depleted foods and disease. Academic Press, London. 1985. 35.
9. N-G. Asp and C-G. Johansson. Reviews in Clinical Nutrition 1984. 54 735.
10. H. Englyst, H.S. Wiggins and J.H. Cummings. Analyst. 1982. 107. 307.
11. L. Prosky, N-G. Asp, I. Furda, J. De Vries, T.F. Schweizer and B. Harland J. Assoc. off Agric. Chem. 1984 67, 1044.
12. S.M. Pilch. editor Physiological, effects and health consequences of Dietary Fibre. Life Science Research Office, Bethesda 1987.
13. P.J. Van Soest Federation Proceedings. 1973. 23, 1804.
14. I.T. Johnson and J.M. Gee. Gut. 1981, 22 398.
15. G.B. Haber, K.W. Heaton, D. Murphy and L.F. Burroughs Lancet 1977 2. 679.
16. R.R. Selvendran Am. J. Clin. Nutr. 1984. 39 320.
17. H.N. Englyst and J.H. Cummings. In. I.D. Morton (editor) Cereals in a European context. Ellis Harwood, Chichester 1987, 221.
18. A.S. Sandberg, H. Andersson, B. Halgren, K. Hasselblad, B. Isaksson and L. Hutter. Br. J. Nutr. 1981, 45, 283.
19. D.A.T. Southgate. In D.F. Hollingsworth and M. Russel editors. Nutritional problems in a changing world. Applied Science Publishers, London. 1973. 190.
20. A.M. Stephen and J.H. Cummings. Nature. 1980. 284.
21. National Research Council. Diet and Health Implications for reducing chronic disease risk. NRC. Washington. 1989.
22. Wahlguist, G.P. Jones, J. Hansky, S.D. Duncan, I. Coles-Rutishauser and G.O. Littlejohn. Food Technology in Australia. 1981. 35. 51.

WORKSHOP REPORT: RECOMMENDATIONS TO THE CONSUMER

M. Ashwell

British Nutrition Foundation
15 Belgrave Square
London SW1X 8PS

1. Is there any evidence that existing 'fibre' recommendations have had any effect up to now?

Minimal changes despite widespread, consistent, advice by expert committees.
BUT REMEMBER
(i) that total energy intake has declined during this time - so there has been a relative increase
(ii) the types of food eaten have changed in the direction of 'fibre' rich alternatives e.g. wholemeal and bran-enriched breads, high fibre cereals

SO

Our gradual change in eating habits and the food industry's response to them, might eventually show up in National statistics.

2. Does the scientific evidence justify giving any 'fibre' recommendations to the consumer at all? If so, for what?

(a) There are three expected benefits from consuming 'fibre' rich foods:
(i) improved large bowel function
(ii) a strategy for lowering dietary fat intake
(iii) slower utilisation of sugar and fat.

(b) The very best evidence i.e. from epidemiology, human trials and animal studies for 'fibre' alone is for the prevention and treatment of constipation. Its role in the management of diabetes is fairly firm but epidemiological data are lacking. The role in the prevention of bowel cancer is better shown epidemiologically than in trials. Its role in cardiovascular disease is suggestive, but not proven.

3. <u>Does the scientific evidence justify giving a numerical target for desirable 'fibre' intake? If so, what should it be? How could it best be used?</u>

(a) Meta-analysis of stool weights in relation to constipation, bowel cancer and diverticular disease would suggest that target stool weight should be 150g/day with a minimum of about 130g/day. This would correspond to a dietary 'fibre' intake of about 30g/day (or 18g NSP). Next step? Relate to energy intake?

(b) Recommendations with even 'soft' quantitation (e.g. eat foods with adequate 'fibre') are too ambiguous and advice should be based on more food orientated guidelines e.g. choose a diet with plenty of vegetables, fruits and grain products.

(c) Setting a quantitative target can only really be useful as 'part and parcel' of a major national campaign monitored with feedback to the consumer as part of the education plan.

(d) Quantitative targets are more useful to the educators and policy makers than the consumer (cf. RDAs).

4. <u>Are fibre recommendations applicable to all sections of the population?</u>

(a) Vulnerable groups are those with high energy requirements (e.g. children, pregnant women) or low energy intakes (e.g. the elderly, the slimmers) or those where plant materials form the major part of the diet. These groups are rarely addressed in standard recommendations and might be least receptive to the message anyway.

(b) Micronutrients most at risk are calcium, iron, zinc and vitamin D because of 'fibre'-associated antinutrients such as phytate.

BUT

These concerns could be counterbalanced by:
(i) the levels of many micronutrients 'per se' are higher anyway in 'fibre'-rich foods
(ii) the body's ability to adapt to the situation

e.g. long-term vegetarians?

and (iii) many of the 'in vitro' effects of 'fibre' on micronutrients cannot be shown 'in vivo'.

so

(i) Ensure that 'vulnerable groups' are given advice in terms of real foods not 'ad hoc' supplements (e.g. not sprinkling bran on everything!)

and (ii) Encourage, but don't force, children, to eat 'fibre'-rich foods at an early age.

5. Should 'fibre' be given as a single value in nutrition labelling on food products? If not, what should appear and how should it be defined?

NO, it should not be given as a single value because:

(i) there are many different types of 'fibre' which can have such a wide variety of properties (e.g. regulates bowel function, controls blood glucose levels, induces satiety, etc.). Different consumers have different needs.

(ii) there are such difficulties with defining and analysing fibre in a consistent manner.

(iii) it could be compared with lumping all the vitamins together as a 'vitamin index' and not acknowledging that different vitamins have different roles.

(iv) it is open to abuse by unscrupulous manufacturers who want to 'sell' useless 'fibre'.

YES, it should be given as a single value because:

(i) consumers could not cope with the concept of different types of fibre
e.g. bowel regulatory fibre
cholesterol lowering fibre
blood glucose controlling fibre

(ii) too much detailed information on the food label could cause problems if it has to be changed at a later date. Scientific credibility could be lost.

FINAL DECISION

Leave as single value (reluctantly) until research is more advanced.

6. Should 'fibre' health messages be allowed? If so, where should the line be drawn? How can health messages be regulated and monitored?

(a) YES, they should be allowed because

(i) they inform the consumer, whose awareness of

diet and health has never been higher

(ii) they ensure the development of 'healthy' products

(iii) they reach a much wider audience than is possible by any other means.

(b) NO, they should not be allowed because:

(i) they can create a false distinction between 'good' and 'bad' foods and negate the importance of diet as a whole.

(ii) they have the potential to mislead in the hands of less reputable manufacturers.

(iii) they give a distorted image if only applied to packaged foods.

(c) Ideally, health messages would be more acceptable to the scientific community if they concentrated on specific physiological functions, e.g. the percentage decrease in the glycaemic index or the absolute increase in stool weight. Not surprisingly, industry and, we suspect, consumers would not be too enthusiastic!

(d) All agree the importance of obtaining 'peer' review and approval before media publicity or pressure groups make demands, but admit that this would be amazingly difficult to achieve!

ACKNOWLEDGEMENTS

My grateful thanks to all the Workshop Participants. Special thanks to Dr. Sheila Bingham, Dr. Martin Eastwood, Dr. Sue Fairweather-Tait, Dr. Kenneth Heaton, Mr. Alan Howells, Dr. D. Kritchevsky, Ms. Janet Lewis, Dr. Barbara Schneeman, Dr. David Topping and Mr. Kevin Yates.

diet and health has never been higher

(ii) they help the development of healthy products

(iii) they reach a much wider audience than is possible by any other means.

Chemistry of Dietary Fibre

THE CHEMISTRY OF DIETARY FIBRE - AN HOLISTIC VIEW OF THE CELL WALL MATRIX

R. R. Selvendran and J. A. Robertson

AFRC Institute of Food Research
Norwich Laboratory
Colney Lane
Norwich NR4 7UA (U.K.)

1 INTRODUCTION

The role of dietary fibre (DF) in human nutrition remains a topical problem. DF was initially defined as the skeletal remains of plant cells that are resistant to the digestive enzymes of man[1] and was later extended to include polysaccharides used as food additives. This made the definition, 'all the polysaccharides and lignin in the diet that are not digested by the endogenous secretions of the human digestive tract'[2]. Accordingly, for analytical purposes, the term DF refers mainly to the non-starch polysaccharides (NSP) and lignin in the diet and methods have been developed to quantify these components in foods,[3-5] for use in food labelling and food composition tables. However, less attention has been given to the dynamics of the DF matrix as a structural and functional entity during gut transit, although this is an integral part of the DF hypothesis. The fibre hypothesis in outline states that, 'a diet rich in foods which contain plant cell walls (eg high extraction cereals, fruits and vegetables) is protective against a range of clinical disorders, in particular those prevalent in western society, and in some instances a diet low in sources of fibre is a causative factor in the aetiology of the disease and in other cases provides the conditions under which other aetiological factors are more active'[6]. The implication is that the protective effects of DF are due to the amount of cell wall material (CWM) in the diet and the protection is derived from the physicochemical properties of the fibre. Thus, methods which measure only the amount of DF in foodstuffs do not adequately explain the functional role of fibre in the diet since no account is taken of how the fibre matrix is affected by processing and has its properties altered by the digestive environment during gut transit[7,8]. For example, in the proximal gastrointestinal tract (stomach and small intestine), although fibre is resistant to digestion by endogenous secretions it may be susceptible to changes in solubility. In the large intestine a significant proportion of the fibre may be fermented and fermentation may be influenced by fibre 'conditioning' in the proximal intestine[9]. Thus, it is apparent that an understanding of the chemical composition of DF in relation to the structure and function of the component biopolymers in the cell wall of the various tissue types common to plant foods is required for a better appreciation of the role of DF in the gastrointestinal tract. The factors involved in the structure and function relationships in DF are outlined in Figure 1 and will be elaborated in the text. Particular attention is drawn to these factors which have been previously only superficially or inadequately considered.

Tissue Structure

Vegetable, fruits, legume seeds and cereals are all sources of DF and are each composed of different tissue types, as illustrated for de-seeded mature runner bean pod and the outer layers of the wheat grain (mainly bran) in Figure 2. The figure also indicates the major biopolymer types that have been isolated from the cell walls of the different tissues by detailed chemical fractionation and illustrates the distinct differences between parenchymal and lignified tissues. Cell wall biopolymers are predominantly NSP but also contain significant amounts of glycoproteins and phenolics, including lignin in tissues where secondary thickening has occurred. The chemical composition and major structural features of cell wall polysaccharides used as sources of DF have been discussed extensively[4,8] but less attention has been given to the types of bond and degree of cross linking between constituent biopolymers in the DF matrix. Each source of DF can be considered to comprise epidermal, parenchymal, vascular and occasionally sclerenchyma or supporting tissues. Parenchymal tissues will predominate in most sources of DF and only in cereal bran products will lignified tissues be important as a foodstuff.

Detailed tissue separation and cell wall fractionation techniques being developed now help us to appreciate the complexity of events involved in understanding the chemistry of DF and also to help target chemical behaviour to dietary effect[8,10,11]. Although mature runner bean pods, as illustrated, are not usually eaten at this developmen-

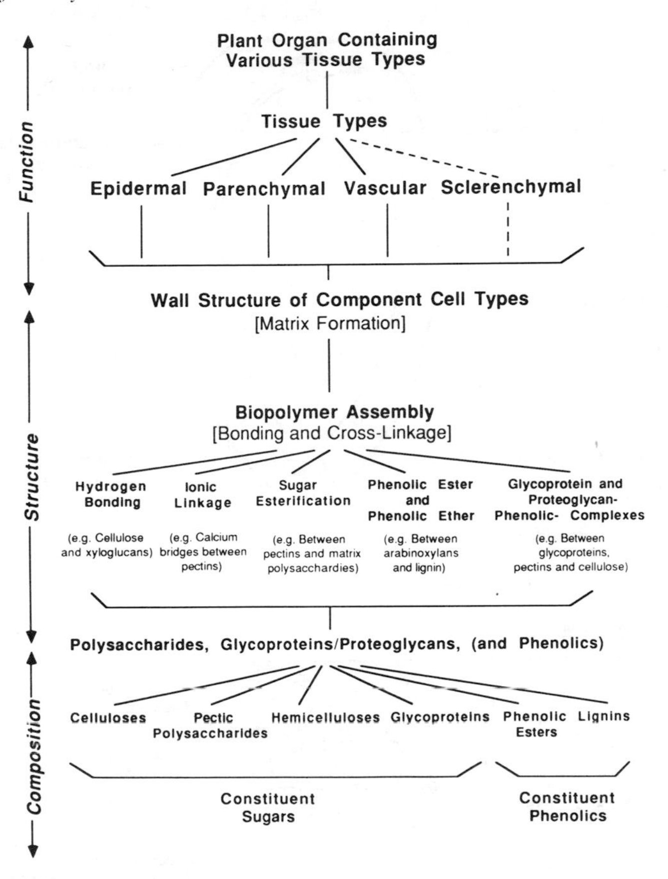

Figure 1 Plant cell wall composition and its relationship to biopolymer assembly, wall structure and cell function within plant tissues

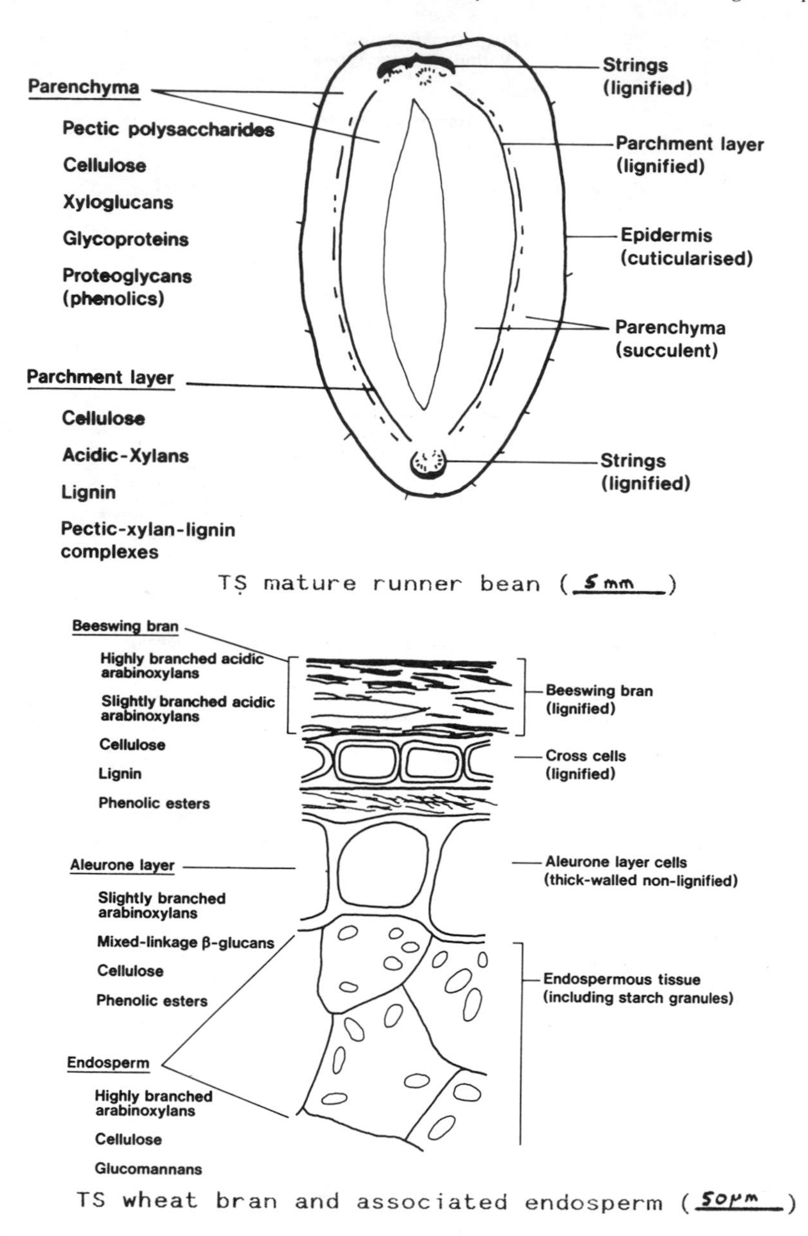

Figure 2 Tissue distribution and compositional characteristics in a vegetable (runner bean) and a cereal (wheat grain) fibre source

tal stage, component tissues can be easily separated for detailed fractionation and chemical analysis (Table 1). In immature beans the cells corresponding to the parchment layers and vascular tissues can also be easily identified and their walls shown to contain cellulose, pectic polysaccharides and acidic xylan-pectic complexes, these complexes corresponding to the acidic xylans in the lignified mature tissues. The deposition of acidic-xylans and lignin is coincident with rapid enlargement of the pod and seed development. This can be inferred from the data on runner bean pod cell wall tissue sugar composition and also composition at different stages of maturity shown in Table 1. The major polysaccharides of runner bean parenchyma that can be inferred from sugar analysis alone, in decreasing order of abundance are : pectic polysaccharides (Gal*p*A, Ara, Gal), cellulose (Glc released on Saeman hydrolysis) and xyloglucans (Glc and Xyl released by M H_2SO_4 hydrolysis). From the composition of the parchment layers it can be inferred that the major polysaccharides present are cellulose and acidic-xylans (Xyl released by M H_2SO_4 hydrolysis). The Klason lignin content of the parchment layer is around 20%. This information in conjunction with detailed fractionation studies can be used to infer the changes in the cell wall polysaccharides which occur during maturation of the pods (de-seeded). Thus, the very immature pod (prior to significant development) contains in decreasing amounts ; pectic polysaccharides, cellulose and xyloglucans and in Stage 2 (when used as a vegetable) rapid pod elongation has occurred but with little lignification or change in polysaccharide type. During Stage 3 the maturation of the pod results in extensive lignification of the 'parchment' layers and 'strings' and corresponding changes in the polysaccharide components (note the increase in the proportion of xylose and glucose). The relatively small increase in xylose content of pods during Stage 2 indicates little secondary thickening at this stage and by more detailed monitoring of bean pod development it can be shown that after this stage rapid secondary thickening occurs in parallel with seed development.

The major cell wall polysaccharides of potato parenchyma and pea cotyledons inferred from compositional data in Table 1 are comparable to those of the runner bean parenchyma, except that the pectic polysaccharides in potato are richer in galactose compared to runner bean and in the case of pea cotyledons pectic arabinans predominate. Pea hulls contain significant amounts of acidic-xylans in addition to pectic polysaccharides, cellulose and xyloglucans. In beeswing wheat bran the major cell wall polymers inferred are (highly branched) acidic

Table 1 Carbohydrate composition of DF from various vegetable and cereal sources and at different stages of vegetable development

Sugar composition (mg/g cell wall material)

Source	Deoxy[1]	Ara	Xyl	Man	Gal	Glc[2]	Uronic[3] Acid
Potato[29]	14	75	17	7	288	339	240
Runner Bean;							
-parenchyma[17]	35	160	48	50	134	358	196
-parchment[18]	4	5	288	4	4	386	77
-Pod#;							
Stage 1	18	37	22	21	81	269	307
Stage 2	19	27	28	24	93	300	309
Stage 3	21	28	70	28	75	326	277
Pea*;							
-hull[30]	16	27	118	3	11	570	154
-cotyledon[30]	9	132	12	5	28	58	60
Wheat bran[28]	4	176	286	8	12	195	63
Beeswing bran[14]	1	306	295	5	24	335	<30
-endosperm[12]	-	340	535	70	25	30	Tr
-aleurone[13]	-	170	480	10	30	310	Tr

[1]Deoxy = Rha + Fuc ; [2]Sugars are based on results of Saeman hydrolysis and hence Glc would be derived from cellulose and non cellulosic polysaccharides (~10% of total Glc is non-cellulosic) ; [3]Uronic acid is mainly galacturonic acid, except in parchment layers and wheat tissues where glucuronic acid or its 4-OMe derivative predominate ; #(R.R. Selvendran and J.A. Robertson, unpublished results) Stage 1 = pod at start of elongation phase ; Stage 2 = pod at stage of vegetable use ; Stage 3 = pod during seed development and deposition of parchment tissue but before pod maturity ; *carbohydrate composition determined on starch depleted alcohol insoluble residue in pea tissues. Tr = Trace.

arabinoxylans, cellulose and lignin, whereas in wheat endosperm cell walls (highly branched) neutral arabinoxylans predominate and cellulose content is very low. The aleurone layer cells of bran contain (slightly branched) neutral arabinoxylans and some β-D-glucans, with small amounts of cellulose. The polysaccharides present in the different tissues of the wheat grain have been deduced from detailed fractionation studies[12-14].

The behaviour of the DF matrix will depend on the ability of the constituent polysaccharides to cross link and the maintenance of these cross links under different physiological conditions. The persistance of cross links for example will have an important influence on fibre solubility and ultimately on resistance to microbial degradation. The different types of cross link that prevail within the DF matrix will be discussed briefly:

Cellulose. Both crystalline and amorphous cellulose can exist in the cell wall and the latter appears to predominate in parenchymatous tissues. Inter (and intra) molecular hydrogen bonding between adjacent β-(1→4)-linked glucans which give rise to the cellulose microfibrils account for the insolubility of cellulose in water. Sulphonation of hydroxyl groups of C-6 disrupts H-bonding and hence accounts for the solubility of cellulose in 12M H_2SO_4. Similarly, converting cellulose to carboxymethyl cellulose shows how cellulose can be rendered soluble.

Xyloglucans. Two main types of xyloglucan, slightly branched and highly branched, can be isolated from parenchymatous cell walls. The slightly branched xyloglucans are the more strongly H-bonded to cellulose microfibrils and require strong alkali (4M KOH) to be solubilised. In primary cell walls this 'coating' of xyloglucan helps to disperse the insoluble cellulose microfibrils within the wall matrix. Xyloglucans and galactomannans are also present as the predominant polysaccharides in some seed cell walls, e.g. tamarind and guar respectively, where they act as food reserves for the seed. H-bonding involving the more highly branched storage xyloglucans and galactomannans is more susceptible to disruption than for the slightly branched xyloglucans and a large proportion of these storage polysaccharides can be solubilised from de-hulled seeds with hot water, and find application as food gums.

Pectins. The pectins are a complex group of acidic polysaccharides, characterised by a backbone containing α-(1→4)-linked-D-galacturonic acid residues with varying degrees of methyl esterification interspersed by (1→2)- and (1→2,4)-linked L-rhamnose. Associated oligosaccharide side chains contain mainly D-galactose and L-arabinose linked mainly to C-4 of rhamnose. The side chains may be involved in ester cross linking within the wall matrix, through ferulic acid linked to terminal Gal*p* or Ara*f* residues to form phenolic esters. These are common features in cell walls of sugar beet, and account for the autofluorescence of the cell walls. Calcium bridges between non-esterified rhamnogalacturonan regions of pectic polysaccharides are also an important cross linking mechanism between molecules and these cross links tend to predominate in the middle lamella region (between adjacent cells). Removal of wall calcium by chelation can result in solubilisation of middle lamella pectins and hence cell separation. This phenomenon occurs during cooking, where intracellular chelating agents, e.g. citrate, diffuse into the cell wall, complex with the calcium and effect tissue softening through cell loosening/separation[15]. The effect is enhanced through trans-eliminative degradation (β elimination) of methyl esterified pectins at cooking temperatures and/or mild alkaline conditions. Here the cooking conditions disrupt chemical bonding adjacent to esterified galacturonic acid residues and results in molecular fragmentation of the pectic material with solubilisation of pectic polysaccharides of reduced molecular size.

Wall glycoproteins and proteoglycans. Glycoproteins present in cell walls can be classified as hydroxyproline rich (Hyp- rich) or Hyp- poor. Hyp- poor glycoproteins are mainly cell wall related enzymes whereas the Hyp- rich glycoproteins appear to have a structural role within the wall matrix. Hyp- rich glycoproteins which have been found in relatively high concentration in runner bean parenchyma, are strongly associated with cellulose and a range of glycoprotein-pectic complexes and proteoglycans have also been isolated from a range of parenchymatous tissues[16,17]. Available evidence now strongly suggests that Hyp- rich glycoproteins can form cross links between polysaccharides and are an important component in the determination of cell wall structure.

Acidic xylans. Acidic xylans and associated acidic xylan-pectic complexes are found mainly in dicotyledonous plants and consist of β-(1→4)- linked xylose residues substituted at C-2 with 4-OMe glucuronic acid or glucuronic acid residues (Glc*p*A). Ester linkage between phenolics and Glc*p*A or 4-OMe Glc*p*A is possible and results in xylan-lignin and xylan-pectic-lignin complexes, as have been isolated from runner bean parchment tissues[18].

Acidic arabinoxylans. Acidic arabinoxylans are associated mainly with cereal bran and, as in the acidic xylans, contain Glc*p*A or its 4-OMe derivative covalently linked to the β-(1→4)- linked xylan backbone of the arabinoxylans. The Glc*p*A residues are cross linked to lignin and in addition the terminal arabinose residues of side chains have associated phenolic residues, e.g. ferulic acid, which can be either ester cross linked or ether cross linked to the terminal Ara*f* of adjacent arbinoxylans within the wall matrix.

Lignin. Lignins are composed of phenyl propane monomers covalently linked to form a network through walls which have undergone secondary thickening. Lignins may be linked to polysaccharides through ester cross links and through phenolic ethers. Lignins vary with plant source as well as with developmental age but information on lignin estimation and its chemical characterisation is limited in comparison with other cell wall components[19].

2 THE DIETARY FIBRE CONTENT OF PLANT FOODS - AS NSP

The NSP are the major constituents of the cell wall of fruits and vegetables and comprise between 1.5 - 2.5% of the plant fresh weight in most commonly used fruits and vegetables[8]. This is typical for the cell walls of parenchymatous tissues. The water content of the cell wall also varies with tissue type/fibre source and stage of maturity. This important cell wall component can represent up to 90% by weight of the cell wall in parenchymatous tissue and contributes to the succulence of fresh fruit and vegetables. In mature tissues the water content of the cell wall is much lower and is very low in 'dry' seed products, e.g. legume seeds. Legume seeds have a variable NSP content, e.g. dried peas 18.6% ; haricot beans 17.1% ; chick peas 9.9% ; butter beans 15.9% measured on a dry weight basis[3,4].

Cell wall composition in fruits and vegetables (% dry weight) is typically ; cellulose 35% ; hemicellulose 15% ; pectins 40% ; and proteins and phenolics each 5%. This is comparable with the composition of cotyledon cell walls in legume seeds. Legume seed hulls, however, are enriched with acidic xylans. The NSP content of cereal grains is variable, for example ; brown rice 2.1% ; porridge oats 7.1% ; pearl barley 7.8% ; and whole wheat 10.4%. In cereal-based foods the extent of cereal grain refinement is important. Thus, high extraction wheat products contain more DF than low extraction products, for example ; wheat bran 41.7% ; whole wheat flour 10.4% ; 72% extraction flour 3.3%. The differences in DF content reflects the differential extraction of grain tissues during milling and this also affects the DF composition of the final product. Thus, white flour (72% extraction) contains only endospermous tissue and has DF composition ;

cellulose 3% ; hemicellulose 85% ; protein 7% and phenolics 5%, but in wheat bran the composition is ; cellulose 30% ; hemicellulose 50% ; protein 8% and phenolics/lignin 12%. For further details of composition and fractionation see Refs 10 & 20. Pectic polysaccharides are absent in most cereal product, one major exception being rice.

3 ANALYSIS OF DIETARY FIBRE AS NSP[4,19]

Analysis of DF involves 3 major steps :

1 Preparation of an extractive free residue (usually alcohol insoluble residue (AIR)).
2 Removal of starch from the residue by gelatinisation and treatment with amylolytic enzymes.
3 Analysis of the de-starched residue for component neutral sugars, usually as alditol acetates, and uronic acids by colorimetry or by decarboxylation.

Minor variations have been developed for the preparation of extractive free residues in different foods, usually depending on the lipid content of the food. For starch removal two major procedures have been developed.

1 Starch in the AIR is gelatinised in buffer at 85°C (30 min), followed by amylolysis with Termamyl also at 85°C for up to 45 minutes. Subsequently the residue, (preferably after precipitation in alcohol to separate low molecular weight starch degradation products), is treated with amyloglucosidase at 60°C (16h) to complete starch digestion and provide the material for detailed fibre analysis[5,21,22].

2 The AIR is finely milled prior to starch gelatinisation in acetate buffer at 100°C (1h). The resultant suspension is treated with α-amylase and pullulanase at 42°C (16h)[23]. (Digestion at 42°C is preferable to lower temperatures to avoid the association and precipitation of partially degraded starch - fragments containing ~30-40 Glc residues). This procedure removes all but the 'resistant starch' and a subsequent modification to disperse the AIR in dimethyl sulphoxide prior to gelatinisation can effect complete starch removal by the enzyme treatment[3].

In both procedures the analytical scheme has also been used to provide an estimate of the 'soluble' and 'insoluble' fibre components in the food, by analysing separately the NSP in the insoluble residue and in the supernatant after amylolytic treatment and alcohol precipitation. However, traditional methods used to fractionate cell wall polysaccharides, using hot water and oxalate to solubilise pectic polysaccharides, have been shown to result in a significant breakdown of these polysaccharides. Analysis of pectic material solubilised from potato cell walls using hot water and oxalate revealed material rich in galactose[24]. More recent fractionation using improved and non-degradative techniques has shown that the bulk of the galactose rich pectic polysaccharides remain associated with the α-cellulose residue and relatively little galactose rich material is solublised using chelators or mild alkaline non-degradative conditions[17] (Table 2). Hence, the use of hot water and oxalate must result in substantial degradation of pectic material, by β-elimination. Cooking may mimic this eliminative degradation of pectins and result in fibre solubilisation or indeed fibre loss in the cooking liquor. The methods developed for DF analysis which require high temperature to gelatinise starch will result in the partial degradation and subsequent solubilisation of pectic polysaccharides and hence could give misleading results in terms of soluble and insoluble fibre contribution from the original food under more physiological conditions. In terms of soluble and insoluble fibre it should be emphasised that cell structure, the profile of cross linking between constituent biopolymers and their ability to persist in the cell wall will be paramount in determining fibre solubility and hence the behaviour of the matrix under the physiological conditions encountered during digestion.

Fractionation of cell walls under non-degradative conditions

Cell wall fractionation involves solubilisation of polysaccharides under non-degradative conditions to determine how biopolymers are cross linked within the cell wall matrix, as outlined in Table 2 for fruit and vegetable parenchyma cell walls. The fractionation protocol can be considered in relation to 4 diverse tissue types :

1 Potato parenchyma cell walls to represent succulent vegetable tissue.
2 Runner bean parchment layer to represent secondary thickened vegetable tissue.
3 Wheat grain endosperm cell walls to represent cereal parenchyma.
4 Beeswing wheat bran to represent tissue with wall polymers cross linked by both phenolic esters and lignin.

Table 2 Scheme for the sequential extraction of polysaccharides from the plant cell wall (Fruits and Vegetables)[17,29]

Sequential Treatment	Effect	Polysaccharide Solubilised
CDTA ; 50mM 2o°C (twice)	Abstraction of calcium from pectins and disruption of ionic cross links. May result in cell separation.	Slightly branched polysaccharides (mostly from the middle lamellae)
Na_2CO_3;50mM (twice) (+20mM $NaBH_4$) 1st Extraction 1°C 2nd Extraction 20°C	Abstraction of calcium within the wall matrix and disruption of ionic cross links. Cleavage of some sugar-ester links	Highly branched pectic polysaccharides from the middle lamella and primary cell walls
KOH ; 1M (+20mM $NaBH_4$)	Cleavage of phenol-sugar ester and protein-carbohydrate bonds	Small amounts of highly branched pectic polysaccharides, glycoproteins and proteoglycans
	Disruption of some H-bonds	Highly branched xyloglucans
KOH ; 4M (+20mM $NaBH_4$)	Swelling of cellulose and disruption of H-bonds	Mainly slightly branched xyloglucans
[RESIDUE]	-	**[INSOLUBLE]**
α-Cellulose	-	[Mainly cellulose and cross linked highly branched pectic polysaccharides and Hyp-rich glycoproteins]
Chlorite/Acetic acid (0.3% w/v: 0.12% v/v) @ 70°C	Disruption of phenolic cross links. - mild delignification	Hyp-rich glycoproteins and some highly branched pectic polysaccharides

The solubility of potato parenchyma cell wall biopolymers and types of cross link disrupted to effect solublisation are outlined in Table 2. The scheme involves use of a powerful chelating agent, such as CDTA at 20°C, and alkali of successively increasing strength to disrupt sugar ester, phenolic ester and H-bonding between polysaccharides. In potato the α-cellulose residue contains little Hyp-rich glycoprotein but in runner bean parenchyma a significant amount of Hyp-rich glycoprotein is found. This requires mild treatment with chlorite/acetic acid for solubilisation and probably depends on disruption of phenolic cross links involving the Hyp-rich glycoproteins[17]. Fractionation of runner bean parchment layers according to the outlined scheme results in negligible polysaccharide solubilisation and delignification (chlorite/acetic acid) is required, i.e. production of holocellulose, prior to fractionation. Little solubilisation of material from the holocellulose occurs before exposure to M or 4M KOH and at this stage significant amounts of hemicelluloses (acidic xylans) are released. This is indicative of acidic xylans in the holocellulose linked through phenolic ester bonds (to degraded lignin) and in the original parchment tissue. The acidic xylans are linked to lignin through glucuronic acid esters[18].

In the case of wheat grain tissues a significant proportion of endosperm cell walls, mainly neutral arabinoxylans,can be solubilised with hot water and the remainder solubilised in M KOH to leave only a trace of α-cellulose. This is analogous to the solubilisation of β-glucans from oat bran using hot water and yielding a viscous solution containing β-glucans. However, in beeswing bran significant solubilisation of polysaccharides (mainly acidic arabinoxylans) only occurs when M or 4M KOH is used and, with delignification after alkali treatment, further polysaccharide extraction with alkali is possible[14]. This indicates that the cross links in the endospermous cell walls are much weaker compared with those in beeswing bran. In beeswing bran phenolic ester (e.g. ferulic acid type) cross links predominate and these are much more persistant than the sugar ester cross links found in potato parenchyma, but the former are more alkali labile than the lignin ester cross links in runner bean parchment tissue.

Through the development of these non-degradative fractionation schemes it is possible to understand better the associations between cell wall biopolymers[10]. They also allow the interpretation of analytical data in terms of cell wall matrix structure rather than just composition. Hence functional models of the cell wall can be formulated as a necessary stage in understanding the behaviour of the fibre matrix during gastrointestinal transit.

4 DIETARY FIBRE BEHAVIOUR DURING GASTROINTESTINAL TRANSIT IN RELATION TO CELL WALL STRUCTURE AND ITS FRACTIONATION

During transit DF is exposed to different environments, notably stomach acid, mild alkaline conditions in the small intestine and fermentation in the hind gut. In the stomach the acid (pH ≈ 2.5) might be expected to displace some of the calcium from the cell wall matrix and this may influence the solubility of pectic polysacharides and hence cell loosening. In the small intestine the alkaline endogenous secretions of bile salts and pancreatic fluid will neutralise excess stomach acid but result in mild alkaline conditions in the ileum. At body temperature these conditions may be sufficient to result in some β eliminative degradation of pectic polysaccharides with their concomitant solubilisation and tissue softening. Degradation of pectic polysaccharides during cooking and possibly during transit in the upper gut also presents the possibility that the physiological effects of 'pectin' may be a consequence of their eliminative degradation in addition to the effect of 'intact' polysaccharides or their fermentation products. Tissue softening whether by cooking or conditioning in the upper gut will enhance the contribution of soluble fibre to the diet and the generation of soluble material will weaken cell structure such that mechanical action in the gut will lead to tissue disintegration (particle size reduction).

The more severe conditions used for cell wall fractionation, strong alkali, will not apply in the gut but an appreciation of how matrix polysaccharides behave using strong alkali also helps in understanding how the matrix might behave under fermentative conditions. To be fermented the polysaccharide must first be solubilised and weaker cross links will be preferentially attacked by microorganisms to effect solubilisation. Thus lignified tissues persist under fermentative conditions as well as during wall fractionation.

Evidence for polysaccharide solubilisation and tissue degradation other than by fermentation has been noted from experiments using the pig as an experimental model to investigate fibre behaviour during gut transit[9]. When fed a minced vegetable (swede) diet or cereal bran diet a major reduction in median particle size was found on the swede diet anterior to the terminal ileum[11]. The reduction was not due to mastication (stomach particle size distribution being similar to the feed) but occurred in the small intestine and was associated with a reduction in the proportion of insoluble uronic acid in the digesta recovered from the terminal ileum. It should be noted that in the case of 'wholefoods' mastication will be important to determine particle size distribution in the stomach. A further small reduction in particle size occurred in the hind gut, along with the rapid disappearance of cell wall material (87.8% apparent digestibility). No corresponding reduction in the median particle size of bran was found although some solubilisation of arabinoxylans may occur in the small intestine and a large proportion of polysaccharide (mainly arabinoxylans) was fermented (41.5% apparent digestibility). However, particle shape did change in bran, from flake to 'coiled cylinder', consistent with the selective fermentation of thick-walled aleurone layer cells[25]. This will alter the mechanical properties of the particle. From particle size analysis it would appear that tissue softening can occur in the stomach and pectic polysaccharides are solubilised in the small intestine. There, mechanical conditions promoting transit leads to tissue breakdown. This will not only increase the proportion of soluble fibre present in the digesta but also the surface area of particulate material. Thus, there exists the possibility of further pectic polysaccharide solubilisation and also an increase in the porosity of the residual cell wall as polysaccharides are removed. Measurement of potential digestibility of insoluble fibre in the pig caecum, using a nylon bag technique to suspend fibre in the caecum[9,26], also showed swede fibre to be fermented almost to completion within 24h and the uronic acid component (representative of pectic polysaccharide) to diminish preferentially. This was in contrast to bran where the uronic acid, as Glc*p*A, is associated with lignification and hence is poorly fermentable. In bran the neutral non-cellulosic sugars, predominantly from arabinoxylans and β-glucans from aleurone cells, were preferentially fermented but potential digestibility was less than 50% for bran even after 48h in the caecum. Similar observations have been made using *in vitro* fermentation of wheat bran with faecal inocula[27,28]. This pattern of fermentation is consistent with the fractionation scheme devised to investigate cell wall cross link structure discussed previously and also illustrates that even after prolonged exposure to fermentation, cross linking within the wall matrix can withstand microbial attack.

Therefore, although the chemistry of DF has been

studied extensively in recent years and methods developed for its quantification in a range of fruits, vegetables, cereals and legume seeds only now are the complexities of the cell wall polysaccharide structure and function beginning to be unravelled and hence the possible significance appreciated in their nutritional context as DF. In developing methods to fractionate CWM for structural analysis it has become apparent that the gastrointestinal tract may also be actively involved in fractionating DF during gut transit and hence it may be more advantageous to investigate DF chemistry in line with this apparent fractionation rather than persist with correlating physiological effects with only quantitative intake as total fibre or perhaps misleadingly as 'soluble' fibre by chemical analysis. This can also be extended to overcome the assumption that all cereal fibres constitute a distinct fibre source and similarly all fruit and vegetable fibres are another distinct source since it is now becoming more apparent that within cereals, as indeed within fruits and vegetables, definite structural and compositional cell wall differences exist. These differences may be as responsible for the reported dietary effects of fibre than the generally assumed differences between subjects.

5. BIBLIOGRAPHY

1. H. Trowell, Am.J.Clin.Nutr., 1976, 29, 417-427.
2. H. Trowell, D.A.T. Southgate, T.M.S. Wolever, A.R. Leeds, M.A. Gassull and D.J.A. Jenkins, Lancet, 1976, 1, 967.
3. H.N. Englyst and J.H. Cummings, Analyst, 1984, 109, 937-942.
4. D.A.T. Southgate and H. Englyst, 'Dietary Fibre, Fibre - Depleted Foods and Disease' [ed. H. Trowell, D. Burkitt and K. Heaton], Academic Press Inc. (London) Ltd., 1985, Chapter 3, pp. 31-55.
5. O. Theander and E. Westerlund, J.Agric.Food Chem., 1986, 34, 330-336.
6. D.A.T. Southgate, 'Dietary Fibre in Health and Disease' [ed. G.V. Vahouney and D. Kritchevsky], Plenum, New York, 1982, Chapter 1, pp. 1-7.
7. R.R. Selvendran, 'Dietary Fibre', [ed. G.G. Birch and K.J. Parker], Applied Science Publishers, London and New York, 1983, Chapter 7, pp. 95-147.
8. R.R. Selvendran, B.J.H. Stevens and M.S. DuPont, Advances in Food Research, 1987, 31, 117-209.
9. J.A. Robertson, S.D. Murison and A. Chesson, J. Nutrition, 1987, 117, 1402-1409.

10. R.R. Selvendran and M.A. O'Neill, 'Methods of Biochemical Analysis' [ed. D. Glick], Wiley (Tube and Science), New York, 1987, 32, pp. 25-153.
11. J.A. Robertson, Proc.Nutrition Soc., 1988, 47, 143-152.
12. D.J. Mares and B.A. Stone, Aust.J.Biol.Sci., 1973, 26, 793-812.
13. A. Bacic and B.A. Stone, Aust.J.Plant Physiol., 1981, 8, 475-495.
14. M.S. DuPont and R.R. Selvendran, Carbohydr.Res., 1987, 163, 99-113.
15. R.R. Selvendran, J.Cell Sci.Suppl., 1985, 2, 51-88.
16. M.A. O'Neill and R.R. Selvendran, Biochem.J., 1980, 187, 53-63.
17. P. Ryden and R.R. Selvendran, Biochem.J., 1990, (In press).
18. R.R. Selvendran and S.E. King, Carbohydr.Res., 1989, 195, 87-99.
19. R.R. Selvendran, A.V.F.V. Verne and R.M. Faulks, 'Modern Methods of Plant Analysis New Series' [ed. H.F. Linskens and J.F. Jackson], Springer-Verlag, Berlin Heidelberg, 1989, 10, pp. 234-259.
20. R.R. Selvendran, Scand.J.Gastroenterol.Suppl., 1987, 22, 33-41.
21. O. Theander and P. Aman, Swed.J.Agric.Res., 1979, 9, 97-106.
22. R.M. Faulks and S.B. Timms, Food Chem., 1985, 17, 273-287.
23. H. Englyst, H.S. Wiggins and J.H. Cummings, Analyst, 1982, 107, 307-318.
24. S.G. Ring and R.R. Selvendran, Phytochemistry, 1981, 20, 2511-2519.
25. R. Moss and D.C. Mugford, J.Cereal Sci., 1986, 4, 171-178.
26. J.A. Robertson, S.D. Murison and A. Chesson, J.Sci. Food Agric., 1986, 37, 359-365.
27. B.J.H. Stevens, R.R. Selvendran, C.E. Bayliss and R. Turner, J.Sci. Food Agric., 1988, 44, 151-166.
28. B.J.H. Stevens and R.R. Selvendran, Carbohydr. Res., 1988, 183, 311-319.
29. P. Ryden and R.R. Selvendran, Carbohydr.Res., 1990, 195, 257-272.
30. R.R. Selvendran, Am.J.Clin.Nutr., 1984, 39, 320-337.

CHANGES IN DIETARY FIBRE POLYMERS DURING STORAGE AND COOKING

K. W. Waldron and R.R. Selvendran

AFRC Institute of Food Research
Norwich
U.K.

1 INTRODUCTION

The major components of dietary fibre (DF) are derived from cell wall polymers (1), and the amounts and relative proportions of these can affect the fibre quality (2). There is little definitive information concerning the changes in such components during maturation, storage and processing of edible plant organs. In this paper, we report changes in DF polymers that occur during (a) maturation and storage of asparagus stems, and (b) cooking of potatoes.

2 MATERIALS AND METHODS

Maturation and Storage of Asparagus Stems

Asparagus stems were harvested and marked, at 5, 10 and 15cm from the apex with indian ink. Spears, in groups of 12, were cut at the marks either immediately or after storage for 3 days at 21°C in darkness in punnets overwrapped with perforated auto-RMFA packaging film. Florets and scale leaves were removed and alcohol insoluble residues (AIRs) were prepared from the cut sections of both fresh and stored stems. The AIRs were analysed for the component neutral sugars and uronic acids as described (3). The alkali-soluble cell wall phenolics were estimated using Staffords (4) adaptation of the method of Bondi and Meyer (5) as described below. This method utilises the pH-dependant absorbance properties of the phenolics. The AIR was extracted 0.5M NaOH for 16h at 70°C, and the extract neutralised. Aliquots were diluted in buffer (pH 7 and pH 12) and the difference between the absorbance at pH 12 and 7 was determined at 340 nm.

Whole asparagus stems (15 cm in length), either fresh or stored for 3 and 7 days under similar conditions, were also extracted for AIR which was analysed similarly. As in the previous experiment, 12 stems were used per sample.

Fibre degradation caused by cooking

Potato parenchyma cell wall material (CWM) was prepared free of intracellular components including starch (6). The pectic polysaccharides were extracted by 2 methods. One, which caused little or no degradation using CDTA and mild alkali, and the other, using hot water and hot oxalate which caused significant degradation of the pectic polysaccharides, particularly the highly branched ones.

3 RESULTS AND DISCUSSION

The results reported in this paper show the same trends in changes of DF and cell wall components as those of preliminary experiments performed during two previous growth seasons. Although the values obtained are from single determinations on pooled samples of 12 comparable stem sections, the consistent changes shown in Figure. 1 and Table 1 allow for comparison in the wall components during maturation and storage. Furthermore, conclusions drawn from this data are supported by the analysis of AIRs from fresh and stored whole asparagus stems shown subsequently in Figure 2 and Table 2.

Changes in DF during maturation of asparagus stems

Since no starch could be detected in the AIRs, levels of dietary fibre are calculated (on a fresh-weight basis) from the values of the cell wall sugars. In fresh asparagus stems dietary fibre decreased with distance from the apex (Figure 1). This was due to a decrease in cell number per unit fresh weight: examination by light microscopy showed that in the immature tissues of the apical 5 cm, the cells were small and closely packed. In the second section, the majority of cells were elongating whilst in the basal section, extension had ceased and some secondary thickening and associated lignification was evident. The carbohydrate composition (Table 1) showed that in immature tissues the main sugars, in decreasing order of abundance, were uronic acids, glucose, arabinose, galactose, mannose and deoxy-sugars. From these values, the main polysaccharides could be inferred to be pectins, cellulose and hemicelluloses. Maturation was accompanied by a decrease in arabinose and an increase in xylose and

Table 1 Changes in the composition of dietary fibre during maturation and storage of asparagus stems

Section	Sugars mol%								g/100g AIR	Phenolic content AOD_{340}/mg
	Rha	Fuc	Ara	Xyl	Man	Gal	Glc	UA		
F1	1	1	18	6	3	12	29	30	33	1.7
F2	1	1	13	7	4	15	33	26	40	2.0
F3	1	1	9	9	3	14	35	27	53	5.5
S1	1	1	17	8	3	9	30	32	35	4.0
S2	1	1	11	9	4	7	36	32	47	5.5
S3	1	1	8	13	3	6	37	31	57	8.0

glucose. There was little change in the levels of other sugars.

Changes in dietary fibre during storage of asparagus stems

The trends shown during maturation in fresh asparagus spears were also shown in those that had been stored for 3 days. However, there were significant differences between the composition of AIRs of stored asparagus stem sections and their corresponding fresh controls. Storage resulted in a decrease in the total dietary fibre level (relative to the fresh 'control' tissue) in the apical tissues, no change in the mid section, and an increase in the basal section (Figure 1). These changes were accompanied by changes in the carbohydrate composition. Storage resulted in a decrease in the relative levels of galactose and arabinose in all sections. In the basal section, storage resulted in an increase in the levels of xylose, glucose

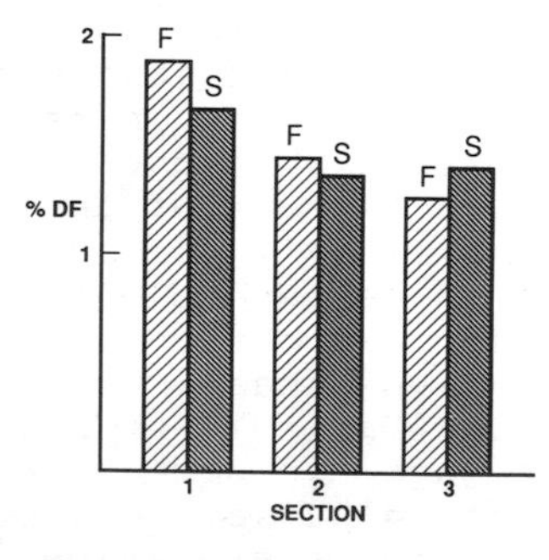

Figure 1 Changes in DF during maturation and storage of asparagus stems.

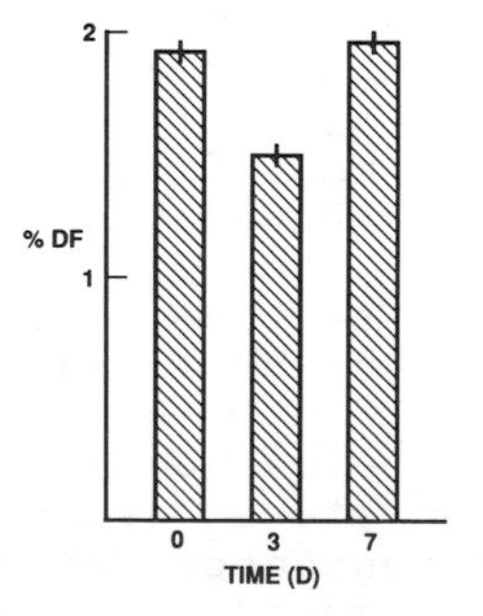

Figure 2 Changes in total asparagus stem DF during storage for up to 7 days

Table 2 Changes in the composition of dietary fibre during storage of asparagus spears

Days of Storage	Rha	Fuc	Ara	Xyl	Man	Gal	Glc	UA	g/100g AIR	Phenolic content AOD340/mg
	Sugars mol%									
0	1	1	14	9	3	12	35	25	42	3.2
3	1	1	13	9	4	7	42	22	38	5.7
7	1	1	9	12	3	5	36	33	41	22.3

and phenolics (Table 1). These changes are in accordance with secondary thickening of the sclerenchyma and vascular tissues, and would account for the increase in DF in the basal section.

These results facilitated the interpretation of changes in DF during prolonged storage of whole asparagus stems for up to 7 days (Figure 2). Over this period, the DF content of the fresh stems, which was 1.9%, decreased to 1.5% after 3 days, and then increased to 2% by day 7. Storage was accompanied by a significant decrease in galactose and arabinose, and a notable increase in phenolics and xylose, particularly by day 7 (Table 2). These changes are probably due to breakdown of pectic polysaccharides in the parenchymatous tissues, and deposition of lignin and xylans, possibly in the sclerenchyma and vascular tissues.

The results above show that pectic polysaccharides can become metabolised and depolymerised during storage. Pectins are also subject to depolymerisation as a result of heat-catalysed degradation. The following section reports on such changes during cooking.

Changes in dietary fibre during cooking

Cooking is usually accompanied by tissue softening and changes in textural characteristics. This occurs as a result of abstraction of calcium from the middle lamella region by suitable chelating agents e.g. citrate (7) and also due to β-eliminative degradation of pectins (Figure 3). Both of these processes encourage cell separation. Degradation of apple pectins by β-elimination has been studied during heating in phosphate buffer at pH 7 (8), and these conditions are similar to those which occur in vegetables during boiling. The above and related studies (9) show that significant breakdown of the pectins can occur during boiling of vegetables.

Table 3 Effects of cooking on the extraction of CW polymers

	Rha	Fuc	Ara	Xyl	Man	Gal	Glc	UA	g/100g
	Component sugars (mol%)								
CWM	2	-	9	2	t	29	35	22	95
Cold treatment									
CDTA x2*	1	2	14	1	1	9	7	66	64
Alkaline solvents to solubilise hemicelluloses									
α-Cellulose	1	t	6	t	1	36	44	11	97
Hot treatment									
Hot water	2	-	14	-	-	53	t	32	98
Hot oxalate	2	-	5	-	-	21	t	72	97
Alkaline solvents to solubilise hemicelluloses									
α-Cellulose	1	-	2	1	2	3	72	18	96

* values refer to the first CDTA estraction.

In order to highlight the effects of β-eliminative degradation on DF polysacharides, the effects of extracting purified potato CWM with hot water and hot oxalate were compared with extracting with CDTA at 20°C. The results (Table 3) show that CDTA-extraction released polysaccharides were rich in uronic acids with some arabinose and galactose. In contrast, the polysaccharides released by both hot water and hot oxalate contained considerable amounts of galactose. The α-cellulose residues remaining after subsequent (non-degradative) extraction with alkali were also compared. Whilst the

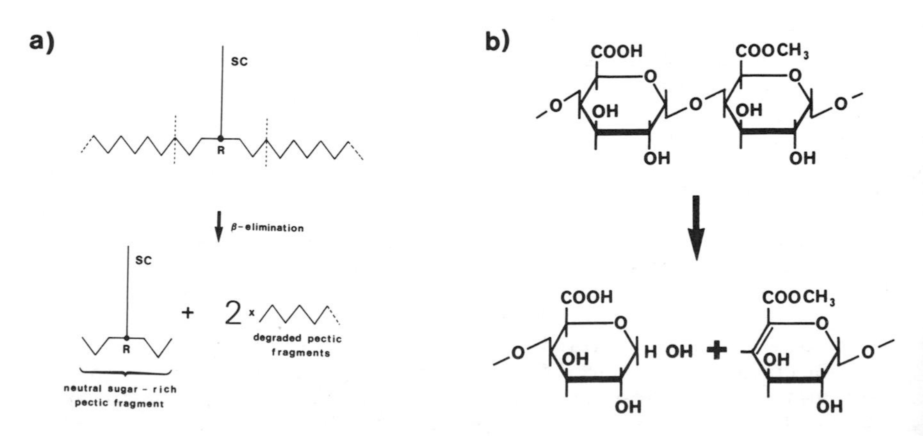

Figure 3 β-eliminative degradation of pectins. (a) diagram of depolymerisation, (b) bond cleavage

α-cellulose from CDTA-extracted cell walls was rich in galactose, little galactose was found in the hot water and hot oxalate-extracted α-cellulose. It can be concluded that the release of galactose by heat was due to depolymerisation, since under non-degradative conditions it remained in the α-cellulose fraction.

4 CONCLUSIONS

This study has focussed attention on the metabolism of cell wall polyaccharides in succulent growing tissues during storage. The changes in pectic polysaccharides, particularly the breakdown of the galactan and arabinan side chains, are comparable to those in ripening tomato (10) and apple (11) although the tissues of ripening fruits are physiologically distinct from those of fast growing stems of asparagus. The increase in xylose-rich hemicelluloses and phenolics are indicative of secondary thickening, and correlate with an increase in DF. Finally, depolymerisation of DF polysaccharides have also been highlighted during cooking. Since the physicochemical characteristics of DF are dependant on the cell wall polymers and their interactions, it is likely that the changes discussed above are likely to affect the DF quality.

5 REFERENCES

1 R.R. Selvendran, B.J.H. Stevens and M.S. DuPont, Adv. Fd. Res., 1987, 31, 117.
2 J.W. Anderson and S.R. Bridges, Am. J. Clin. Nutr., 1988, 47, 440.
3 B.J.H. Stevens and R.R. Selvendran, Phytochem., 1984, 23, 107.
4 H.A. Stafford, Plant Physiol. 1960, 35, 108.
5 A. Bondi and H. Meyer, Biochem. J., 1948, 34, 248
6 P. Ryden and R.R. Selvendran, Carbohydr. Res., 1990, 195, 257.
7 R.R. Selvendran, J. Cell Sci. Suppl. 2, 1985, 51-88
8 A. J. Barrett and D. H. Northcote, Biochem. J., 1965, 96, 617.
9 S.A. Matz, 'Food Texture', The AVI Publishing Company, Inc, Connecticut, 1962.
10 K. Gross and S. Wallner, Plant Physiol., 1979, 63, 117.
11 I.M. Bartley and M. Knee, Food Chem., 1982, 9, 47-58.

HETEROGENEITY OF MONOMERIC COMPOSITION IN GRAMINEAE, MAIZE AND WHEAT, LIGNINS.

B.MONTIES

Laboratoire de Chimie Biologique, I.N.R.A,
Institut National Agronomique Paris-Grignon (CBAI),
Centre de GRIGNON,
78850 THIVERVAL-GRIGNON (France).

I - INTRODUCTION

Even if the definition of crude-, detergent-, or dietary-fibres is not a simple task[1], plant cell wall phenolics such as lignins and linked phenolic acids have a great importance in connection with the indigestibility and thus the definition[2] of these fibres.

Several levels of heterogeneity of lignins, which include molecular, macromolecular and super-molecular levels, have been described in addition to associations with polysaccharides and phenolic acids[3,4]. Furthermore, biochemical variations mainly due to differences in plant species, cytological origin and stage of development could change the structure and the reactivity of these phenolics[5], and thus the digestibility of the corresponding fibres. Heterogeneity and inhomogeneity in angiosperm lignins has been, until now, particularly studied in the case of woods[6,7]. Only few studies have been published about the heterogeneity and the formation of grass lignins[8-10]. Evidences of heterogeneity in monomeric composition in wood lignin have been previously reported by comparison of the composition of lignin fractions isolated from different species[11,12]. However in the case of wheat- and of rice- straw lignins, clear occurrence of heterogeneity in monomeric composition were not found[8,9,13].

Evidence of this phenomenon was however recently obtained, after isolation of lignin fractions from maize culm internodes. A publication being in preparation (TOLLIER, LAPIERRE and MONTIES), these results are summarized here, with emphasis on data previously published in the laboratory.

II - MATERIAL AND METHODS

Extractive free cell wall residues (CWR) were prepared by successive Soxhlet extractions with toluene-ethanol (2-1, vol), ethanol and water from roughly ground (0.1 to 0.5 mm particle size) internodes of straw from wheat (Triticum sp.) and of culm from maize (Zea sp.) grown under field or greenhouse. Milled-stem lignin, LM, and enzyme lignin, LE, were successively isolated from ultraground extractive free CWR, according to standard procedure[7,13]. Alkali lignin, LA, was isolated after N sodium hydroxide extraction at 35°, with stirring under nitrogen[14], with recovery of the insoluble final residue, saponification residue, SR, by filtration and washing in dilute HCl and water[8]. Monomeric composition of these fractions was characterized by nitrobenzene oxidation with reversed phase HPLC identification of p-hydroxybenzaldehyde (h), vanillin (v) and syringaldehyde (s)[7,8] and by thioacidolysis with GC determination of the TMS derivatives of the tri-thioethylether monomeric units of phenyl (H), guaiacyl (g) and syringyl (S) types with identification by mass-spectrometry[7,15]. Ester- and ether- bound p-coumaric-(PC)- and ferulic-(FE)-acids were analyzed after successive alkaline and acid hydrolysis of the CWR and lignin fractions with HPLC determination[8,14]. Total lignin, KL, contents were obtained according to Klason type procedure without prehydrolysis[16]. Mean standard deviations for Klason lignin, nitrobenzene oxidation, thioacidolysis and phenolic acids were respectively, about : 10, 10, 5 and 5 per cent for independent and repeated series of determinations.

III - RESULTS

Dry plant stems were harvested about two weeks after maturity of the grain in the case of wheat, grown in field conditions and in the case of maize, grown in greenhouse. As previously reported[17], a normal line of Maize was compared to the corresponding brown-midrib mutant ($b.m_3$), mainly characterized by a low lignin content and the occurrence of an additional type of monomeric unit with a 5-hydroxyguaiacyl substitution pattern[17,18].

Total lignin contents in wheat straw, in normal- and in mutant-maize lines were respectively : 19.5 - 15.2 and 11.3% CWR. Corresponding lignin contents in saponification residue were respectively: 13.8 , 9.2 and 3.8% CWR. These last results agree with the well established high alkaline solubility of gramineae lignins and indicate a higher solubility in the case of b.m.-mutant. The molecular origin of this higher solubility (etherification by phenolic acids[19] and/or larger content in free phenolic hydroxyl groups[20]) is under study.

According to yields in non condensed monomeric units of phenyl, guaiacyl and syringyl type recovered after nitrobenzene oxidation (b,v,s respectively) and after thioacidolysis (H,G,S respectively), maize culm lignins appeared as typical H-G-S lignins, as shown in table 1.

Table 1 : Monomeric composition of cell wall (CWR)- and saponification (SR)- residues of normal maize stem as shown by nitrobenzene oxidation and thioacidolysis.

$\mu M.g^{-1}KL$	Nitrobenzene b	Nitrobenzene v	Nitrobenzene s	Thioacidolysis H	Thioacidolysis G	Thioacidolysis S
CWR	1300	750	830	21	256	365
SR	459	491	693	trace	188	246

The relative contents in phenyl (b,H) compounds was dramatically decreased after alkali extraction. Such behaviour has been previously reported for other monocots[21], suggesting that a significant part of b and v recovered after nitrobenzene oxidation arose from p-coumaric- and from ferulic-acids and not from lignin. The same type of variations were obtained in the case of wheat straw[15] and b.m maize mutant[17,22]. In each case, thioacidolysis yields were lower than for nitrobenzene, confirming[7,15] that, in addition to aryl-alkyl-ether linkages, other structures, such as diaryl-propane, were degraded. As however the characteristic phenylpropane structure of phenolic acids and of lignin monomers is not degraded by thioacidolysis, thioacidolysis data shown in fig. 1 provide the more exact figure of the monomeric composition of the total stem lignin.

Significant contents of p-coumaric- and ferulic-esters and ethers were found linked to the three types of lignin fractions, obtained from wheat- and maize-stem CWR (data not shown). In all cases, p-coumaric ester contents were about ten times higher than the corresponding ferulic ester content with PC-ester contents ranging from about 150 to 600 $\mu M.g^{-1}$ of LM and LE fractions with relatively lower contents in wheat. Phenolic esters were not significantly found in LA fractions. Again, in each cases, p-coumaric-and ferulic-ether contents were found in the range of 10 to 70 $\mu M.g^{-1}$ of LM, LE and LA fractions with a characteristic relative higher content of ferulic ethers in LA fractions of both wheat and maizes. These data agree with previous results[3,14,15,19], indicating a preferential association of p-coumaric esters to lignin-rich fractions of monocot plant cell wall, and a higher content in phenolic ethers in alkali lignins. Large differences in absolute contents in linked phenolic acids were found between independent experiments confirming the importance of environmental and genetic factors on these contents[17,23].

Very significant differences of contents in non condensed guaiacyl (G) and -syringyl (S) monomeric units were found by thioacidolysis between LM and LE lignin fractions isolated from normal maize, but not from b.m-mutant line (table 2).

Table 2 : Monomeric composition of LM and LE lignin fractions isolated from normal- and mutant-maize shown by thioacidolysis.

μMole.g^{-1}	Normal		Mutant	
	LM	LE	LM	LE
H	12	9	12	9
G	169	144	197	293
S	201	220	90	130
H/G	0.07	0.06	0.06	0.03
S/G	1.19	1.53	0.45	0.44

Differences in monomeric composition, also indicated in table 2 by H/G ratio, appear only in the case of mutant and require confirmation. Further, in agreement with previous data[17], a lower relative content in syringyl units in LM and LE lignin fractions of the mutant, appears associated with a significantly higher content in guaiacyl units. Biosynthetic correlations between these variations are very likely[17,18].

Variations in monomeric composition of lignin fractions isolated from wheat and maizes are shown in table 3, with two independent series of samples[22,24], in the case of normal- and mutant-maize.

Table 3 : Monomeric composition of LM and LE lignin fractions isolated from wheat, normal- and mutant-maïze as shown by thioacidolysis (data : S/G molar ratio with two independant experiment for maïzes : years 1988 and 1989).

	Wheat[a]	Normal-maïze		Mutant-maïze	
		exp.1[b]	exp.2[c]	exp.1[b]	exp.2[c]
LM	0.84	1.19	1.01	0.45	0.46
LE	0.92	1.53	1.42	0.44	0.49
LA	1.04	1.80	1.70	0.47	0.50

a, b, c : according respectively to ref. : 19, 22 and 24

Differences in syringyl to guaiacyl ratio reported in the case of normal-maïze unambiguously shows the occurrence of an heterogeneity of monomeric composition in gramineae lignin which compares to other cases previously reported in angiosperm woods lignin fractions[7,11]. In the case of wheat, weak differences in S/G ratio may thus, tentatively, indicate the occurrence of lignin heterogeneity[12,19]. Furthermore, higher S/G values reported in table 3, for LA lignin fractions, have to be discussed cautiously as some water-soluble-colloïdal lignin fractions may be selectively lost during the isolation steps.

IV - CONCLUSION

Occurrence of heterogeneity of monomeric composition between lignin fractions, isolated from extractive free cell wall residue from internodes of the culm of maïze, has been reported here for the first time. Due to its selectivity, thioacidolysis only allowed an unambiguous evidence of this phenomenon previously reported, in the case of angiosperm woods, for related guaiacyl-syringyl lignin. Gramineae lignin, however, appears as more condensed and intermediate in its molecular structural scheme[4] between angiosperm- and gymnosperm-lignins. Furthermore, lignins of monocotyledons examined by radiotracer methods were not so intensively heterogeneous in structure with regard to corresponding tree lignins[9,10,26]. It is likely that the differences in heterogeneity found between wheat and maïze may be related to the fact that, in the stem of these gramineae, seen in trans-section, widely spaced vascular bundles are not restricted in one circle but appear either in two circles in wheat, or scattered throughout the section in Maize[27,28].

REFERENCES

1. P.J.Van Soest and J.B.Robertson, Nutrition Rev., 1977, 35, 12.
2. D.A.T.Southgate, Nutrition Rev., 1977, 35, 31.
3. B.Monties and C.Lapierre, Physiol. Veg., 1981, 19, 327.
4. B.Monties, "The biochemistry of plant phenolics", C.F.Van Sumere and P.J.Lea edit., Oxford Univ. Press, 1985, Vol.25, Chapter 9, p.161.
5. B.Monties, "Methods in plant biochemistry", P.M.Dey and J.B.Harborne edit., Academic Press, 1989, Vol.1, Chapter 4, p.113.
6. N.Terashima, K.Fukushima and K.Takabe, Holzforschung, 1986, 40 Suppl., 101.
7. M.T.Tollier, B.Monties, C.Lapierre, Holzforschung, 1986, 40 Suppl., 75.
8. U.Sharma, J.M.Brillouet, A.Scalbert and B.Monties, Agronomie, 1986, 6, 265.

9. B.Monties, M.T.Tollier and M.Gaudillere, Proc. 4th Int. Symp. Wood Pulping Chem. (PARIS), Centre Technique du Papier, GRENOBLE, France pub., 1987, Vol.2, p.41.
10. L.He and N.Terashima, Mokuzai Gakk. 1989, 35, 116.
11. C.Lapierre, Thèse de Doctorat d'Etat, Université PARIS-SUD at ORSAY, INRA pub., 1986.
12. C.Lapierre, A.Scalbert, B.Monties and C.Rolando, Bull Groupe Polyphenols, 1986, 13, 128.
13. A.Scalbert, B.Monties and J.Y.Lallemand, Holzforschung, 1986, 40, 119.
14. A.Scalbert, B.Monties, J.Y.Lallemand, E.Guittet and C.Rolando, Phytochemistry, 1985, 24, 1359.
15. C.Lapierre, D.Jouin and B.Monties, "Physico chemical characterisation of plant residues for industrial and feed use", A.Chesson and E.R.Ørskov edit., Elsevier Applied Sc. pub., 1989, Chap.10, p.118.
16. B.Monties, Agronomie, 1984, 4, 387.
17. M.Gaudillere and B.Monties, "Plant cell wall polymers : biogenesis and biodegradation", N.G.Lewis and M.G.Peace edit., Am. Chem. Soc. pub., Vol.399, ACS Symp. Ser., Chap.13, p.182.
18. C.Lapierre, M.T.Tollier and B.Monties, C.R. Acad. Sci. Paris, 1988, 307 : serie III, 723.
19. A.Scalbert, B.Monties, Holzforschung, 1986, 40, 249.
20. C.Lapierre, D.Jouin and B.Monties, Phytochem., 1989, 28, 1401.
21. T.Higuchi, Y.Ito and I.Kawamura, Phytochemistry, 1967, 6, 875.
22. T.Queral, Mémoire de DEA, Institut National Agronomique, INRA pub., 1988.
23. F.Bayet and B.Monties, Hoppe-Seyler's Z. Physiol. Chem. 1977, 358, 1173, cf. also F.Bayet, Thèse de Docteur, Institut National Agronomique, INRA pub., 1979.
24. M.T.Tollier, C.Lapierre and B.Monties, 1989, results to be published : see introduction.
25. C.Lapierre and B.Monties, Proc. 5th Intern. Symp. Wood and Pulping chemistry (RALEIGH, USA) 1989, Vol.1, p.615.
26. L.He and N.Terashima, Proc. 5th Intern. Symp. Wood and Pulping Chemistry (RALEIGH, USA) 1989, Vol.1, p.555.
27. K.Esau, "Anatomy of seed plants", John Willey and Sons pub., 1977, 2 edit., p.260 (fig.16-2) and p.315 (fig.17-17).
28. A.Fahn "Plant anatomy", Pergamon press pub., 1982, 3 edit., p.139 (fig.106-4,5).

STUDIES ON PENTOSANS IN FLOUR, DOUGH AND BREAD FRACTIONS

P. Åman[a,b], E. Westerlund[a] and S. Bengtsson[a,b]

[a]Department of Chemistry and
[b]Department of Animal Nutrition and Management
Swedish University of Agricltural Sciences
S-750 07 Uppsala, Sweden

INTRODUCTION

Rye and wheat are widely grown cereals and the grains are mainly used in bread and other products for human consumption and in animal feeds. The starchy endosperm of common cereals contains predominantly starch and protein but significant amounts of dietary fibres and other components are also present. In wheat and especially rye these fibres are mainly composed of pentosans, including arabinoxylans and arabinogalactan-proteins.[1] Arabinoxylans are partly soluble in water and it has been shown that the average molecular weight of soluble arabinoxylans in rye flour is about 2-3 times greater than that of soluble arabinoxylans isolated from wheat flour.[2]

The influence of pentosans on the milling behaviour of grain,[3] the quality of flour,[4] rheological properties,[5] bread volume and staling[6] have been reported but little is known about the molecular mechanisms of these influences.

This paper will deal with the structure of pentosans in wheat and rye, including methods for their determination and effects of baking on pentosan structure and solubility.

STRUCTURE OF WATER-SOLUBLE PENTOSANS

Soluble arabinoxylans in the cell-walls of starchy endosperm in rye and wheat are composed of 4-, 3,4- and 2,3,4-linked β-D-xylopyranosyl residues together with terminal α-L-arabinofuransoyl residues (Figure 1).[7-9] The major water-soluble arabinoxylan in rye grain contains arabinose and xylose residues in a ratio of about 1:2

together with only traces of other components.[8] Methylation and NMR analysis revealed that this polysaccharide (arabinoxylan I) consists of a main chain of 4-linked β-D-xylopyranosyl residues of which about 50 % are substituted at position 3 by terminal α-L-arabinofuranosyl residues. Small amounts of xylose residues are substituted at both position 2 and 3 also by terminal α-L-arabinofuranosyl residues. However, it is not known if these double-branched residues are a part of arabinoxylan I.

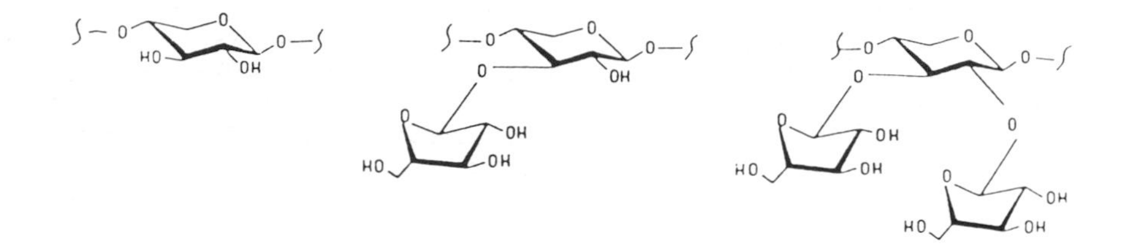

Figure 1 Structural units in soluble rye and wheat arabinoxylans

Xylopyranosyl residues branched at position 3 are stable to periodate oxidation. A sequential oxidation and reduction procedure was used for complete oxidation of the arabinose and unbranched xylose residues in arabinoxylan I.[10] Quantitative analysis of products formed after mild acid hydrolysis revealed the presence of glycerol-xylosides with one, two or three xylose residues in the molar ratio of 1.0; 0.86; 0.02. The xylose residues in these oligosaccharides originated from the branched residues in the main chain of the polysaccharide. It was concluded that these branched residues are predominantly present as isolated units or small blocks of two residues in arabinoxylan I.

Arabinoxylan I

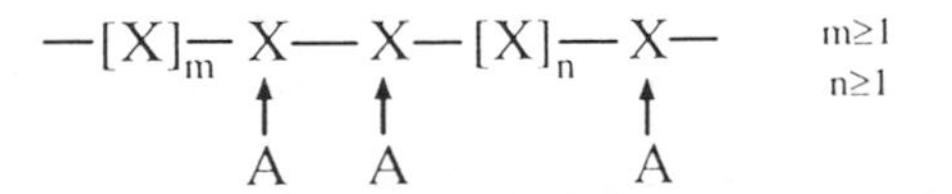

Figure 2 Structural features of arabinoxylan I, the major water-soluble polysaccharide in rye grain

A water-soluble arabinoxylan fraction from rye grain with a structure similar to arabinoxylan I but containing about 4 % of double-branched residues, was hydrolysed with different amounts of a semi-purified xylanase from Trichoderma reesei.[11] The double-branched xylose residues became enriched in the residual polymeric fraction. After extensive hydrolysis with the enzyme, the residual polymeric fraction contained a main chain of 4-linked xylose residues in which about 70 % of the residues were substituted at both 0-2 and 0-3 with terminal arabinose units. It was evident that the semi-purified enzyme contained high xylanase and arabinosidase activities. These activities rapidly degraded un- and monosubstituted xylose residues in arabinoxylan I while the double-substituted xylose residues were very resistant. The results show that a new polysaccharide structure, arabinoxylan II, is present in rye grain as a separate polymer or as a region in a complex macromolecule. Since the double-branched xylose residues were slowly degraded by the enzyme it is evident that arabinoxylan II contains longer regions with double-branched xylose residues.

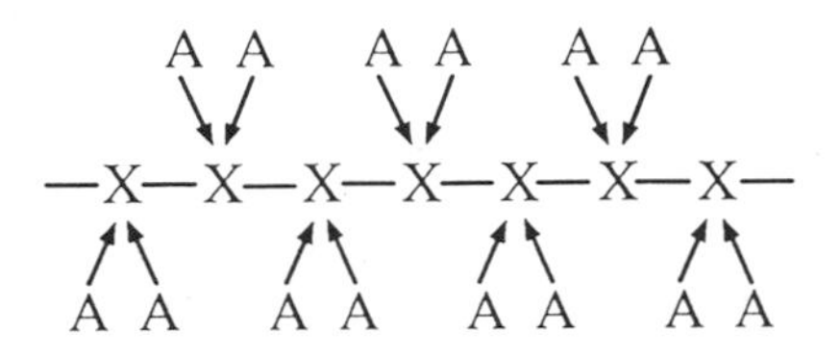

Figure 3 Structural features of arabinoxylan II, a minor water-soluble polysaccharide in rye grain.

Water extracts of wheat flour contain an arabino-galactan-protein with a hydroxyproline-rich polypeptide backbone.[7] This polymer has been reported to contain branched β-galactans substituted with arabinofuranosyl substitutents. Our preliminary results indicate that all three water-soluble pentosan structures are present in both rye and wheat grain, although the contents differ significantly.

EFFECTS OF BAKING ON POLYSACCHARIDES IN WHITE BREAD FRACTIONS

Bread baked with white flour at 210°C for 22 or 35 min was divided into outer crust, inner crust and crumb.[12] The content and composition of water-soluble and insoluble dietary fibre polysaccharides were determined in these fractions and the dough by using a gas-liquid chromatographic method. It was shown that the content of pentosans (arabinose, xylose and galactose residues) decreased during baking, probably due to thermal degradation of sugar residues or losses of low-molecular weight fragments. This trend was particularly noticeable in the outer crust fractions. The content of Na-acetate buffer insoluble arabinoxylans also decreased significantly during baking. The decrease was evidently related to the extent of heat-treatment as it was largest for fractions of outer crust and also became more apparent with the use of longer baking time. It can not be taken for granted, however, that the noted decrease in arabinose residues arises solely from effects on arabinoxylans, since arabinogalactan-proteins also are present in wheat flour. An increasing content of buffer-extractable arabinoxylans in the order crumb to outer crust showed that a redistribution from insoluble to extractable arabinoxylans had occurred. The amount of extractable galactose residues decreased substantially in the order crumb to crust and was also lower in bread fractions than in dough. This finding suggested that a soluble polymer containing galactose residues (probably the arabinogalactan-proteins) was rendered progressively insoluble as a result of more severe thermal treatment during baking. This was also confirmed by a higher level of insoluble galactose residues in the bread fractions compared with dough. It is possible that denaturation of the protein part of arabinogalactan-proteins is responsible for these effects.

Polysaccharides in the white bread fractions were extracted with water and fractionated on DEAE-cellulose.[9] Fractions isolated were characterized by sugar and NMR analysis, revealing the presence of arabinoxylan mixtures with widely different degrees of arabinose substitution. Arabinogalactan-proteins isolated, however, showed little variation in galctose/arabinose ratio (1.5-1.8). Water-soluble arabinoxylans had a higher degree of arabinose substitution in bread fractions than in flour and dough, probably due to increased solubilization of highly substituted arabinoxylans during baking, assuming that originally insoluble arabinoxylans have a more highly substituted structure than their soluble counterparts. It is noteworthy that arabinoxylans of outer crust need stronger borate concentrations for elution, possibly reflecting some modifications in the polysaccharide structure caused by the more severe heat-treatment of this fraction during baking.

From these studies it is evident that the structure and solubility of pentosans are affected by baking. These polysaccharide modifications may have nutritional implications and significant effects on bread properties.

REFERENCES

1. D. Pettersson and P. Åman, Acta Agric. Scand., 1987, 37, 20.
2. F. Meuser and P. Suckow, Spec. Publ. - R. Soc. Chem., 1986, 56, 42.
3. H.J. Moss and N.L. Stenvert, Aust. J. Agric. Res., 1971, 22, 547.
4. R. Tao and Y. Pomeranz, J. Food Sci., 1967, 32, 162.
5. R. Mod, F. Normand, R. Ory and E. Conkerton, J. Food Sci., 1981, 46, 571.
6. S.K. Kim and B.L. D'Applonia, Cereal Chem., 1977, 54, 225.
7. G.B. Fincher and B.A. Stone, Advances in Cereal Science and Technology, 1986, 8, 207.
8. S. Bengtsson and P. Åman, Carbohydr. Polym., 1990, 12, 267.
9. E. Westerlund, R. Andersson, P. Åman and O. Theander, J. Cereal Sci., 1990, in press.
10. P. Åman and S. Bengtsson, Carbohydr. Polym., 1990, Submitted.
11. S. Bengtsson and P. Åman, Carbohydr. Polym., 1990, Submitted.
12. E. Westerlund, O. Theander, R. Andersson and P. Åman, J. Cereal Sci., 1990, 10, 149.

THE EFFECTS OF EXTRUSION COOKING ON THE PHYSICAL AND CHEMICAL PROPERTIES OF DIETARY FIBRE IN SNACKFOODS

P. Dysseler, C. Krebs, C. Van Cappel and D. Hoffem

CERIA-ULB
Food Science and Technology Department IIF-IMC
av. Emile Gryzon 1
B - 1070 Brussels

1. INTRODUCTION

The importance of the intake of dietary fibre in relation to health and disease is sustained by scientific and clinical evidence for a number of physiological effects[1]. This has brought about the interest shown by the food manufacturing industries in the nutritional aspects of dietary fibre.

Today, a number of food manufacturing companies produce DF enriched snackfoods. These snackfoods are obtained by extrusion, i.e. HTST[2] (high temperature short time) processing under high pressure evolving important shear forces.
The impact of high temperature combined to high pressure, which prevails during extrusion, has already been studied for maize starch, the essential component of snackfoods[3]. In the present investigation the impact of these same parameters is examined for DF enriched snackfoods. It is indeed logical to assume that possible structural and/or chemical alteration of DF resulting from the extrusion process might affect the functional properties of the DF and the importance to assess such alterations is self evident.

The evaluation of possible modifications of the dietary fibre is carried out by measuring some physico-chemical properties such as the water and oil absorption capacities as well as the water activity of the samples which are also examined by SEM(scanning electron microscopy). Furthermore, the chemical composition of the DF is also assessed in view to ascertain whether the extrusion process is likely to produce any analytical modifications of the DF conponents.

2. MATERIAL AND METHODS OF PROCESSING AND ANALYSIS

Material

The samples are prepared from pre-gelled maize starch (Cerestar, Belgium) added with fibre-rich products such as hulls from maize (Cerestar, Belgium) or from pea (DDS, Denmark), or sugarbeet fibre (British Sugar, UK). The blends are prepared from 80 % or 90 % maize starch and 20 % or 10 % fibre-rich material.

Prior to extrusion all samples are powders and these are analyzed as such. The snackfood obtained by the extrusion process is suitably prepared for analysis (see hereunder).

Extrusion

Extrusion is carried out in a twin screw extruder CLEXTRAL bivis BC 45, operating under temperatures of 120°, 160°or 200°C with moisture contents of 14 % or 24 %. The rotating speed is 152 RPM (this represents optimum conditions of expansion and viscosity for samples made from maize starch only[3]). The nozzle diameter is 8 mm. The resulting pressures lay between 15 and 22 bars. It should be pointed out that besides the high temperature and high pressure, the high shear forces prevailing during the extrusion process are also likely to contribute to possible modifications of the polymer network of the DF.

Physico-chemical measurements

The absorption capacities of oil and of water by the product are assessed by a centrifuge technique, wherein the water absorption is measured according to the AACC (American Association of Cereal Chemists) method N°88-04, and the groundnut oil absorption is measured according to the method by Caprez et al.[4]. All the samples are ground to less than 0,5 mm prior to analysis.

For SEM the samples prior to extrusion are examined as such. The extruded cylindrical shaped snackfood samples are cut transversally to offer the inner surface to the SEM. Each sample is glued on an aluminium stub with TEMPFIX resin and then sputtered with 20 nm gold-palladium (30 mA, dist. 30 nm for 120 s.).

SEM is carried out with a JEOL 840 A, operating at 15 kW or 20 kW and 10^{-11} to 6.10^{-11} amperes.

Chemical analysis

Total water soluble and water insoluble fibre content is measured according to the AOAC (Association of Official Analytical Chemists) method of Prosky et al.[5].

RESULTS AND DISCUSSION

The following summarizes the results obtained and highlights the most significant features.

Oil absorption

The extrusion process increases the oil absorption capacity (x 2) for each kind of sample (i.e. 10 % or 20 % fibre content, and extruded with 14 % and 24 % moisture contents at 160°C and 200°C).

The experimental date reveal that differences in extrusion temperatures do not affect the increase of the oil absorption capacities for the extruded samples. However, the oil absorption capacities are significantly higher for samples extruded at lower moisture contents (14 % INSTEAD OF 24 %). Fig. 1 shows the results obtained for extrusion with 14 % moisture.

Figure 1 Oil absorption capacities of samples before and after extrusion. Extrusion conditions : 14 % moisture content at 160°C

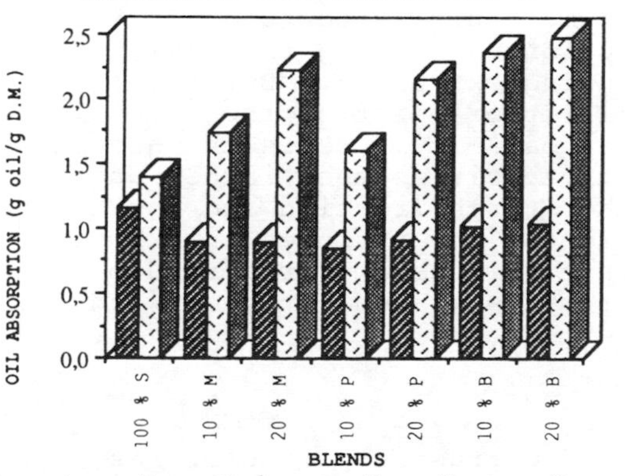

S : Starch ; M : Maize ; P : Pea ; B : Sugarbeet

WATER ABSORPTION

The assessment of the water absorption capacity is carried out on duplicate samples from the same extrusion tests as those used for the oil absorption assessments.
The results show an increase of the water absorption capacity (x 2 to x 6) which is of about the same magnitude as the increases recorded for the oil absorption capacities (see Fig. 2).

Figure 2 Water absorption capacities of samples before and after extrusion. Extrusion conditions : 24 % moisture content at 160°C.

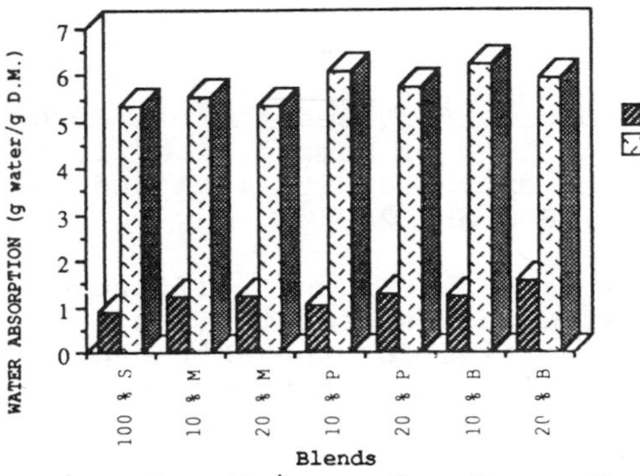

S : Starch ; M : Maize ; P : Pea ; B : Sugarbeet

As for the results of the oil absorption capacity, the extrusion temperature does not influence the water absorption capacity either. On the other hand, however, for all the samples moisture contents of 24 % lead to higher water absorption capacities than moisture contents of 14 % during extrusion. But the fibre content itself has little impact on the results of the water absorption capacity. Hence the alterations in water absorption capacities seem rather due to modifications of the starch fraction and not to those of the fibre fraction of the extruded snackfood.
It is noticeable that Chinnaswamy[3] recorded maximum expansion ratio for maize starch extruded with 13.4 % moisture content.

Hygroscopicity

Only one type of extrusion parameter has been tested : 14 % moisture content at 200°C. As shown in

Fig. 3, the extrusion process lowers the hygroscopicity at 0.75 a_w for all the blends (10 % or 20 % fibre and 90 % or 80 % starch content) with each type of fibre (maize, pea or sugarbeet).

Figure 3 % Moisture content at 0.75 a_w for samples before and after extrusion (a_w : water activity)

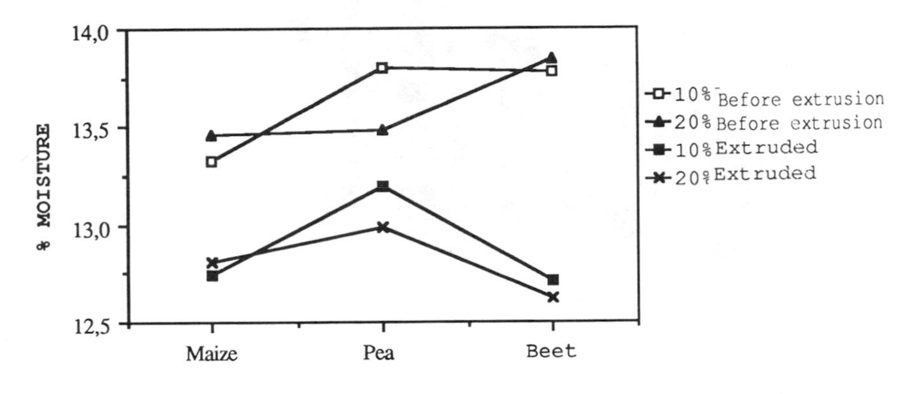

SEM

Under conditions applied SEM has the advantage of great magnification but cannot differentiate between the different fibre components. Thus it is not possible to identify visually cellulose, pectins, etc..., though SEM has been applied not only to the snackfood blends before and after extrusion but also to the fibre residues obtained in the analytical fibre assessment according to Prosky et al. Also, in spite of substantial magnification (x 15.000) SEM reveals no significant alteration of the DF network.

Chemical analysis

As shown in Fig. 4 a loss of 10 % of total DF content is recorded for all the samples extruded with 14 % moisture content at 200°C. Furthermore, as summarized in fig. 5, from the analysis of water soluble and water insoluble DF content, it appears that the ratio soluble/insoluble fibre content changes after extrusion. This seems due more specially to modifications of the insoluble fibre content. It should be kept in mind also that for very small amounts of soluble dietary fibre contents Prosky's analytical method shows poor reproducibility and consequently the differences of soluble DF recorded before and after extrusion are not significant.

Figure 4 Total dietary fiber content (%) for samples before and after extrusion. M : maize, P : Pea

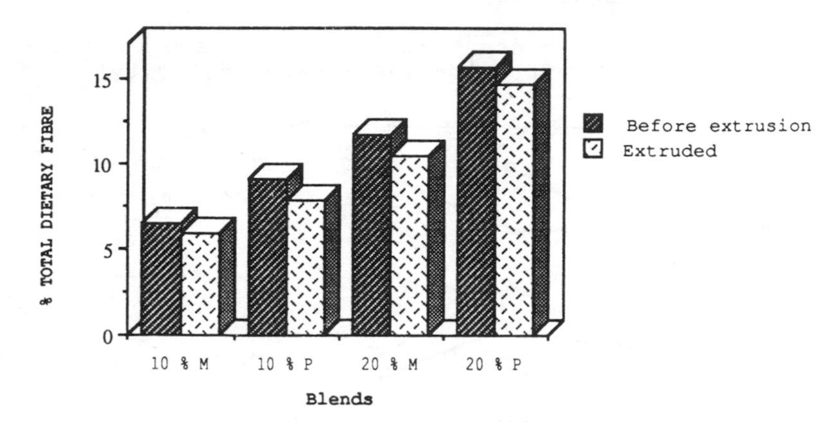

Figure 5 % water soluble and water insoluble dietary fibre contents before (first value) and after (second value) extrusion

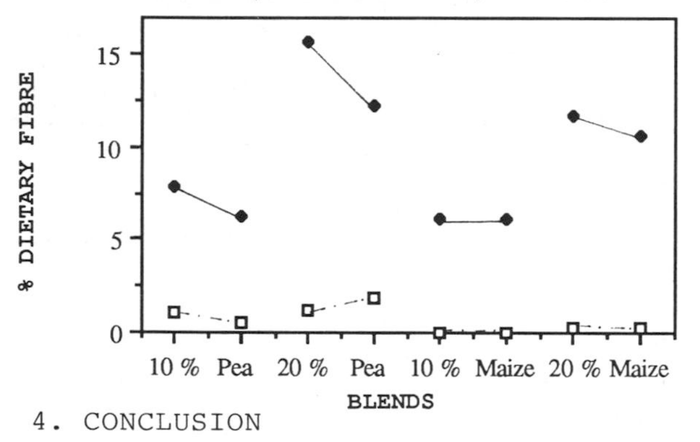

4. CONCLUSION

DF enriched maize starch based snackfoods produced by extrusion process loose 10 % of their total DF content. Water and oil absorption capacities are only slightly modified by the extrusion process.

5. REFERENCES

1. Comité Consultatif d'Experts sur les Fibres Alimentaires, 'Santé et Bien-être social', Canada, 1986, 54 p.
2. C. O'Connor. 'Extrusion technology for the food industry'. Elsevier Applied Science, London and New-York, 1986.
3. R. Chinnaswamy and M. A. Hanna, J. Food Sci., 1988, 53, (3), 834.
4. A. Caprez, E. Arrigoni, R. Amado and H. Neukom, J. Food Sci., 1986, 4, 233.
5. L. Prosky, N. G. Asp, I. Furda, J. W. Devries, T. F. Schweizer and B. F. Harland, J. Assoc. Off. Anal. Chem., 1988, 71, (5), 1017.

CHEMICAL AND PHYSICO-CHEMICAL PROPERTIES OF FIBRES FROM ALGAL EXTRACTION BY-PRODUCTS

M. Lahaye and J.-F. Thibault
INRA, Laboratoire de Biochimie et Technologie des Glucides,
BP 527, Nantes 44026 CEDEX 03 France

INTRODUCTION

Certain species of marine algae are sources for the industrial production of gelling and/or viscosifying polysaccharides used as food additives (alginates, carrageenans, agars). By-products from these extractions are not or poorly valorized and in a project aiming at differentiating the sources and types of dietary fibres such materials were investigated. The present study focus on the chemical and physico-chemical characterization of by-products from *Eucheuma cottonii* used as raw material for the production of the gelling polysaccharide kappa-carrageenan[1] (Fig.1).

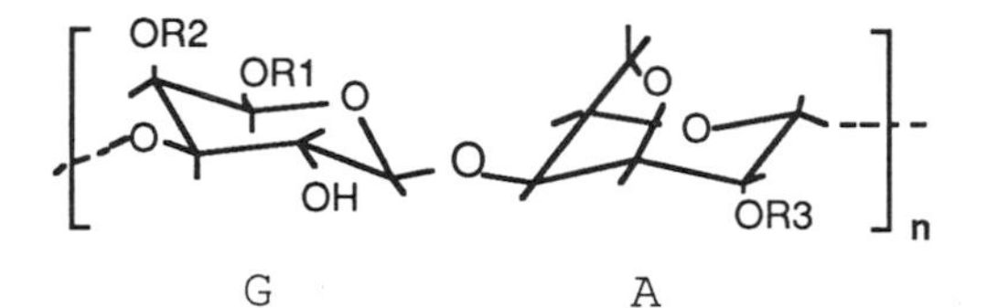

G: ß-D-galactopyranose; A: 3,6-anhydro-α-L-galactopyranose
R1 = H, R2 = OSO3, R3 = H: kappa-carrageenan
R1 = H, R2 = OSO3, R3 = OSO3: iota-carrageenan
R1 = H or CH3, R2 = H or OSO3, R3 = H or OSO3: 'deviant' carrageenan

Figure 1. Chemical structures of carrageenans

MATERIALS AND METHODS

Eucheuma cottonii industrial residues were kindly provided as dried pieces from Copenhagen Pectin Fabrik. These pieces were ground to particle size ≤ 0.5 mm and washed extensively with 80% ethanol (sample EC), or washed with 80% ethanol acidified with HCl (0.01N) followed by 60% ethanol containing NaCl (0.1M), 60%

ethanol and 80% ethanol (sample ECNa). Preparation of the calcium form (ECCa) was obtained by suspending ECNa in 60% ethanol containing $CaCl_2$ (1.0M) followed by 60% and 80% ethanol washes. These preparations were dried by solvent exchange and over P_2O_5 at 40°C *in vacuo*. All results are expressed on a dry weight basis.

Dietary fibre yields are obtained gravimetrically[2]. Soluble fibres were precipitated in 4 vol. ethanol from the hot buffer used for enzymatic digestions and after subsequent re-extractions with water at 90-95°C. Hot-water insoluble residues represent the insoluble fibres.

Chemical determinations. Neutral sugar contents were determined by gas liquid chromatography of their alditol acetates obtained after sulfuric acid hydrolysis[3]. Anhydrogalactose (36AG) and sulfate (SO_4) were determined by colorimetry[4] and turbidimetry[5], respectively. Ashes were measured gravimetrically after mineralization. Protein concentrations were determined by the Kejdhal method (N x 6.25).

Physico-chemical determinations. Infrared transmission spectra were recorded on films prepared as described[6] using a Fourier transform Nicolet 10-DX.

Water-absorption was determined using a Bauman apparatus consisting of a fritted glass covering a thermostated (25°C) solvent reservoir to which a graduated pipette is connected. The volume of solvent absorbed by 50-100 mg sample on the fritted glass is read on the pipette after 30 minutes.

Water-retention is the moisture content (1h at 120°C) of the fibre pellet recovered after overnight agitation in the solvent at room temperature, 1h centrifugation (14 000g), and removal of excess solvent over fritted glass G1 for 2 h.

Water-swelling was determined after 16 or 48 hours incubation at room temperature of samples in solvent using a graduated cylinder.

RESULTS AND DISCUSSION

Eucheuma cottonii extraction by-products (EC) are rich in soluble (57%, ECsol) and insoluble fibres (32%, ECins), or 41.5% soluble and 29.2% insoluble dietary fibres (after subtracting ash and protein contents).

Chemical characteristics of the polysaccharides found in the soluble and insoluble fractions are shown in Table 1. The water soluble polysaccharides are rich in anhydrogalactose, galactose and sulfate, whereas galactose and glucose are the major sugars of the insoluble fractions. Anhydrogalactose content in the latter fraction could not be determined by colorimetry due to the presence of interfering sugars. From these results a carrageenan structure for the soluble polysaccharides and a mixed cellulose-galactans population in the insoluble fibres are proposed. The infrared spectrum of the soluble fibres allowed for the location of the sulfate and confirmed the carrageenan structure of these polysaccharides[6] (Fig. 2). From the sulfate content and the lower absorbance observed at 820 cm^{-1} as compared to iota-carrageenan (theoretical 37.3% sulfate), it can be concluded that this fraction consists either of a mixed population of iota- and kappa-carrageenan and/or 'deviant' (hybrid) iota-carrageenan molecules.

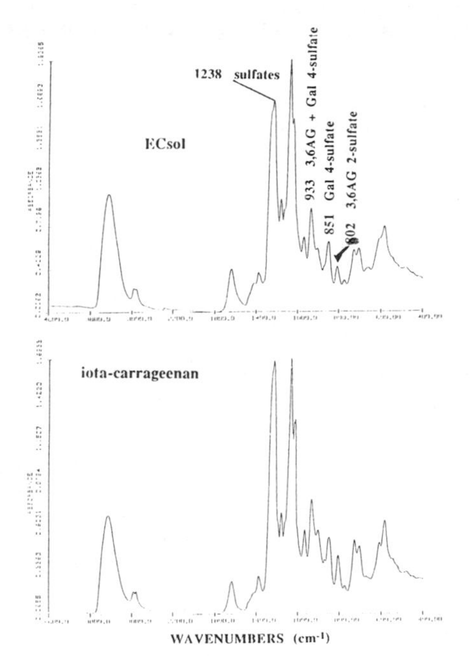

Figure 2. Infrared spectrum of ECsol and iota-carrageenan

The physico-chemical properties of *E. cottonii* fibres reflect the ionic nature and the high proportion of the water-soluble fibres. Kappa- and iota-carrageenan are known for their selectivity for potassium and calcium, respectively, in their physico-chemical and rheological characteristics. All sodium forms of carrageenan are soluble in cold water whereas their calcium form will swell as a function of the type and concentration of cations present in the solution and temperature[7]. Being a mixed population and/or hybrids of both carrageenan types, the soluble fibres demonstrated differences in physico-chemical properties in relation with the salt solution and their ionic form.

Water-absorption values for EC, ECsol, and ECins with regard to the type of ions present in the solution are shown in Table 2. Although absorption obtained with water was higher in cases of EC and ECsol, reproducible results were only obtained with salt solutions. Water-absorption is higher with NaCl than with KCl and is very close for the EC and ECsol fractions. For these latter samples, water-absorption is related to the ionic strength of the solution (Fig. 3). No such marked variation is observed for ECins nor for commercial iota- and kappa-carrageenan Other chemical and/or physical particularities of EC and ECsol have therefore to be taken into account to explain this phenomenon.

The ionic form of fibres affected water-absorption, water-retention, and water-swelling values (Table 3). Thus, fibres in sodium form absorb more water than those in calcium form at both ionic strength and for the three salt solutions investigated. At low ionic strength, absorption of both forms of fibres increases in the order K > Ca > Na, whereas at high ionic strength only the calcium form show a small increase in absorption in the order Ca > K > Na.

Water-retention values are very low and relate most probably to the insoluble fibre fraction as most of the soluble fibres were expected to be dissolved in the solvents. For this reason, the values are very close for both ECCa and ECNa and do not show marked variations between salt solutions. They show slightly higher retention at low ionic strength than at I = 0.5.

Water-swelling is higher for both fibre forms at low ionic strength than at I = 0.5 and values are close for both forms of fibres. Measurements were complicated for fibres in sodium form by the soluble polysaccharides that formed viscous solutions retarding the settling of the insoluble fibres. Water-swelling values increase in the order Ca > K > Na for fibres in calcium form at both ionic strength. Fibres in sodium form swell less in calcium solutions as compared to sodium or potassium solutions.

Table 1. Composition of fibre fractions from *E.cottonii* by-products (% weight)

	Rha	Fuc	Ara	Xyl	MeGal[1]	Man	Gal	Glc	36AG[2]	SO_4[3]
ECsol				2.0	0.4	0.3	30.6	1.7	38.3	27.5
ECins	0.3	0.1	0.6	2.1		3.3	10.9	55.2		6.1

[1] 6-O-methyl galactose, [2] 3,6-anhydrogalactose, [3] sulfate

Table 2. Water-absorption of fibre fractions from *E. cottonii* by-products (g / g)

	H_2O	NaCl (100mM)	KCl (100mM)
EC	20.0	17.2	14.1
ECsol	24.8	18.5	14.1
ECins	8.1	10.7	6.5

Table 3. Physico-chemical characteristics of fibres in the sodium and calcium form of *E. cottonii* by-products

FORM		SOLUTION					
		NaCl		KCl		$CaCl_2$	
	Ionic strength	0.01	0.5	0.01	0.5	0.01	0.5
Na	Absorption (g/g)	17.8	9.4	13.4	9.9	18.3	9.5
	Retention (g/g)	2.0	1.3	2.1	1.1	2.3	1.0
	Swelling (ml/g)	22.5	19.1	24.4	17.3	18.8	15.2
Ca	Absorption (g/g)	15.2	9.2	8.4	8.5	12.9	7.3
	Retention (g/g)	2.0	1.3	1.8	0.9	1.8	0.8
	Swelling (ml/g)	27.4	18.2	25.5	15.3	21.2	13.7

CONCLUSION

Extraction by-products of *E.cottonii* are potentially useful as they are specially rich in soluble dietary fibres. Being essentially composed of ionic polysaccharides, the physico-chemical properties of these fibres can be tailored by changing their ionic form and ionic environment.

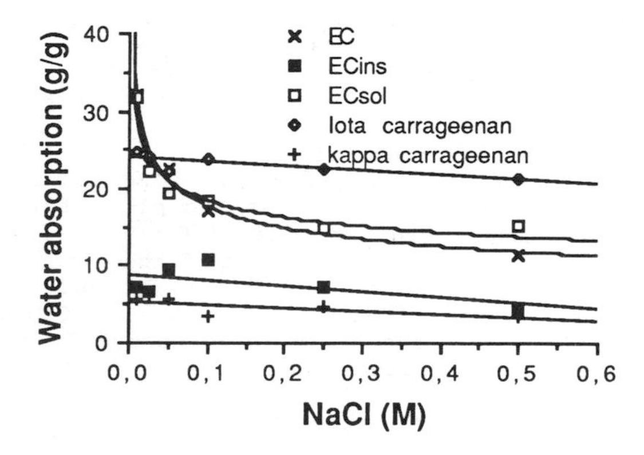

Figure 3 Effect of ionic strength on water-absorption

[1] M. Glicksman, Hydrobiologia, 1987, 151/152, 31.

[2] L. Prosky, N.G. Asp, T.F. Schweizer, J. DeVries, I. Furda, J. Assoc. Off., Anal. Chem., 1988, 71, 1017.

[3] A.B. Blakeney, P.J. Harris, R.J. Henry, B.A. Stone, Carbohydr. Res., 1983, 113, 291.

[4] W. Yaphe, G.P. Arsenault, Anal. Biochem., 1965, 13, 143.

[5] J.S. Craigie, Z.C. Wen, J.P. Van der Meer, Bot. Mar., 1984, 27, 55.

[6] C. Rochas, M. Lahaye, W. Yaphe, Bot. Mar., 1986, 29, 335.

[7] Marine Colloids Division, 'Carrageenan', FMC Corporation, Marine Colloids Division, Springfield, N.J. USA, 1977, Monograph number one, p.14.

THE CONTRIBUTION OF MICROBIAL NON-STARCH POLYSACCHARIDES (NSP) TO THE TOTAL NSP CONTENT OF FAECES

A. C. Longland and A. G. Low

AFRC Institute for Grassland and Animal Production
Shinfield, Reading RG2 9AQ, Berkshire

1 INTRODUCTION

There is increasing interest in the use of alternatives to cereals as energy sources for pigs, because cereals are relatively expensive and can be consumed directly by man. There are, by contrast, many foods which are unpalatable for humans but which are readily eaten by pigs. These are often by-products of the food or feed industries and often have a high content of non-starch polysaccharides (NSP), which are fermented to varying degrees by the gut microflora. Thus, measurements of the digestibility of the NSP fraction are of importance when evaluating their potential as energy sources. Digestibility values are usually obtained by determining the difference between NSP intake in the feed and NSP output in the faeces. However, in addition to undigested diet residues, faeces contain gut microbes and endogenous secretions, and therefore NSP present in these fractions of non-dietary origin will contribute to the total faecal NSP content. This will result in an under-estimate of the digestibility of the NSP in the diet, the extent to which will clearly depend on the amount of NSP of non-dietary origin in the faeces. Such digestibility measurements are therefore approximate and are referred to as the 'apparent' digestibility of a diet. However, correction for the contribution made by non-dietary components will give a more accurate 'true' digestibility value.

It is known that microbial components in the faeces of pigs fed conventional diets may account for up to 75% of faecal dry matter (DM). As faecal microbes contain some of the same NSP constituents as those in diets, it is important to estimate the contribution of microbial NSP to the faecal total, because NSP from such a major non-dietary fraction of the faeces could cause substantial under-estimates of the digestibility of NSP in a feedstuff.

The aim of this study was to determine both the types and amounts of NSP in cultures of faecal microbes, so that the extent of their contribution to apparent digestibility measurements could be assessed. This will be discussed in conjunction with an estimate of the contribution of NSPs present in endogenous material to faecal NSP.

2 MATERIALS AND METHODS

A mixed microbial population was obtained from a 36 kg Large White x Landrace boar pig, which had been fed a diet containing 32% cereals, 45% molassed sugar beet feed, 19% soya bean meal, 2.5% fishmeal and a vitamin and mineral premix.

One gramme of fresh faeces was introduced into a 250 ml screw-capped bottle, containing 100 ml of yeast-dextrose broth (YDB). After a 24 h incubation at 37°C without agitation, 1 ml of the culture fluid, free from non-microbial debris, was sub-cultured into each of 10, 250 ml bottles containing YDB and incubated for a further 24 h. The resultant cultures were centrifuged at 3,000 x g for 20 min., and the supernatants discarded. The pellets were then resuspended in 100 ml saline and centrifuged as before. This washing procedure was repeated twice, and the washed pellets were pooled and lyophilised.

The Englyst and Cummings[1] method was used to measure the constituent NSP monomers in feeds and faeces. This required boiling of samples in dimethylsulphoxide, followed by enzymic de-starching and washing in 85% ethanol. The NSP residue was then hydrolysed, and alditol acetates were prepared from the hydrolysates for analysis of neutral sugars by gas liquid chromatography. A colorimetric technique was used to quantify the uronic acid content of hydrolysates. These methods were also used to determine the NSP content of the lyophilized microbial material.

3 RESULTS

NSP accounted for 4.33% microbial dry matter. The constituent monomers were, in descending order of importance, glucose, galactose, rhamnose and uronic acids (Table 1).

4 DISCUSSION

Dietary NSP is largely derived from the structural components of plant cell walls. Although plant cell walls differ in their composition depending upon their botanical source, those used in pig diets typically contain little rhamnose, some mannose and galactose, large amounts of glucose, and intermediate levels of arabinose or xylose. Quantities of uronic acids, however, vary greatly depending on their botanical origin.[2,3,4]

In this study, faecal microbes contained low levels of rhamnose, galactose, glucose and uronic acids. Theoretically, therefore, these individual NSP components of microbial origin could result in slight under-estimation of the digestibility values for their dietary counterparts. However, dry pig faeces frequently contain in excess of 35% NSP, and even if faecal microbes represented all of the remaining faecal DM, the maximum possible contribution of microbial NSP to the faecal total would be very small (<2.9%, i.e. 1.6%, 0.6%, 0.4% and 0.19% of DM in the form of glucose, galactose, rhamnose and uronic acids respectively).

A further potential source of non-dietary NSP is the glycoprotein in endogenous secretions. However, the only NSP constituent found in mucus from the mouth, stomach and colon of the pig was a small amount of galactose.[5] As galactose is usually well digested by pigs from a variety of diets[2,3,6] any effect of endogenous galactose on overall galactose digestibility is likely to be small.

Values obtained for the excretion of non-dietary non-cellulose NSP by pigs fed diets with cellulose as the sole NSP source are given in Table 2 (Longland et al., unpublished). These values are considerably less than the low, maximum theoretical values for microbial NSP quoted above. In either case, such small amounts of non-dietary NSP are unlikely to have much effect on NSP digestibility measurements. Therefore the 'apparent' digestibilities obtained for NSP monomers by pigs can probably be regarded as near to 'true' digestibility values. This is certainly the case for the pentosans arabinose and xylose which have not been detected in either faecal microbes or endogenous secretions.

In the light of the above results, it would seem

likely that the only occasion when a non-dietary NSP sugar monomer could cause a major under-estimate of its apparent digestibility would be when it was a very minor component of the diet.

Table 1 The NSP content of the lyophilised faecal microbes (% of dry matter)

Rhamnose	Arabinose	Xylose	Mannose	Galactose	Glucose	Uronic Acids	Total NSP
0.71	0	0	0	0.96	2.36	0.30	4.33

Table 2 NSP constituents of non-dietary origin in the faeces of pigs fed a semi-purified diet with cellulose* as the sole NSP source

		NSP (% faecal DM)			
Rhamnose	Arabinose	Mannose	Galactose	Uronic Acids	Total NSP
0.02	0	0	0.12	0.20	0.34

* contained glucose and xylose
(Longland et al., unpublished results)

5 REFERENCES

1. H.N. Englyst and J.H. Cummings, Analyst, 1984, 9, 937.
2. H. Graham, K. Hesselman and P. Åman, J. Nutr., 1986, 116, 242.
3. H. Graham and P. Åman, Anim. Feed Sci. Technol., 1987, 17, 33.
4. A.C. Longland and A.G. Low, Anim. Feed Sci. Technol., 1989, 23, 67.
5. A. Allen, Br. Med. Bull., 1978, 34, 28.
6. A.C. Longland and A.G. Low, Proceedings of the Nutrition Society, 1988, 47, 104A.

DISTRIBUTION OF SOLUBLE FIBER IN VARIOUS MILLSTREAMS OF WHEAT

G. S. Ranhotra, J. A. Gelroth and K. Astroth

American Institute of Baking
Manhattan, KS 66502 U.S.A.

E. S. Posner
Kansas State University
Manhattan, KS 66506 U.S.A.

1 INTRODUCTION

Fiber, described today as Total Dietary Fiber (TDF), can be broadly divided into water Insoluble Fiber (IF) and Soluble Fiber (SF). Whereas IF appears to be helpful in various disorders of the intestinal tract, SF has variously been reported[1,2] to lower elevated blood Cholesterol (CH) levels, a risk factor in heart disease.

Cereal-based foods made with white flour, although not high in TDF, do contain a good portion of the fiber as SF. This fiber (source: bread products) has been shown to lower blood CH in animals[3]. A recent study with hypercholesterolemic men has also shown a CH-lowering effect of SF present in refined cereal-based products[4]; cereal-based products tested in this study included white and variety bread, hamburger buns, saltine crackers, biscuits and waffles.

White flour contains about 2.5% TDF; two-fifths or more of this is SF. The study reported here was undertaken to both develop information on the distribution of SF in various millstreams of wheat and to identify a stream(s) that might be a more significant source of SF than white flour itself, and thus, could be used in food products to add more SF to the diet.

2 MATERIALS AND METHODS

A properly tempered hard red winter wheat was milled in the pilot mill of Kansas State Univeristy. The procedure used 19 grinding steps and yielded 23 flour streams and some by-products (bran, shorts, red dog and the resultant millfeed stream). Patent and clear flours are made up of blends from selected flour streams; straight grade flour is the blend of all flour streams.

All millstreams, including wholewheat flour, were analyzed for fiber -- IF and SF[5]. Samples were also analyzed for other components (information not included here).

3 RESULTS AND DISCUSSION

In the milling procedure followed, combining all flour streams yielded 76.4% straight grade flour (66.4% patent flour and 10% clear flour). The germ fraction accounted for 0.5% of the yield. The millfeed stream accounted for the remaining 23.1%.

Although millstreams have been extensively characterized by chemical and physical measurements, this characterization has not included an assessment of the distribution of IF and SF. The results of this study show that bran contained more SF (content, 2.1%) than any other millstream except the red dog fraction, a minor fraction of wheat milling. However, by virtue of its high IF content (41.9%), bran remains primarily a source of IF. Wholewheat flour and the germ fraction each contained about 10% TDF; however, only about one-tenth of this amount is SF. SF in flour streams that make up white flour ranged between 1 and 1.9%; IF in these streams ranged between 0.9 and 4.9%. These ranges would be much narrower, and IF and SF values much closer, if a few minor streams (bran and shorts duster, sixth middling, etc.) were excluded from consideration.

No one flour stream emerged as exceptionally high in SF; the fraction sixth middling is a little higher (SF content, 1.9%), but it is also higher in IF (content, 4.9%). A fraction exceptionally high in SF could be manipulated, through modification in the milling procedure, to increase its yield. In the absence of such a revelation, patent and straight flours remain the flours of choice to be used in products to simultaneously increase our intake of both IF and SF. Patent flour contained 1.3% SF and 1.2% IF while straight grade flour contained 1.1% SF and 1.4% IF. Clear flour contained about the same amount (content, 1.4%) of SF as the other two flours, but it was substantially higher in IF (content, 3.3%).

Products made with 150 g of patent flour, the current level of flour consumption in the U.S.A., would add about 2 g SF (4 g TDF) to our diet. A 50% increase in flour consumption would add another 1 g SF; this will also help increase the level of utilizable complex carbohydrates in the diet to the levels now recommended[6]. Other plant-derived foods can easily add another 4 to 5 g SF to our diet. This way SF would represent about one-third of the recommended intake of 20 to 35 g fiber a day in the U.S. diet. A more dramatic change in food habit (including specially formulated foods, for example) would be required to increase SF intake well above 8 g a day.

In conclusion, no single flour stream from the traditional wheat milling procedure could be identified as being exceptionally high in SF. Studies are in progress, however, to air-classify white flour which may yield a subfraction noticeable higher in SF.

4 REFERENCES

1. J. W. Anderson, Am. J. Gastroenter., 1986, 81, 892.
2. B. P. Kinosian and J. M. Eisenberg. J. Am. Med. Assoc., 1988, 269, 2249.
3. G. Ranhotra, J. Gelroth and H. Bright, J. Food Sci., 1987, 52, 1420.
4. J. W. Anderson, S. Riddell-Lawrence, T. L. Floore and D. W. Dillon. Am. Coll. Nutr. (30th Annual Meeting), 1989, Norfolk, VA.
5. L. Prosky, N-G. Asp, T. F. Schweizer, J. W. DeVries and I. Furda, J. Assoc. Off. Anal. Chem., 1988, 71, 1017.
6. F. J. Cronin and A. M. Shah, Nutr. Today, 1988, 23(6), 26.

INFLUENCE OF AUTOCLAVING ON PHYSICOCHEMICAL PROPERTIES OF THE BEET PULP FIBRES AND ON SEVERAL OF THEIR PHYSIOLOGICAL EFFECTS

M. Champ, F. Guillon and C. Gourgue.

Lab. de Technologie Appliquée à la Nutrition, I.N.R.A., BP. 527.
44026 NANTES CEDEX 03. France.

1 INTRODUCTION

Beet pulp fibres represent an important by-product of the sugar industry which has been used up to now in animal feeding. Their physico-chemical properties suggest their possible utilization as dietary fibres in human nutrition and possibly in drug industry. Technological treatments have been shown to be able to optimize several of the properties of the fibres and thus, may be used to improve their physiological effects. For instance, extrusion-cooking of wheat bran is able to increase its soluble fibre content and its water absorption capacity [1]. This same process has been used to induce physico-chemical transformations in the fibres of sugar-beet pulp, citrus peels and apple marcs [2].
In the present work, the physico-chemical characteristics of these fibres have been compared to those of soluble (apple pectins) or mainly insoluble (wheat bran) fibres. Some physiological effects of these products have been evaluated on rats.

2 MATERIAL AND METHODS

Material: Sugar beet fibres, wheat bran and apple pectin (Rapid set pomme) were respectively purchased from S.R.D. Society (Sucre-Recherche-Développement, Compiègne, France), Breteau-Aubert (LaVarenne, France) and Unipectine (Redon, France).

2 Kg of beet fibres (average particle size: 570 µm) were swollen during 1 hour in a sodium acetate buffer (0.03 M) pH4. Half of the fibres were then autoclaved at 126°C during 1 hour. Both lots of fibres were freeze-dried without elimination of the unabsorbed solution and coarsely grounded using a Braun mixer.

Analytical methods and water holding capacity determination: Analytical methods for crude protein, neutral sugars, uronic acids, volatile fatty acids and chromic oxide were previously described [3]. Hydrogen and methane were analysed as described by Champ et al. [4]. Dietary fibre content was determined gravimetrically after enzymatic removal of starch and proteins [5].

The water holding capacity was determined by the centrifugation method according to the procedure of Mc Connell et al. [6] and results expressed as g of solution (0.15M NaCl) per 100 g of dry residue.

In vivo experiments

Animals: 4 male Wistar rats (mean weight: 320 g) were placed individually in separate metabolic cages. The animals were successively adapted to each experimental diet during 9 days; the different nutritional and physiological parameters were determined at the end of each adaptation period. The animals were allowed free access to water and feed. Their feed intake was controlled and the animals were weighed 4 times during each experimental period. Transit time measurement has been estimated on a 48 hours period. Feed containing 2% chromic oxide was distributed during 2 hours (from 3:00 to 5:00 p.m.); food without chromic oxide was then distributed (*ad libitum*). Faeces were hourly collected during 48 hours and were pooled during the next 20 hours. The faeces were freeze-dried and kept at room temperature until analysed for chromic oxide. The procedure for quantification of excreted gases using a respiration chamber has been described previously [4]. At the end of the experiment, the rats were killed at 11:00 a.m. by intra peritoneal injection of Nembutal, the caecum was quickly collected; its content was controlled for pH, frozen at -20°C and kept for volatile fatty acids analysis.

Diets: A casein-DL-methionine mixture (about 50:1, w/w) was used as protein source, supplying, with the fibre sources, 20 g (N*6.25)/100 g, dwb, in the diets. All diets contained, 4.5 % mineral mixture, 0.5 % vitamin mixture and 5 % maize oil. The composition of the mineral and vitamin mixtures has been given in full previously [3]. Autoclaved (BFa), unautoclaved (BFua) beet pulp fibres, wheat bran (WB) and apple pectins (AP) were respectively incorporated to BFa, BFua, WB and AP diets at the following levels: 9.1, 8.8, 10.2 and 6.8 %. These diets contained respectively 58.6, 58.9, 59.0 and 60.6 % instant mashed potatoes (dwb).

Statistical evaluation: The *in vivo* data were expressed as the mean ± the standard deviation. Student's t-test was used to determine the significance of the differences between groups.

3 RESULTS AND DISCUSSION

Analytical and physicochemical characteristics of the fibre sources

Table 1 Chemical and physical characteristics of the fibre sources

Fibre	BFa	BFua	WB	AP
Soluble fibre[1]	27.4	12.0	n.d.	92.7
Insoluble fibre[1]	57.0	80.3	n.d.	0
Total fibre[1]	84.4	92.4	44.0	92.7
Sugars (%/DM)[2]				
Rha	0.98	0.98	0.00	0.63
Ara	17.51	18.60	6.49	5.41
Xyl	1.30	1.20	12.35	1.01
Man	1.33	1.35	0.43	0.26
Gal	4.38	4.35	0.63	4.53
Glu	20.01	20.15	8.43	16.76
Uronic ac.(%/DM)[3]	20.41	21.60	2.70	59.30
W.H.C.(g. H_2O/g.MS)				
1h	11.15	10.71	4.25	nd
24h	12.71	11.23	4.67	nd

[1] Prosky et al.[5]; [2] Hoebler et al.[7]; [3] Thibault[8].

Dietary fibre content (total, soluble and insoluble) of the different fibres sources is reported in Table 1. The total beet dietary fibre content, was relatively high (92.4-84.4%) within the range reported by other authors [9-13]. A shift from insoluble to soluble fibres was observed when autoclaved. Since the autoclave conditions were severe, some of the dietary fibres were made soluble and degraded; therefore, they did not precipitate in alcohol. That explained the slight decrease in total dietary fibre for the BFa. This phenomenon was also reported by Guillon [13]. She had shown that soluble fibres were mainly constituted of pectic material and arabinans. Apple pectins were soluble fibres, mainly composed of galacturonic acid (59.3%), glucose (16.8%), arabinose (5.4%) and galactose (4.5%) as neutral sugars. Wheat bran contained 44.4% of total dietary fibres; these data are in agreement with the values reported in the literature [1,9]. Xylose, glucose and arabinose were the major sugars of these fibres. The water holding capacity obtained for wheat bran were lower than those of the beet fibres. These values are difficult to compare with data of the literature because they depend on the technique, the experimental conditions and the preparation of the substrates [6,14,15]. However, values obtained on several industrial by-products [16], (vegetables) [17] are higher

(2.6-35.1 g.g-1) than for cereals (1.5-3.0 g.g-1). These differences could be related to the individual properties and structures of the various components of the substrates. Cereals often contained some residual starch in a crystalline form and lignin, an hydrophobic substance, while non cereal products are rich in pectins which are very absorbent. Values for autoclaved and initial fibres were not significantly different and did not vary with the duration of the hydratation. They are lower than those reported by other authors [10,11,13]. Pectins act as a cement between the different cell wall components; their removal should have caused a decrease in the water holding capacity.

Nutritional and physiological properties of the fibre sources (Table 2)

Food consumption and weight gain were slightly lower (NS) for AP diet whereas the best food efficiency is obtained for WB diet. Total transit time is not deeply affected by the source of fibres. However, if the parameter "excretion of 80% of the marker" is considered, the tendancy appeared to be in agreement with the literature [18]: their is a positive correlation between the soluble content of the fibre and the transit time. Faecal excretion was lower with the AP diet which contained highly fermentable pectins; the same observation has been made by Cummings et al.[18]. Slightly higher dry matter of the faeces and caecal content observed in rats fed wheat bran may be explained by a higher excretion of dietary residue and/or by a smaller bacterial mass [20]. Most of the characteristics of the caecal content agreed with the data of the literature [3,21]. Caecal pH was higher for AP group and the VFA profile of this same group revealed a low formation of butyric acid. Both diets containing beet fibres had intermediate VFA profiles between WB and AP diets. Hydrogen and methane excretions determined in a respiration chamber indicated a poor production of methane for the WB group compared to the other groups. Diets with PFa, BFua and AP induced a high excretion of methane which may be due to the methyl groups of their highly methylated pectins.

4 CONCLUSION

Physico-chemical characteristics of beet pulp fibres (with or without autoclaving) have been compared to those of soluble (apple pectins) or mainly insoluble fibres (wheat bran). These fibres have then been incorporated to a semi-synthetic diet to evaluate their effect on food consumption, transit, faecal excretion and fermentability in the large intestine (gases and short-chain fatty acids formation) in the rat.

It appears from this study that beet pulp fibres have physiological effects which are due to their soluble fibre fraction (hydratation of the faeces, flatus), or to their insoluble fibres fraction (faecal excretion). However, these effects are not influenced by an hydrothermic treatment as it has been applied in the present study (autoclaving in a buffered medium, pH 4).

Table 2 Nutritional and physiological effects of the fibres (mean ± SD)

Diet	BFa	BFua	WB	AP
Food consumption[1]				
(g DM/day)	22.0±3.2[a3]	21.6±1.4[a]	21.3±5.0[a]	18.3±4.2[a]
Weight gain[2]				
(g/day)	2.16±0.59[a]	2.06±2.21[a]	2.66±2.52[a]	1.79±3.25[a]
Transit time				
(hr:min)				
maximal conc.				
of the marker	13:15±2:04[a]	14:00±8:21[a]	14:00±8:26[a]	13:53±7:12[a]
excretion of 80%				
of the marker	27:50±5:17[a]	26:22±2:11[a]	24:46±10:29[a]	28:56±1:43[a]
Fecal excretion				
g DM/day	1.99±0.20[a]	1.89±0.18[a]	2.02±0.71[a]	1.41±0.10[a]
% DM	42.6±2.9[a]	42.0±1.1[a]	46.5±3.0[a]	42.0±2.8[a]
Caecal content characteristics[4]				
caecal wght/				
body wght (%)	0.95-0.61[ab]	0.93-0.86[a]	0.71-0.67[b]	0.66-0.13[ab]
% DM	15.2-14.4[a]	12.8-14.5[a]	22.1-21.6[a]	18.1-10.0[a]
pH	6.60-6.36[b]	7.03-6.19[ab]	6.61-6.70[b]	7.32-7.33[a]
VFA				
µmoles/200g rat	260-136[a]	196-263[a]	164-145[a]	117-28[a]
molar ratio (%)				
C2	65.4-75.6[ab]	62.5-64.6[a]	59.0-56.8[b]	65.9-76.9[ab]
C3	18.1-14.8[a]	12.0-22.4[a]	16.8-13.8[a]	23.6-15.7[a]
C4	16.5-9.6[b]	25.5-13.1[ab]	24.2-29.4[a]	10.4-7.4[b]
H_2 and CH_4 excretion				
(cm^3/24hr/g. ing. fibre)				
H_2	4.80±5.47[a]	3.23±0.91[a]	4.57±3.81[a]	5.53±6.79[a]
CH_4	13.78±5.79[a]	18.07±6.04[a]	7.34±2.92[b]	12.09±8.21[ab]
total	18.58±7.13[ab]	21.31±5.15[a]	11.92±5.60[b]	17.63±3.67[ab]

[1] during the last 3 days of the adaptation period; [2] during the 14 days of the adaptation period; [3] means in the same line bearing common superscripts are not significantly different ($p \leq 0.05$);
[4] data obtained from 2 rats

REFERENCES

1. M.C. Ralet, J.F. Thibault, G. Della Valle, J. of Cereal Sci., 1990, in press.
2. J.F. Thibault, G. Della Valle, M.C. Ralet, French Patent n°88 11601.
3. M. Champ, J.L. Barry, C. Hoebler, J. Delort-Laval, Anim. Feed Sci. Technol., 1989, 23, 195.
4. M. Champ, J.L. Barry, C. Bonnet, S. Berot, J. Delort-Laval, Sci. Aliments, 1990, in press.
5. L. Prosky, N.G. Asp, I. Furda, J.W. De Vries, T.F. Schweizer, B.F. Harland, J. Assoc. Off. Anal. Chem., 1984, 67, 1044.
6. A.A. Mc Connell, M.A. Eastwood, W.D. Mitchell, J. Sci.Food Agric., 1974, 25, 1457.
7. C. Hoebler, J.L. Barry, A. David, J. Delort-Laval, J. Agric. Food Chem., 1989, 37, 360.
8. J.F. Thibault, Lebens. Wiss. Technol., 1979, 12, 247.
9. T.F. Schweizer, P. Würsch, J. Sci.Food Agric., 1979, 30, 613
10. C. Bertin, X. Rouau, J.F. Thibault, J. Sci.Food Agric., 1988, 44, 15.
11. F. Michel, J.F. Thibault, J.L. Barry, R. De Baynast, J. Sci.Food Agric., 1988, 42, 77.
12. C. Leclere, C. Cherbut, F. Guillon, M. Champ, Sci. Aliments, 1990, in press.
13. F. Guillon, 6ème Journées Sciences des Aliments de l'Association Française de Nutrition, Nantes, 1989, Abstract.
14. V.F. Rasper, 'Food Carbohydrate', Lineback, D.R. and Inglett, G.E., Eds, AVI, Wesport (Conn.), 1982, p 333.
15. M.A. Eastwood, W.D. Mitchell, 'Fiber in human nutrition', Spiller, G.A. and Amen, R.J.,Eds, Plenum Press, New York, 1976, p. 109.
16. F. W. Sosulski, A.M. Cadden, J. Food Sci., 1982, 47, 1472.
17. W.D. Holloway, R.I. Greig, J. Food Sci., 1984, 49, 1632.
18. K.I. Wrick, J.B. Robertson, P.J. Van Soest, B.A. Lewis, J.M. Rivers, D.A. Roe, L.R. Hackler, J. Nutr., 1983, 113, 1464.
19. J.H. Cummings, W. Branch, D.J.A. Jenkins, D.A.T. Southgate, H. Houston, W.P.T. James, Lancet, 1978, 1, 5.
20. F.W. Sosulski, A.M. Cadden, J. Food Sci., 1982, 47, 1472.
21. C. Remesy, Thesis University of Clermont II, France, 1982, 285 p.

BIOSYNTHESIS OF THE HEMICELLULOSE, GLUCURONOXYLAN, IN PLANT FIBRE

C.T. Brett

Plant Molecular Science Group
Department of Botany
University of Glasgow
Glasgow G12 8QQ

K.W. Waldron

AFRC Institute of Food Research
Norwich Laboratory
Colney Lane
Norwich NR4 7UA

1 INTRODUCTION

Glucuronoxylans are the main hemicelluloses of secondary walls of dicots, and hence are major constituents of fibre derived from these plants. They contain a backbone of β(1-4)-linked xylose residues, to about 10% of which 4-O-methylglucuronic acid residues are linked by α(1-2)-linkages. About half the xylose residues are acetylated on C-2 or C-3, and some arabinose residues may also be present as side chains[1](Fig.1). Similar molecules, with the same backbone but with a higher proportion of arabinose side-chains, are found in the primary walls of dicots and in monocot hemicellulose.

The biosynthesis of this group of polysaccharides is carried out by glycosyltransferases, which are probably located in the Golgi apparatus[2]. Methyl- and acetyl-transferases are also thought to be involved. This paper summarises what is known about the biosynthetic system, and focusses on the functional interactions of the enzymes.

2 MATERIALS AND METHODS

Epicotyls (50g) from 7-day-old etiolated pea (*Pisum sativum*, var. Alaska) seedlings were homogenised with 150ml Tris/HCl (10mM, pH 7.5) at 4°C in a Polytron homogeniser. The homogenate was filtered through muslin and centrifuged at 97,000*g* for 30 min. The pellets were re-suspended in 50mM Tris/HCl, pH 7.5, to form the enzyme preparation.

Incubations were carried out at 25°C with 50µl enzyme

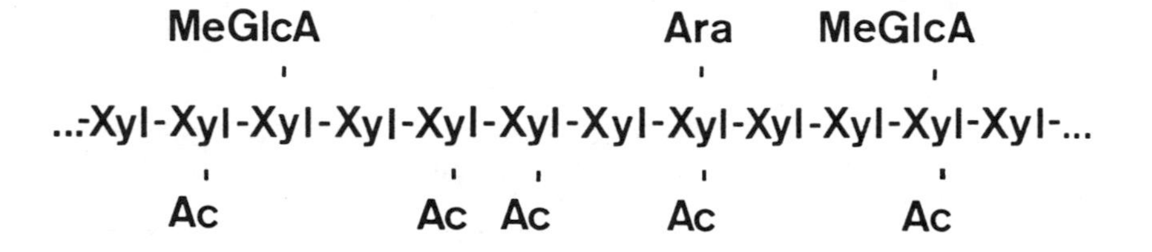

Figure 1: Structure of glucuronoxylan

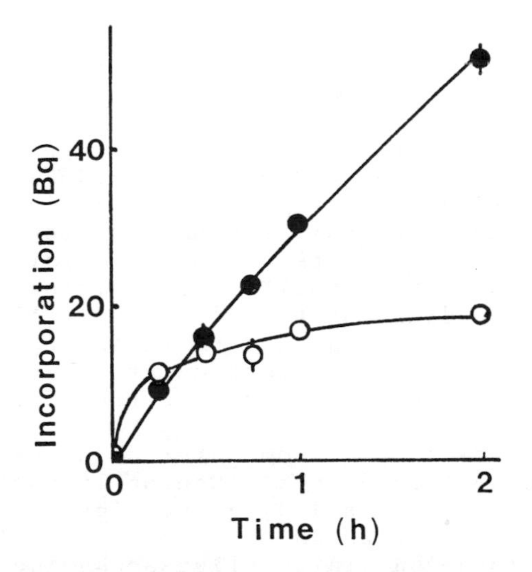

Figure 2: Incorporation of (^{14}C)GlcA into hemicellulose in the presence (●) and absence (○) of UDPXyl.

preparation in a total volume of 100µl. The standard incubation media were as follows. For glucuronyltransferase: UDP-(^{14}C)glucuronic acid (460 Bq, 0.5µM), $MnCl_2$(10mM), ± UDP-xylose (UDPXyl) (1mM). For xylosyltransferase: UDP-(^{14}C)xylose (8.3 kBq, 1mM); $MnCl_2$(10mM). For methyltransferase: S-adenosyl-L-(methyl-^{14}C) methionine (6.5k Bq, 0.3µM); $CoCl_2$(1mM). All incubations were terminated by addition of 70% ethanol. Particulate material was washed twice with 70% ethanol, extracted twice with EDTA-NaH_2PO_4 buffer (50mM, pH 6.8) at 100°C for 15 min., and washed with water. The radioactivity in the insoluble, polymeric material was then estimated by scintillation counting.

3 RESULTS AND DISCUSSION

The xylosyltransferase is found in the particulate fraction sedimenting at 97,000g[1]. Under our assay conditions, 84% of the radioactivity in the polymeric product was present as xylose residues. The xylose-containing product is β (1-4)xylan, as judged by its degradation by β(1-4)xylanase but not by xyloglucanase. The remaining 16% is composed of radioactive arabinose, as a result of epimerisation to UDP-arabinose followed by the action of an arabinosyltransferase. It is not clear whether the arabinose is part of the same polymer as the xylose, or whether it is part of another polymer, such as arabinogalactan.

The glucuronyltransferase is present in the same membrane fraction as the xylosyltransferase. Only very limited incorporation of glucuronic acid into polymeric material occurs when UDP-glucuronic acid (UDPGlcA) is the only sugar nucleotide present. However, in the presence of UDPXyl, sustained incorporation of glucuronic acid can occur for at least eight hours, though the initial rate is lower as a result of competitive inhibition by UDPXyl (Fig.2). This dependance of the glucuronyltransferase on the presence of UDPXyl is thought to be due to the need for xylan to be formed in the particulate preparation, onto which glucuronic acid can then be transferred[3].

To look more closely at the interaction between the two enzymes, the ability of the glucuronyltransferase to transfer GlcA onto pre-formed xylan was investigated. First, it was shown that transfer onto added xylan or short, β(1-4)-xylose-containing oligosaccharides did not occur. Secondly, a preincubation experiment was carried out. The particulate enzyme preparation was preincubated either with non-radioactive UDPXyl (1mM) or with no added sugar nucleotide. It was washed free of added sugar

Table 1 Effect of preincubation of the enzyme preparation with UDPXyl on incorporation of (^{14}C)GlcA from UDP(^{14}C)GlcA into glucuronoxylan.

Preincubation		Incubation with UDP-(^{14}C)GlcA		Incorporation of (^{14}C)GlcA (Bq)
Length	UDPXyl	Length	UDPXyl-	
30 min	+	4h	+	29.1 ± 2.0
"	+	"	−	9.9 ± 0.9
"	−	"	+	32.3 ± 1.2
"	−	"	−	5.1 ± 0.4
2h	+	"	+	22.2 ± 0.1
"	+	"	−	7.4 ± 0.2
"	−	"	+	22.7 ± 0.1
"	−	"	−	8.2 ± 2.0

nucleotides by centrifugation at 97,000*g*, and then incubated with UDP-(^{14}C)GlcA for 4 hours, either with or without non-radioactive UDPXyl(1mM). The results are shown in Table 1; each result is the mean of duplicate incubations.

The incorporation can be seen to be dependent only on the presence or absence of UDPXyl during the incubation in which UDP-(^{14}C)GlcA was present. Neither the presence of UDPXyl in the preincubation, nor the length of the preincubation, had any significant effect.

Thus it appears that GlcA cannot be added to preformed xylan. Instead, the GlcA must be added to the xylan backbone either at the same time as, or shortly after, that part of the backbone is formed. Since xylan is formed in the absence of UDPGlcA[1], it would appear that the proportions of Xyl and GlcA in the polymer are dependent on the relative concentrations of the two substances present at the time of backbone elongation. This implies a tightly regulated supply of substrates to the biosynthetic system *in vivo*, since the ratio of xylose to glucuronic acid is thought to be fairly constant in cell-wall glucuronoxylan, and pure xylan is only found in specialised tissues such as seed endosperms[4].

In contrast to the addition of GlcA, the addition of methyl groups to the 4-hydroxyl of GlcA appears to take place after formation of the polymer. Incorporation of (^{14}C)methyl groups from S-adenosyl-(^{14}C)methionine into glucuronoxylan continues for at least 8 hours in the absence of added sugar nucleotides. If UDPXyl and UDPGlcA are added, the rate of incorporation of (^{14}C)methyl groups is doubled. So the methyltransferase appears to be able to transfer methyl groups onto both pre-formed and newly-formed polymers[5].

The enzymes which add arabinose and acetyl groups to the polymer have yet to be identified. The donors are probably UDP-arabinose and acetyl-CoA.

The glucuronyl- and xylosyltransferases in pea epicotyls appear to be involved in the synthesis of both primary and secondary walls. The glucuronyltransferase is present throughout the length of the 7-day-old epicotyls, including both the youngest, rapidly-expanding region at the top (primary wall synthesis) and the oldest region (where probably only secondary wall synthesis is occurring). Stimulation of the glucuronyltransferase by UDPXyl also occurs throughout the epicotyl, indicating the presence of the xylosyltransferase. Glucuronyltransferase has been found to increase during xylem differentiation in bean tissue culture[6], and xylosyltransferase also increases[7,8].

Our recent work is concerned chiefly with the solubilisation and purification of the xylosyl-, glucuronyl- and methyltransferases. All three enzymes are soluble in non-ionic detergents such as Triton X-100[5,9]. The functional interactions between the enzymes will be investigated further using the solubilised and purified enzymes.

4 REFERENCES

1. E.A-H. Baydoun, K.W. Waldron and C.T. Brett, Biochem. J. 1989, 257, 853.
2. K.W. Waldron and C.T. Brett, Plant Science, 1987, 49, 1.
3. K.W. Waldron and C.T. Brett, Biochem. J., 1983, 213, 115.
4. C.T. Brett and K.W. Waldron, 'Physiology and Biochemistry of Plant Cell Walls', Unwin Hyman, London, 1990.
5. E.A-H. Baydoun, J.A-R. Usta, K.W. Waldron and C.T. Brett, J. Plant Physiol., 1989, 135, 81.
6. C.M. Cumming, K.W. Waldron, C.T. Brett and J.R. Hillman unpublished work.
7. G. Dalessandro and D.H. Northcote, Planta, 1981, 151, 61.
8. P. Bolwell and D.H. Northcote, Planta, 1981, 152, 225.
9. K.W. Waldron, E.A-H. Baydoun and C.T. Brett, Biochem. J., 1989, 264, 643.

PHYSICAL PROPERTIES OF DIETARY FIBRE IN RELATION TO BIOLOGICAL FUNCTION

Edwin R. Morris

Department of Food Research and Technology,
Cranfield Institute of Technology,
Silsoe College, Silsoe,
Bedford MK45 4DT, UK.

1 INTRODUCTION

The 'non-starch polysaccharides' which constitute dietary fibre are also used widely for control of product texture in the food and related industries, and their physical properties have therefore been studied extensively. The aim of this article is to give a brief overview of the molecular understanding that has resulted from such research and to consider the possible implications for physiological activity as dietary fibre.

The first crucial distinction to be drawn is between insoluble materials (the traditional 'roughage'), whose main importance is likely to be for lower-bowel function, and soluble polysaccharides, which may be particularly significant in controlling absorption of nutrients from the small intestine. Between these two extremes, polysaccharides often exist as hydrated networks in, for example, soft plant tissues and many food products. The factors determining which of these states will be adopted, and the properties of each,[1] are outlined below.

2 ORDER AND DISORDER

In solution, polysaccharides normally exist as expanded, fluctuating coils. Assembly into compact, unhydrated forms, such as cellulose fibrils or starch granules, by contrast, requires the chains to adopt regular, ordered shapes which can pack together efficiently. Whether such ordered assemblies remain stable on exposure to excess water or dissociate into individual hydrated chains depends on the interplay of a number of different factors.

The most obvious is charge. Many polysaccharides have charged groups (e.g. COO^- or SO_3^-) which repel one another, thus favouring expanded coil geometry in solution, rather than compact ordering. Less obviously, the presence of small co-solutes, such as sugars or salts, can alter the solvent-quality of water and shift the balance between polymer-polymer interactions and polymer-solvent interactions, normally in the direction of decreased solubility.

Adoption of an ordered structure also entails considerable loss of mobility (i.e. entropy), which must be more than offset by favourable non-covalent interactions between the participating chains. Non-covalent bonds (such as hydrogen bonds, electrostatic attractions and van der Waals forces) are individually weak and are effective in stabilising polymer assemblies only when they act *co-operatively* in extended arrays, so that ordered structures have a minimum critical length, typically of about 10 residues in each chain. Thus changes in primary sequence (such as branching, irregularly-spaced sidechains or anomalous residues in the polymer backbone) can be an important drive to solubility if they occur at spacings less than the minimum sequence length for ordered association. The presence of sidechains can, in itself, promote solubility by providing an additional entropic contribution from freedom of rotation about the sidechain-mainchain linkage. Similarly, polysaccharides with flexible linkages in the polymer backbone are less likely to form stable, solvent-resistant assemblies than those that are inherently stiff.

Finally, temperature is an important determinant of solubility. The overall change in free energy (ΔG) on going from a compact, ordered structure to a disordered coil is related to the enthalpy change (ΔH) from loss of favourable non-covalent interactions and the entropy change (ΔS) from increased conformational mobility by:

$$\Delta G = \Delta H - T\Delta S \tag{1}$$

Thus entropic effects, favouring solubility, become increasingly important with increasing temperature (T). Because of the simultaneous, co-operative involvement of large numbers of weak bonds, the formation and melting of polysaccharide ordered structures usually occur as sharp processes (in many ways analogous to phase transitions of small molecules), in response to comparatively small changes in temperature (or other external variables such as ionic strength or pH).

3 INTER-RESIDUE LINKAGE PATTERNS

A major difference between polysaccharides and most other types of polymer is that the same monomer unit can be linked together in different ways to give different macromolecular structures. Figure 1 shows four polymers of D-glucose (cellulose, amylose, laminarin and dextran). Despite being built up from the same monomer, these materials all have very different physical properties. Conversely, different sugars linked together in the same way often give polysaccharides with closely similar properties. For example cellulose, chitin, mannan and xylan all form mechanically-strong, solvent-resistant matrices, but are built up from different monomer units (glucose, *N*-acetylglucosamine, mannose and xylose). In each case, however, the component residues are linked together through equatorial bonds diagonally opposite each other across the sugar ring, so that the bonds to and from each residue are parallel and almost co-linear. This linkage pattern (Figure 2a) gives flat, ribbon-like structures that can pack into tough, fibrillar assemblies.

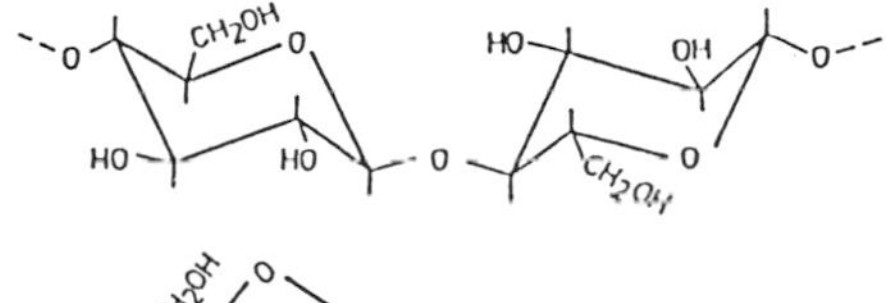

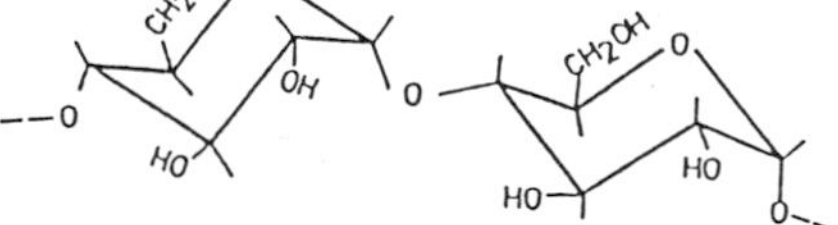

LAMINARIN

DEXTRAN

Figure 1 Polymers of D-glucose, linked β-1,4 (cellulose), α-1,4 (amylose), β-1,3 (laminarin) and α-1,6 (dextran).

Other types of bonding pattern also give rise to characteristic ordered structures, with associated characteristic physical properties, largely independent of the nature of the component residues (unless gross changes, such as the presence or absence of charge, are also involved). For example, when the bonds to and from each residue are once more directly opposite one another but axial rather than equatorial, so that, although still parallel, they are now offset by the full width of the sugar ring, the resulting ordered structures are again ribbon-like, but highly buckled (Figure 2b) and pack together with cavities which, when the chains are charged, can accommodate site-bound counterions.

When the bonds to and from each residue are no longer parallel (Figure 2c), a systematic 'twist' in chain direction is introduced, giving helical ordered structures which are usually stabilised by packing together co-axially. This situation can arise from bonds which are no longer diagonally opposite each other (as in the laminarin structure shown in Figure 1) or which are diagonally opposite, but with one axial and one equatorial (as in amylose). The ordered structures of amylose and laminarin are double and triple stranded helices, respectively. Linkage outside the sugar ring (as in the dextran structure in Figure 1) joins the component residues by three rather than two covalent bonds, giving additional entropic stability to the disordered form.

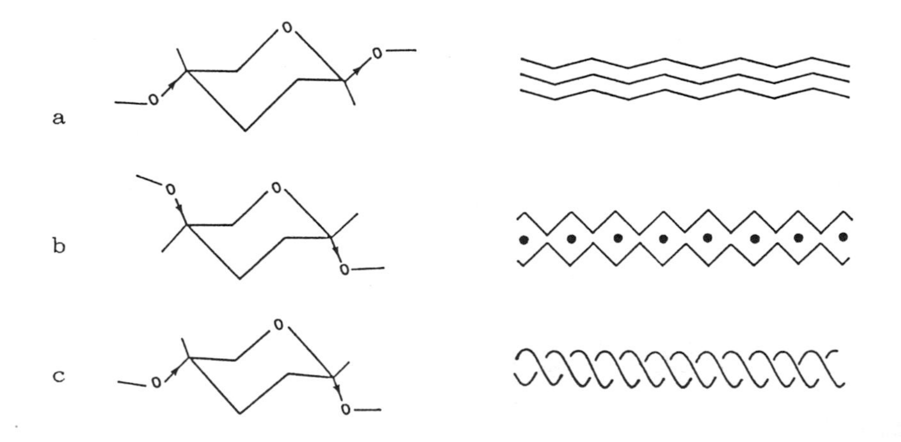

Figure 2 Inter-residue linkage patterns giving: a) flat ribbons, b) buckled ribbons, c) hollow helices.

4 DISORDERED COILS IN SOLUTION

In contrast to the ordered structures of polysaccharides which, as outlined above, are critically sensitive to detailed primary structure, the properties of disordered chains in solution are dependent predominantly on molecular size,[2] which in turn is therefore likely to be a crucial determinant of biological function.

The most direct index[3] of the volume occupied by individual coils is the 'limiting viscosity number' or 'intrinsic viscosity', $[\eta]$. This is the fractional increase in viscosity per unit concentration of polymer, under conditions of extreme dilution (where there are no interactions between chains), and has units of reciprocal concentration (rather than units of viscosity). Intrinsic viscosity is related to molecular weight (M) by the Mark-Houwink relationship:

$$[\eta] = KM^{\alpha} \quad (2)$$

Theoretically, α has a value of 2 for fully-extended rigid rods and 0.5 for freely-jointed coils. For most real disordered polysaccharides α is close to 1, so that intrinsic viscosity (i.e. coil volume) is roughly proportional to molecular weight.

The constant of proportionality (K) is largely dependent on the nature of the linkages between adjacent residues in the polysaccharide backbone. Linkage patterns (Fig. 2) giving extended, ribbon-like ordered structures (flat or buckled) also give rise to expanded coil dimensions in solution (i.e. high values of K), whereas the non-parallel linkages that promote helical ordered structures give small, compact coils (low values of K).

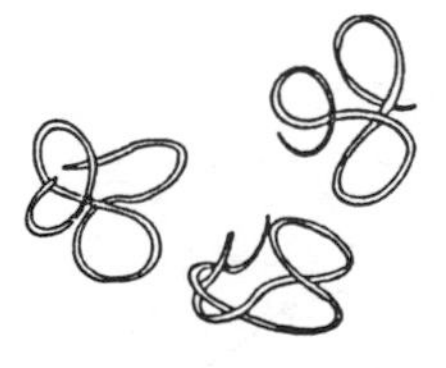
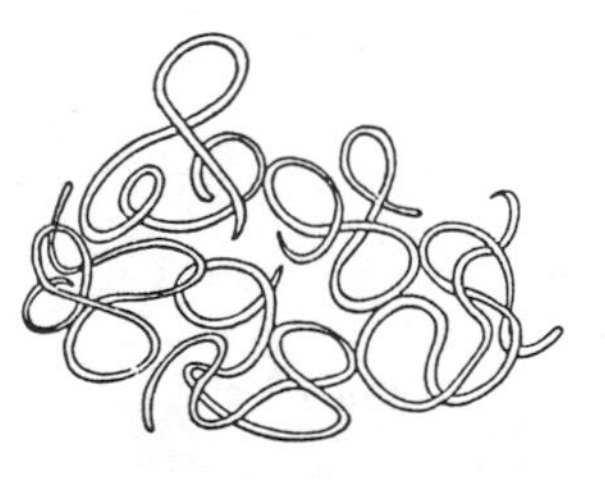

Figure 3 Space-occupancy by disordered polysaccharide coils. Left: $c < c^*$ Centre: $c = c^*$ Right: $c > c^*$.

Short sidechains (such as the single-residue galactose substituents on plant galactomannans) have little, if any, effect on coil dimensions; the intrinsic viscosity of such materials is determined almost entirely by the length of the polymer backbone.[4] Extensive branching (as in amylopectin), however, gives rise to very compact coils (i.e. extremely low values of K) and to a decreased dependence of coil volume on molecular weight ($\alpha < 0.5$).

As the concentration of a polymer solution is increased, a stage is reached at which the individual coils are forced to interpenetrate one another to form an entangled network, as shown schematically in Figure 3. The concentration at which this occurs is known as c^*, and is inversely proportional to coil volume, as characterised by intrinsic viscosity (i.e. the larger the individual coils, the fewer are required to fully occupy the space available). In general, $c^*[\eta] \approx 4$.

Below c^*, where individual coils are free to move through the solvent, with little mutual interference, viscosity is virtually independent of shear rate (i.e. 'Newtonian' behaviour). Above c^*, where chains can move only by the much more difficult process of 'wriggling' (reptating) through the entangled network of neighbouring chains, viscosity (η) becomes highly dependent on shear rate ($\dot{\gamma}$). At low shear rates, where there is sufficient time for entanglements pulled apart by the flow of the solution to be replaced by new entanglements between different chains, with no net change in 'crosslink density', η remains constant at the maximum 'zero shear' value, η_0. With increasing shear rate, however, the rate of re-entanglement falls behind the rate of forced disentanglement, and viscosity falls (Figure 4).

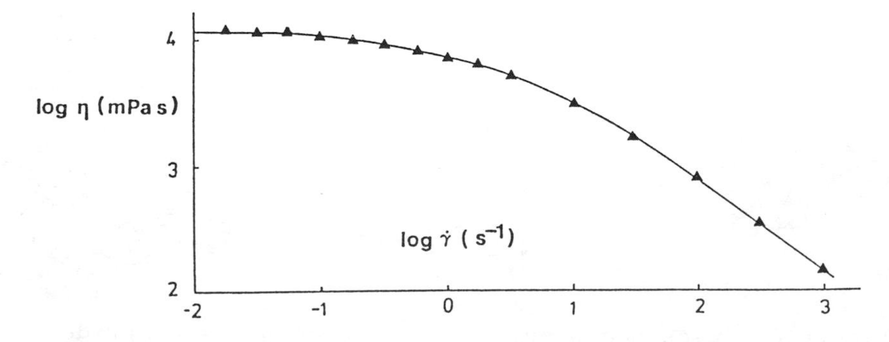

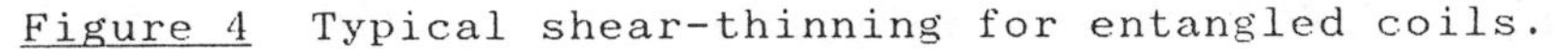
Figure 4 Typical shear-thinning for entangled coils.

The reduction in η with increasing $\dot{\gamma}$ is known as 'shear thinning' and can be very large, with viscosities often dropping by a factor of 100 or more over the shear-rate range available on most commercial viscometers.

The maximum viscosity (η_0) for disordered polysaccharides varies in a general, predictable way with concentration (c), irrespective of primary structure or molecular weight. At concentrations below c*, the increase in viscosity above that of the solvent (i.e. $\eta_0 - \eta_s$) is proportional to $\sim c^{1.3}$ (i.e. doubling concentration increases the 'excess' viscosity by a factor of $\sim$2.5). Above c*, the concentration-dependence increases to $\sim c^{3.3}$, so that doubling concentration gives about a 10-fold increase in viscosity. The viscosity at c* is usually about 10 times that of water, a difference that would be difficult to detect visually. Thus almost all practical situations where solutions are noticeably 'thicker' than water involve the 'entangled' regime.

For samples with a high polydispersity of chainlength (which is invariably the case for food polysaccharides) at concentrations above c*, the form of shear-thinning is entirely general and can be fitted[5] with good precision by:

$$\eta = \eta_0/[1 + (\dot{\gamma}/\dot{\gamma}_{\frac{1}{2}})^p] \qquad (3)$$

where $\dot{\gamma}_{\frac{1}{2}}$ is the shear-rate required to reduce viscosity to $\eta_0/2$, and p = 0.76 (the absolute value of the slope of log η *vs.* log $\dot{\gamma}$ at high shear rate). Rearrangement of Equation 3 gives:

$$\eta = \eta_0 - (1/\dot{\gamma}_{\frac{1}{2}})^p \eta \dot{\gamma}^p \qquad (4)$$

Thus plotting η against $\eta\dot{\gamma}^{0.76}$ gives a straight line of intercept η_0 and gradient $-(1/\dot{\gamma}_{\frac{1}{2}})^{0.76}$. For any given solution, viscosity ( ) at any shear-rate ($\dot{\gamma}$) can therefore be defined completely (Equation 3) by two parameters, η_0 and $\dot{\gamma}_{\frac{1}{2}}$, both of which can be derived from a simple linear plot. It should be noted that the common practice of fitting shear-thinning data to a 'power-law' of the form:

$$\eta = k\dot{\gamma}^{(n-1)} \qquad (5)$$

where n is the so-called 'pseudoplasticity index', is equivalent to treating double-logarithmic plots of η *vs.* $\dot{\gamma}$ (such as that shown in Figure 4) as straight lines, and is therefore totally invalid for entangled coils.

5 HYDRATED NETWORKS

One of the most characteristic features of many polysaccharides *in vitro* is their ability to form cohesive gels at low concentrations (typically ∿1% w/v, although some polysaccharides gel at concentrations of 0.1% or less). The same behaviour often occurs *in vivo* in formation of hydrated tissue structure. Crosslinking of such networks (Figure 5) involves extended 'junction zones' in which the participating chains pack together into ordered assemblies, as in the solid state (Figure 2). Formation of a hydrated network rather than an insoluble precipitate, however, requires the junctions to be linked by interconnecting sequences that are conformationally-disordered (as in solution), providing an entropic drive to solubilise the network. Co-existence of ordered and disordered regions within the same polysaccharide system can occur in a number of ways. As a general mechanism, the presence of a small amount of residual disorder in an otherwise fully ordered polymer can lower the overall free-energy by an entropic contribution that more than outweighs the enthalpic advantage of further ordering. In most polysaccharides, however, ordered association is limited by primary structure.[1]

Covalent Features Limiting Ordered Association

The simplest situation is where the polymer has a block structure, with regions that are soluble under the prevailing solvent conditions interspersed by regions capable of forming stable associations. A well characterised example is the calcium alginate gelling system, where sequences of one type (polyguluronate) adopt a 'buckled-ribbon' conformation and form junctions stabilised by site-bound calcium ions (Figure 2b) and

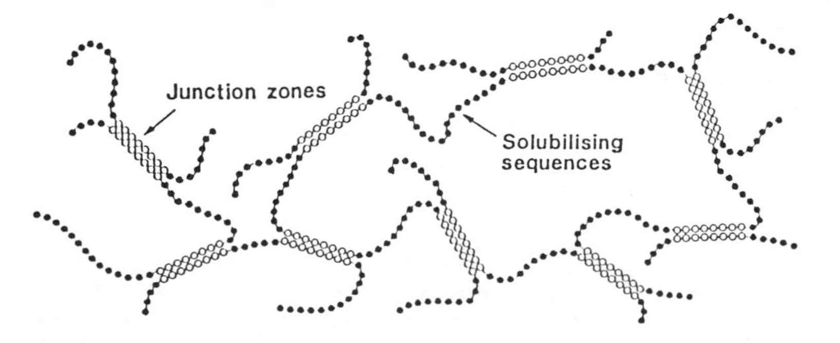

Figure 5 Hydrated polysaccharide network (schematic).

other sequences which cannot bind calcium in this way confer solubility (polymannuronate and heteropolymeric blocks incorporating both mannuronate and guluronate). In other systems, such as the plant galactomannans (which include guar gum and locust bean gum), pendant sidechains limit association of the polymer backbone.

Interchain junctions can also be terminated by the occurrence in the primary sequence of anomalous residues that are geometrically incompatible with incorporation in the ordered structure. The structural properties of agar and carrageenan polysaccharides in algal tissue are regulated in this way. 'Kinking' residues of rhamnose have a similar role in pectin, and are spaced such that the polygalacturonate sequences of the polymer chain have a regular length of ∿25 residues. These sequences form calcium-mediated junctions in the same way as the polyguluronate regions of alginate, but the extent and stability of association may be modified by the occurrence of a proportion of the galacturonate residues as the methyl ester.

Polysaccharide 'Weak Gels'

Certain polysaccharides, notably the bacterial exopolysaccharide xanthan, have properties intermediate between those of solutions and gels. The origin of this behaviour is that xanthan exists in solution in a rigid, ordered conformation and forms a tenuous three-dimensional network by weak, side-by-side association of the ordered chains.[6] Other polysaccharides with soluble ordered structures behave similarly. Such 'weak gel' networks are often sufficiently cohesive to hold particles in suspension or stabilise emulsions over long periods of time. They are not, however, strong enough to support their own weight under gravity, and therefore flow like normal polymer solutions.

The shear-thinning behaviour of 'weak gels' is quite different from that of entangled coils. In particular, since a finite stress is required to rupture the network, the viscosity (defined as the ratio of stress applied to shear-rate generated) increases progressively with decreasing shear rate, rather than reaching a maximum equilibrium value (as in Figure 4 for entangled coils). Indeed, for most 'weak gel' systems, double-logarithmic plots of η *vs.* $\dot{\gamma}$ are linear (so that in this case the 'power law' analysis of Equation 5 is valid), and have a slope greater than the maximum value of -0.76 for disordered polysaccharides.

6 IMPLICATIONS FOR STUDIES OF BIOLOGICAL FUNCTION

Since the physical properties of polysaccharides are determined more by the way in which the component residues are linked together than by the nature of the residues, analysis of dietary fibre for the types and proportions of sugars present is unlikely, in itself, to give much indication of function. Perhaps the only useful generality is that charged polysaccharides are usually more readily soluble than neutral chains, although even here the presence of a sufficient concentration of appropriate counterions can promote stable association of polyelectrolytes into solvent-resistant assemblies.

Determination of the linkages between adjacent residues in the polysaccharide repeating sequence is, of course, a substantial improvement, but is still an incomplete description since, as detailed above, minor departures from structural regularity (e.g. branching, anomalous residues, sidechains) can have a major influence on physical properties. When the spacing of these interruptions is larger than the critical sequence length for ordered association, their effect is to promote formation of hydrated networks rather than insoluble aggregates. At shorter spacings they may completely abolish interchain association and solubilise an otherwise insoluble primary sequence.

Characterisation of the distribution of structural irregularities along the polymer chain (i.e. determination of 'fine structure') has been achieved for some polysaccharide systems, by selective chemical or enzymic degradation, but this is a highly skilled, lengthy undertaking that could certainly not be adopted for routine analysis of dietary fibre. Progress in understanding the mechanism of action of insoluble fibre or swollen, hydrated networks is therefore more likely to come from relating physiological effects to the macroscopic properties of these materials rather than to their chemical composition.

An obvious exception to this rather sweeping assertion is the susceptibility of fibre to bacterial degradation in the lower bowel. However, although the potential susceptibility of polysaccharides to cleavage by specific enzymes is, of course, dependent on their primary sequence, the extent of hydrolysis which occurs in practice can be drastically reduced by conformational ordering and packing (as in 'resistant starch').

Physiological Effects of Soluble Fibre

The ability of soluble fibre to retard postprandial release of nutrients from the gut almost certainly arises from increased viscosity reducing rates of mixing (rather than, for example, to any further suppression of much slower transport processes such as diffusion). As detailed previously, the viscosity generated by disordered coils is dependent on the degree of space-occupancy by the polymer, characterised by the (dimensionless) product of concentration (proportional to the number of chains present) and intrinsic viscosity (proportional to the volume which each occupies). It is therefore just as essential when documenting studies of physiological activity to give an indication of molecular size as it is to report concentration.[7] A particular danger is that apparent differences in the effectiveness of different soluble fibres may be due simply to differences in the molecular weights of the particular samples used, rather than reflecting a genuine dependence of physiological activity or chemical composition.

The most direct index of molecular size is intrinsic viscosity, whose determination is described in detail elsewhere.[3] An acceptable alternative would be to measure viscosities at a few concentrations in the range of practical usage. In view of the extreme shear-thinning of polysaccharide solutions, it is virtually useless to report viscosity at a single shear rate (particularly if the rate used is not specified). Measurements at a few different shear rates, however, can be used to construct a linear plot of η *vs*. $\eta\dot{\gamma}^{0.76}$ and hence derive the values of η_0 and $\dot{\gamma}_{\frac{1}{2}}$ which define completely[5] the viscosity at all shear rates. For 'weak gel' systems the equivalent parameters are k and n (Equation 5) from a linear plot of log η *vs*. log $\dot{\gamma}$.

A complication in interpreting results of clinical studies of soluble fibre is that concentrations in the lumen may be quite different from those ingested. In particular, recent studies using a pig model[8] have shown a substantial increase in the total volume of digesta passing through the small intestine with increasing concentration and molecular weight of soluble polysaccharide in the feed (thus partially offsetting the initial differences in viscosity). The mechanism of this feedback process is not yet fully understood, but it is clearly of importance in defining optimum levels and types of soluble fibre for clinical use.

A further obvious consideration is that polysaccharides can confer viscosity only once they have dissolved. This may explain conflicting reports on the effectiveness of guar gum in reducing postprandial hyperglycaemia, since different commercial guar granulates dissolve at quite different rates.[9]

Finally, increased dietary intake of charged polysaccharides may have implications for the availability of essential mineral nutrients. Tight site-binding of counterions within the ordered structure of charged polysaccharides such as pectin has been detailed above. Weaker 'atmospheric' binding can occur around disordered polyelectrolyte chains in solution or at the surface of insoluble fibres. In all cases, electrostatic binding will, of course, occur only when the ionic substituents of the polysaccharide retain their charge. Sulphate groups (as in carrageenan) remain charged to very low pH, but the pKa of carboxyl groups in, for example, pectin and alginate is around 3.4. Thus at gastric pH these materials may become partially un-ionised and shed their bound counterions, while at higher pH in the small intestine they will regain their charge and once more participate fully in electrostatic binding, perhaps with different ions from those with which they were initially associated.

REFERENCES

1. D.A. Rees, E.R. Morris, D. Thom and J.K. Madden, in 'The Polysaccharides', (Ed. G.O. Aspinall), Academic Press, New York, 1982. Vol. 1, Chapter 5, p. 195.
2. E.R. Morris, A.N. Cutler, S.B. Ross-Murphy, D.A. Rees AND J. Price, Carbohydr. Polym., 1981, 1, 5.
3. E.R. Morris, in 'Gums and Stabilisers for the Food Industry 2', (Eds. G.O. Phillips, D.J. Wedlock and P.A. Williams), Pergamon Press, Oxford, 1984, p. 57.
4. B.V. McCleary, R. Amado, R. Waibel and H. Neukom, Carbohydr. Res., 1981, 92, 269.
5. E. R. Morris, Carbohydr. Polym., 1990, in press.
6. I.T. Norton, D.M. Goodall, S.A. Frangou, E.R. Morris and D.A. Rees, J. Mol. Biol., 1984, 175, 371.
7. P.R. Ellis, E.R. Morris and A.G. Low, Diabetic Medicine, 1986, 3, 490.
8. F.G. Roberts, H.A. Smith, A.G. Low, P.R. Ellis, E.R. Morris and I.E. Sambrook, Proc. Nutr. Soc., 1990, in press.
9. D.B. Peterson and P.R. Ellis, Practical Diabetes, 1988, 5, 133.

CHARACTERIZATION OF SOME PHYSICAL PARAMETERS OF SOLUBLE FIBRES

E. Brosio, M. Delfini, A. Di Nola, A. D'Ubaldo and C. Lintas*

Department of Chemistry
University of Rome "La Sapienza"
* National Institute of Nutrition
Rome

1 INTRODUCTION

Gels are an intermediate state of matter between solid and liquid. The gelation process of a polymer solution consists of the formation of cross-linkages involving extensive segments of polymer molecules - junction zones - to form a three-dimensional network holding water. As a consequence of the gelation process, the polymer solution becomes a viscoelastic medium in which the mobility of the different components (gelling molecules and water molecules) is considerably reduced.

Nuclear Magnetic Resonance (NMR) is a suitable technique for studying both gelation process and gel structure. Investigation on gels can be based on measurements of longitudinal (T_1) and transverse (T_2) relaxation times, these parameters being related to the dynamic state of a system. Since T_1 and T_2 are inversely related to the correlation time of a molecule, the decrease in the mobility of the gel components leads to a corresponding decrease in T_1 and T_2 values. Consequently, lower T_1 and T_2 values can be observed in relation to more extended junction zones and/or to a better efficiency of the intermolecular forces.

In this paper we report an NMR study of high-methoxyl pectin gels. High-methoxyl pectins are linear polymers of

D-galacturonic acid in which more than 50% of the acidic groups are esterified with methanol. At pH values below 3.5 and in the presence of a cosolute, such pectins form gels. The structure of pectin gels is stabilized by both hydrogen bonding and hydrophobic interactions between the ester methyl groups.[1] The nature and the concentration of the cosolute affect the contribution of hydrophobic interactions to the stability of gels.[2]

Aim of the present study was to characterize pectin gels obtained with different cosolutes - sucrose, ethanol, tert-butanol and dioxane - by measuring the NMR parameters T_1 and T_2.

2 MATERIALS AND METHODS

Pectin from apples, 70–75% degree of methylation (Fluka), was used without further treatment.

The dry pectin was mixed first with the cosolute and thoroughly dispersed. The appropriate weight of water was added and the pH adjusted to about 3.5 with 0.1 M citric acid. The mixture was placed in a water bath (95°C for samples containing sucrose as cosolute; 60°C for samples containing ethanol, tert-butanol and dioxane as cosolutes) for 1 hr under stirring.

Gel preparation was carried out directly in NMR glass tubes (18 mm external diameter). Concentrations were measured as % (w/w). Gel samples were stored at 25°C for 24 hr before measurement.

The empirical gel threshold was determined by inversion of the NMR tubes: a sample that did not flow was considered to have gelled.

NMR measurements were carried out on a pulse low resolution spectrometer Minispec pc120 (Bruker Spectrospin, Germany) operating at a frequency of 20 MHz for protons.

The spin-lattice relaxation times were measured by the inversion recovery technique,[3] sampling 10 points with delays spanning between 0.002 and 6 s and using a repetition rate in the range between 1 and 8 s (in any case larger than 5 T_1). Signal enhancement was achieved by averaging over 4 acquisitions using a repetition delay ranging between 1 and 8 s.

The spin-spin relaxation times were measured by the Carr-Purcell-Meiboom-Gill (CPMG) technique,[3] sampling up to 500 echoes; 4 FIDs were detected, using a repetition delay in the range between 1 and 8 s, and the signal averaged in order to obtain signal enhancement.

Samples were thermostated for 30 min at the temperature of measurement (40°C).

3 RESULTS AND DISCUSSION

The variation of T_1 and T_2 values with pectin concentration in gels containing 55% sucrose as cosolute is reported in Figure 1. The figure points out a decrease in both T_1 and T_2 values as pectin concentration increases, as expected since T_1 and T_2 values are inversely related to solution viscosity.

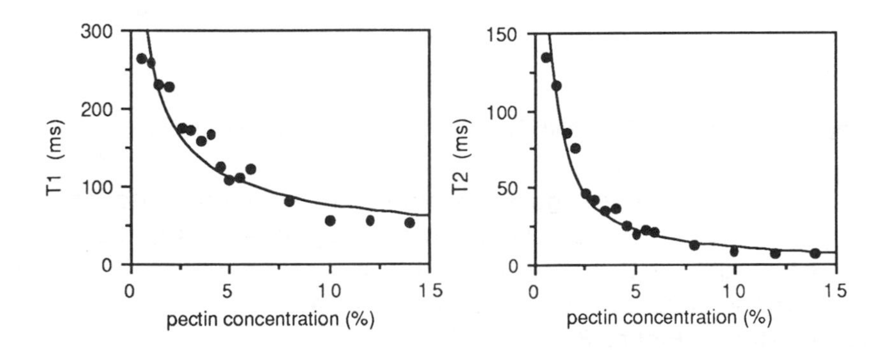

Figure 1 Variation of T_1 and T_2 with pectin concentration in gels containing 55% of sucrose as cosolute.

At the pectin concentration at which the gel is formed - evaluated as described in the previous section - T_1 and T_2 become constant and their absolute values depend on the rigidity of the molecular network. The measurement of physical- chemical parameters can thus replace an empirical evaluation in the determination of the gel threshold.

The variation of T_1 and T_2 values with sucrose concentration in 5% pectin gels is reported in Figure 2. A behaviour of the relaxation times curves analogous to the previous one can be observed, so confirming the reliability of relaxation times measurements to determine the gel threshold for gelling agent as well as for cosolute.

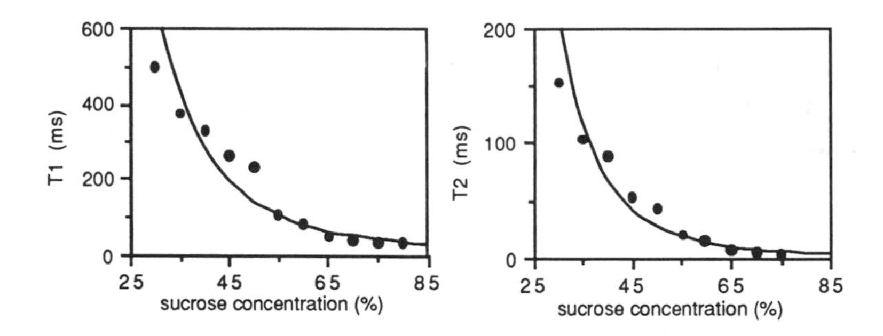

Figure 2 Variation of T_1 and T_2 with sucrose concentration in 5% pectin gels.

In order to investigate the role of hydrophobic interactions in gel formation, sucrose was replaced by other cosolutes, such as ethanol, tert- butanol and dioxane. The variation of relaxation times with cosolute concentration in 5% pectin gels is reported in Figure 3.

It is evident that the behaviour of relaxation times is different when using cosolutes other than sucrose. In fact, in the case of ethanol, tert-butanol or dioxane the T_1 and T_2 curves pass through a minimum which corresponds to the cosolute concentration at which the gel is formed. That

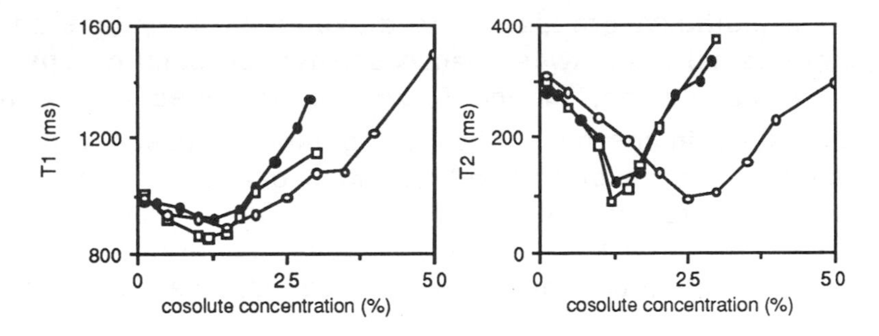

Figure 3 Variation of T_1 and T_2 with cosolute concentration in 5% pectin gels: ● - ethanol; □ - tert-butanol; ○ - dioxane.

indicates a critical cosolute concentration in the gel formation which depends on the nature of the cosolute. The T_2 curve has a more pronounced minimum than the T_1 curve, thus indicating that T_2 relaxation time is a better parameter to describe gel properties. That follows from the fact that spin-spin relaxation time is sensitive to low-frequency molecular motions while spin-lattice relaxation time depends on high-frequency molecular motions.

The relaxation time curves measured for pectin gels prepared with ethanol, tert-butanol and dioxane closely match the curves for rupture strength (RS) versus cosolute concentration of similar gels.[4] These authors find that RS passes through a maximum at the cosolute concentration at which the gel is formed. These results can be explained by considering the effect of the different cosolutes on the interactions that stabilize gel structure. Solutions of 55% sucrose, 15% ethanol and 14% tert-butanol (these percentages corresponding to cosolute concentrations at the minimum in T_2 curves) have a dielectric constant of 57.5, 65.9 and 61.5, respectively, as calculated by interpolating data reported by Åkerlof et al.[5] A solution of 30% dioxane has a dielectric constant of 49.0. Thus, sucrose, ethanol and tert-butanol seem to stabilize hydrophobic interactions between pectin methoxyl groups, while dioxane is likely to stabilize polar interactions between pectin unesterified polar groups.

In conclusion, T_1 and T_2 relaxation times are reliable parameters in the description of gelation process and gel structure; moreover, relaxation times can be related to the rheological properties of gels. Thus, besides being non-invasive and rapid, the NMR technique provides very reproducible parameters, that can favourably replace empirical evaluations.

REFERENCES

1. M. D. Walkinshaw and S. Arnott, J. Mol. Biol., 1981, 153, 1075.
2. D. Oakenfull and D. E. Fenwick, J. Chem. Soc. Faraday Trans. I, 1979, 75, 636.
3. T. C. Farrar and E. D. Becker, "Pulse and Fourier Transform NMR", Academic Press, New York and London, 1971, Chapter 2, p. 18.
4. D. Oakenfull and A. Scott, J. Food Sci., 1984, 49, 1093.
5. G. Åkerlof, J. Am. Chem. Soc., 1932, 54, 4125.

CONDENSED TANNINS AND RESISTANT PROTEIN AS DIETARY FIBRE CONSTITUENTS

F. Saura-Calixto, I. Goñi, E. Mañas and R. Abia

Instituto de Nutrición y Bromatología (CSIC-UCM)
Facultad de Farmacia. Ciudad Universitaria
28040 Madrid

1 INTRODUCTION

The term dietary fibre (DF) includes all the plant polysaccharides and lignin which are resistant to enzyme digestion[1]. Several enzymatic-gravimetric methods are given to determine DF as a whole, e.g. the new AOAC method[2]. Other insoluble substances not included in the DF definition may be found in the DF residues[3].

Cell wall protein could be regarded as part of DF, because of lower apparent digestibility. On the other hand, many vegetable foods contain phenolics which may inhibit proteases and form complexes with protein[4,5] thereby increasing the amount of indigestible protein.

Recently, Trowell[6] suggested that resistant protein must be considered in the DF definition.

The presence of resistant protein and condensed tannins (polyphenols) in DF residues is considered in the present article.

2 MATERIALS AND METHODS

White and red grapes (Airen and Garnacha varieties) were used. The juice was extracted and the remaining pulp was freeze-dried and milled to pass 0.5 mm sieve.

Insoluble dietary fibre (IDF) was determined by the AOAC method[2]. One gram of sample was successively treated with heat-stable α-amylase (100º C, pH=6, 30 min), protease (60º C, pH=7.5, 1 hr) and α-amyloglucosidase (60º C, pH=4.5, 30 min).

Then, filtration and washing with destilled water, 95% ethanol and acetone were carried out. The residue corresponded to IDF.

The residue obtained after acid hydrolysis of IDF residues (12M H_2SO_4, 20ºC, 3 hrs; 1M H_2SO_4 refluxing 2 hrs) was determined as Klason lignin (KL).

Klason Lignin (KL) residues were treated with HCL-activated triethyleneglycol at 125º C, 1 h[7]. The solubilized lignin was measured by the Morrison procedure[8]. Kjeldalh and Biuret determinations of protein were made.

Condensed tannins were spectrophotometrically quantified in the anthocyanidin solutions obtained by sample treatment with HCl-BuOH[9].

Other samples rich in condensed tannins have previously been studied with similar procedures and purpose: carob pods[9], apples[10] and Spanish sainfoin[11].

3 RESULTS AND DISCUSION

The samples studied are rich in condensed tannins (proanthocyanidin polymers). The contents of CT and total protein are listed in Table 1.

The IDF values shown in Table 2 correspond to the gravimetric residues obtained by the AOAC method, after substracting the ash content. Significant amounts of protein were found in these residues representing between 25 and 80% of total protein in the original samples. These amounts must include both cell-wall and intracellular protein.Nevertheless, after the protease treatment usually only very low percentages of protein remained in the IDF residues. The high retention of protein in these samples is related with the presence of CT. A double effect of CT on protein was reported: inhibition of protease and formation of tannin-protein complexes[8]. Both

TABLE 1.- PROTEIN AND CONDENSED TANNINS (CT) IN THE SAMPLES (% DRY MATTER)

SAMPLES	PROTEIN	CT
Carob pod	2.7 ± 0.1	17.9 ± 0.4
Cider wastes	5.4 ± 0.2	3.1 ± 0.1
Spanish sainfoin	8.2 ± 0.4	7.0 ± 0.6
White grape pulp	10.7 ± 0.2	14.5 ± 0.5
Red grape pulp	13.8 ± 0.3	36.4 ± 0.9

Mean value ± standard deviation

TABLE 2.- INSOLUBLE DIETARY FIBRE (IDF) AND PROTEIN IN IDF (% DRY MATTER OF ORIGINAL SAMPLES)

SAMPLES	IDF RESIDUE	PROTEIN IN IDF
Carob pod	32.8 ± 0.6	1.8 ± 0.1
Cider wastes	51.3 ± 0.4	3.0 ± 0.1
Spanish sainfoin	54.6 ± 0.4	2.1 ± 0.1
White grape pulp	62.5 ± 1.1	7.5 ± 0.2
Red grape pulp	65.7 ± 0.6	12.2 ± 0.3

Mean value ± standard deviation

factors contribute to the non-digestion of protein (resistant protein).

Klason lignin was determined in IDF residues and the results are indicated in Table 3. A significant part of the resistant protein is not solubilized by the strong acid treatment performed to obtain KL. The KL residues include lignin, protein and CT.

Condensed tannins resist acid hydrolisis, yielding a red solution of anthocyanidins and remaining an amorphous residue of phlobaphenes[12]. Klason Lignin treatment was performed on CT standard and only 9% was dissolved.

Finally, the KL of grapes and lignin standard residues were treated with HCL/triethyleneglycol[7]. A lignin standard was completely dissolved while only a partial solution of KL was observed. The composition of the supernatant, expressed as percentage of original dry matter was: 2.6% lignin and 7.6% CT in red grape pulp and 2.5% lignin and 2.2% CT in white grape pulp. Protein was not detected in the supernatants.

The presence of protein and tannin in IDF and KL residues may indicate these compounds are quite resistant to enzymatic

TABLE 3.- KLASON LIGNIN (KL) AND PROTEIN IN KLASON LIGNIN (% DRY MATTER OF ORIGINAL SAMPLE)

SAMPLES	KL	PROTEIN KL
Carob pod	15.7 ± 0.1	0.8 ± 0.05
Cider wastes	20.0 ± 1.0	1.9 ± 0.1
Spanish sainfoin	n.d.	n.d.
White grape pulp	41.2 ± 1.2	2.6 ± 0.1
Red grape pulp	39.7 ± 0.6	2.8 ± 0.05

Mean value ± standard deviation
n.d.= not determined

and acid hydrolysis, and subsequently indigestibles. Both substances, condensed tannins (polyphenolic polymers) and resistant protein could be considered constituents of DF. Condensed tannins are common in many vegetables of the diet.

REFERENCES

1. H. Trowell, D.A.T. Southgate, T.M.S. Wwolever, A.R. Leeds, M.A. Gassull and D.J.A. Jenkins. Lancet, 1976, 1, 967.
2. L. Prosky, N.G. Asp, T.F. Schweizer, J.W. DE Vries and I. Furda. J. Assoc. Off. Anal. Chem., 1988, 71, 5.
3. D.A.T. Southgate. In: "Handbook of dietary Fiber in Human Nutrition". Chapter 3. CRC press, Spiller, G.A. 1, 1988.
4. R. Kimar and M.J. Singh. Agric. Food Chem., 1984, 32, 447.
5. M. Tamir and E. Alumot. J. Sci. Food Agric., 1984, 20, 199.
6. H. Trowell. Am. J. Clin. Nutr., 1988, 48, 1079.
7. C.S. Edwards. J. Sci. Food Agric., 1973, 24, 381.
8. I.M. Morrison. J. Sci. Food Agric., 1972, 23, 455.
9. F. Saura Calixto. J. Food Sci., 1988, 53, 6, 1769.
10. I. Goñi, M. Torre and F. Saura-Calixto. Food. Chem., 1989, 33, 51.
11. F. Saura-Calixto. Food Chem., 1987, 23, 95.
12. E. Haslam. "Chemistry of vegetable tannins". Academic Press. London, 1966.

WORKSHOP REPORT: PLANT SUBSTANCES ASSOCIATED WITH DIETARY FIBRE

D. Topping

CSIRO Division of Human Nutrition
Glenthorne Laboratory
O'Halloran Hill
SA5158
Australia

The workshop addressed 8 broad questions while recognising that there were very many important issues that could not be accommodated because of time:

Question 1 What (a) makes a fibre polysaccharide and (b) how does the plant produce them?

(a) Fibre polysaccharides are essentially carbohydrate chains with various carbohydrate and non-carbohydrate side chains. (b) There is no template, control of synthesis is vested in the enzymes that add on or split off units. We concluded that compounds shorter than 10 polymer units were not fibre polysaccharides.

Question 2. What is lignin?

Lignin is a 3-dimensional phenolic network associated with different compounds. Although lignin consumption is only some 200-300mg/person/day in Britain, these small quantities are important because they modify the effects of other food components, e.g. phenolics. Lignification can also affect the mechanical properties of food.

Question 3. Is phytate important?

Mineral availability can be greatly increased by phytase although some foods (e.g. legumes) lack this enzyme and in oats phytase is destroyed by processing. Because increased consumption of plant foods raises mineral supply, phytate is most probably not a problem in advanced countries.

Question 4. Protease and other inhibitors.

Trypsin inhibitors in ingested soya greatly stimulate pancreatic flow and (acting through cholescystokinin) clearly affect gut function. Cooking destroys a large fraction of the inhibitor but the residue may still be significant. Lipase inhibitors (e.g. in wheat) and amylase inhibitors may also be beneficial by slowing digestion. This area was identified as requiring further research.

Question 5. What are the effects of Maillard Products?

Maillard products at low levels have pleasant taste properties and may be functionally similar to fibre. Although laboratory Maillard products bound bile acids *in vitro* they had no effect on cholesterol *in vivo*. Because some products are potentially carcinogenic more research is urgently required.

Question 6. Starch

Under some circumstances starch may behave like fibre although quantities of dietary resistant starch are probably small. Processing may increase this fraction especially through heating. The original source of raw material is important in this respect as starch digestibility can be influenced by environment and genetics. Good analysis for starch is essential and the quantities entering the large bowel needs to be established by experimentation. There was no consensus on a definition of resistant starch except that it was a functional property of food.

Question 7. Are oligosaccharides fibre?

Although certain di- and oligo- saccharides enter the large bowel in significant quantities and are fermented, they are not fibre. The term 'fibre' should be reserved strictly for plant cell wall and (possibly) other related structural materials. Oligosaccharides are fermented rapidly and give widely different products depending on structure. Their effects and those of sugar alcohols are urgently in need of investigation.

Question 8. How should we regard antinutrients and related compounds?

Numerically these greatly exceed synthetic additives. They have been investigated to a very limited extent and many studies are of dubious significance as purified compounds only have been used. Vegetarians have exposures higher than average but their health status seems unimpaired and resistance to adverse effects is probably well established. Exposures in the general population are set to increase due to nutritional and agronomic trends. Concern was expressed that plant genetic manipulation may increase adversely the level of some components. In people with eccentric practices (e.g. eating potato peel or drinking certain herbal infusions) toxic reactions are possible. More research and monitoring was recommended.

Analytical Techniques

EVALUATION OF DIETARY FIBER METHODS AND THE DISTRIBUTION OF ß-GLUCAN AMONG VARIOUS FIBER FRACTIONS

Joseph L. Jeraci[a], James M. Carr[b], Betty A. Lewis[b], and Peter J. Van Soest[b]. Syracuse Research Corporation Merrill Lane, Syracuse, NY 13210[a] and a Division of Nutritional Sciences, Savage Hall, Cornell University, Ithaca, NY 14583[b].

1 SYNOPSIS

Total dietary fiber (TDF) values for fruits, vegetables, processed foods and cereal grains were determined by the urea enzymatic dialysis (UED) method. For 12 oat-based cereal products, the ß-glucan levels of the TDF residues were determined by a NaOH enzymatic assay. The insoluble dietary fiber and soluble dietary fiber values were determined for an oatmeal cereal, an oat bran cereal and rolled oats. The insoluble and soluble dietary fiber residues were analyzed for ß-glucan. The results obtained using the UED method were compared with the AOAC procedure and the neutral-detergent fiber method. Crude protein and ash contamination was usually lower with the UED method compared with the AOAC method. The UED method was effective in removing starch even at the relatively low temperatures of the assay (50°C). The TDF values were comparable using the UED method and the AOAC method, but these methods gave significantly different insoluble and soluble dietary fiber values. The insoluble dietary fiber values determined by the AOAC method were larger than values obtained using the UED method and the neutral detergent fiber procedure. In the evaluation, the UED method quantitatively recovered, with less variation, more of the ß-glucan in the insoluble and soluble dietary fiber residues. The data indicate that the UED method is more precise and accurate than the AOAC method.

2 MATERIALS AND METHODS

Locally purchased cereal-based food products representing national brands and in-store bakery products were used. These products were air-dried, ground to pass a 20 mesh sieve and then stored at -20°C in polyethylene bottles until needed for analyses.

The water-soluble ß-glucan and total ß-glucan content were determined for the cereal-based products using a ß-glucan enzyme assay (Carr et al., in press).

Total dietary fiber (TDF), soluble (SF), and insoluble dietary fiber (IF) contents were analyzed by the modified AOAC method[1] and by the urea enzymatic dialysis (UED) method[2]. All dietary fiber residues were analyzed and corrected for protein (Kjeldahl N x 6.25) and ash (ignited at 525 °C for 8 hours).

Neutral detergent fiber (NDF) content was analyzed by the modified method described by Jeraci et al.[3]. All NDF residues were analyzed and corrected for protein (Kjeldahl N x 6.25) and ash (ignited at 525 °C for 8 hours).

Fiber residues from each of the three runs were used to determine the ß-glucan content of the various fiber fractions. These fiber fractions generated were collected on Whatman No. 54 filter paper and air dried. A filter paper blank was carried through the analysis for each set of samples and consistently yielded a value of 0.0 mg ß-glucan.

3 RESULTS AND DISCUSSION

The cereal-based products exhibits a wide variation in their composition which is attributable to the proportion of ingredients and to the processing procedures used in the manufacture of the product[4].

In terms of TDF for the food products analyzed, there was close agreement between the two TDF methods. Advantages of the UED method have been discussed[2] (Table 1). These advantages include (1) removal of essentially all of the starch, (2) generation of smaller crude protein and ash correction factors, (3) better quantitative recovery of semi-purified dietary fiber products, (4) more efficient in terms of labor, equipment and assay time, (5) milder assay conditions ($\leq$50°C) and (6) fewer enzymes and reagents.

The NDF values were consistently lower than the TDF values. The NDF method determines an "insoluble" dietary

Table1. Composition of Various Cereal-Based Food Products (% dry weight)[a].

Method:	NaOH[b]	Urea Enzymatic Dialysis[c]			AOAC-Prosky[d]		
Food Product[e]	ß-Glucan,%	TDF,%	± SD	ß-Glucan,%[f]	TDF,%	± SD	ß-Glucan,%[f]
"Quick" rolled oats	4.3	11.2	1.8	3.3	11.3	1.1	3.1
Rolled oats	4.6	9.5	1.6	3.1	12.6	1.2	3.5
Oat RTE cereal	2.5	8.4	1.2	2.2	9.0	0.5	1.7
Oat bran RTE cereal	1.9	17.4	0.7	1.6	17.2	1.5	1.6
Oat bran flaked RTE Cereal	2.1	8.4	1.4	2.1	9.4	2.7	1.6
Oat bran hot cereal	8.9	19.3	2.1	6.9	21.3	2.6	7.1
Oat bran	7.3	18.8	2.7	5.6	21.1	1.2	5.8
Granola bar	1.4	8.5	1.8	1.1	6.8	2.8	1.2
Oat bran muffins	0.6	5.7	0.8	0.4	7.3	1.7	0.2
Oatmeal bread A	0.9	5.5	0.7	0.5	7.4	1.7	0.6
Oatmeal bread B	1.1	7.2	0.6	1.0	7.9	1.1	0.4
White bread	0.4	3.4	0.8	0.9	5.4	0.5	0.48

[a] Values corrected for water content. Total dietary fiber (TDF) values are corrected for ash (525°C, for 8 hr) and protein (Kjeldahl nitrogen x 6.25).

[b] 1.0 N NaOH extracted ß-Glucan values adapted from Carr et al. (in press).

[c] TDF determined by the method described by Jeraci et al. (in press). Mean and standard deviation (SD) for TDF from two replicated experiments, n=6.

[d] TDF determined by the method described by Prosky et al. (1988). Mean and standard deviation (SD) for TDF from two replicated experiments, n=6.

[e] RTE: ready to eat.

[f] ß-Glucan content of TDF residues expressed as percent of the original sample dry matter using a modification of the NaOH method described by Carr et al. (in press). Mean for ß-Glucan from two replicated experiments, n=2.

Table 2. Composition (Insoluble Fiber, and Soluble Fiber) for Three Ready to Eat Cereals Using the Urea Enzymatic Dialysis (UED) Method, the AOAC Total Dietary Fiber Method, and the Neutral-Detergent Fiber Method [a,b,c,d].

	% Dietary Fiber ± SD				
Method:	UED		AOAC		NDF
Cereals	Insoluble	Soluble	Insoluble	Soluble	Insoluble
Oat bran hot	9.6 ± 0.0	10.2 ± 0.7	23.8 ± 0.2	1.6 ± 0.9	14.0 ± 0.3
Oat bran	16.8 ± 1.1	2.9 ± 0.7	18.2 ± 1.5	4.2 ± 2.1	14.9 ± 0.2
Rolled oats	3.5 ± 0.4	6.8 ± 1.2	12.8 ± 2.1	1.0 ± 1.0	4.9 ± 0.8

a Corrected for water.
b The AOAC Total Dietary Fiber Method (Prosky et. al., 1).
c The Neutral-detergent fiber described by Jeraci and Van Soest (5) modified to calculate insoluble fiber corrected for ash and crude protein.

Table 3. Composition (Insoluble Fiber, and Soluble Fiber) for Three Ready to Eat Cereals Using the Urea Enzymatic Dialysis (UED) Method and the AOAC Total Dietary Fiber Method[a,b].

	% ß-Glucan					
Method:	UED		AOAC		UED	AOAC
Cereals	Insoluble	Soluble	Insoluble	Soluble	Sum[c]	
Oat bran hot	0.71	3.50	1.42	0.75	4.21	2.17
Oat bran RTE	0.32	1.04	0.28	0.12	1.36	0.40
Rolled oats	0.32	1.65P	1.42	0.75	1.97	0.99

a Corrected for water.
b The AOAC Total Dietary Fiber Method (Prosky et. al., 1).
c Sum of ß-glucan in the insoluble (I) and soluble (S) dietary fiber fractions.

fiber. With the samples used in this study, high NDF values relative to TDF values may reflect elevated NDF values caused by difficult filtration or less ß-glucanase activity in the amylase use in this modified NDF method[3,4,5]. Previous reports that NDF does not recover ß-glucan[6] are not confirmed by the present study and the amount of ß-glucan (0.2-7.0% ß-glucan) in NDF may be due to the heat stable α-amylase used in the present study.

The UED method had relatively better recovery of ß-glucan for the majority of the products than the 1988 modification of AOAC method for TDF. Both of the TDF methods, however, only recovered from 39 to 89% of the ß-glucan (Table 1).

The content of soluble and insoluble dietary fiber for three cereal-based products was determined (Table 2) by the modified AOAC method[1] and the UED method[2]. There was better agreement between the determined and calculated (sum) values for TDF using the UED method than the TDF values using the modified AOAC method. There is a large difference in the distribution of the ß-glucan between soluble and insoluble fiber fractions as measured by the two methods. The UED method gave much higher proportions of soluble fiber than the modified AOAC method for two of the three products. Both of these methods had lower ß-glucan recoveries in the soluble and insoluble dietary fiber fractions than in the TDF (Table 3). The UED method, however, was significantly better than the modified AOAC method in the recovery of ß-glucan for the soluble and insoluble dietary fiber fractions (Table 5).

REFERENCES

1. L. Prosky, N.-G. Asp, T.F. Schweizer, J.W. Devries, and I. Furda, J. Assoc. Off. Anal. Chem., 1988, 71, 1017.
2. J.L. Jeraci, B.A. Lewis, P.J. Van Soest, and J.B. Robertson, J. Assoc. Off. Analyt. Chem., 1989, 72, 677.
3. J.L. Jeraci, T. Hernandez, B.A. Lewis, P.J. Van Soest, and J.B. Robertson, J. Assoc. Off. Analyt. Chem., in press.
4. J.M. Carr, S. Glatter, J.L. Jeraci, and B.A. Lewis, Cereal Chem., in press.
5. J.L. Jeraci and P.J. Van Soest, Am. Chem. Soc. Symp. Ser., in press.
6. C.J. Patrow and J.A. Marlett, J. Amer. Diet. Assoc., 1986, 86, 794.

SOME PROBLEMS ASSOCIATED WITH THE DETERMINATION OF DIETARY FIBRE IN CITRUS BY THE A.O.A.C. METHOD

E. Mañas, R. Abia and F. Saura-Calixto

Instituto de Nutrición y Bromatología (CSIC-UCM)
Facultad de Farmacia. Ciudad Universitaria
28040 - MADRID
SPAIN

1 INTRODUCTION

The official A.O.A.C. method for dietary fibre (DF) determination was recently modified[1,2]. The authors reported that "the precision of the method for the individual fractions especially for SDF, is not yet satisfactory" and "further studies of fruits, vegetables, and leguminous seeds are needed".

On this basis, citrus samples were chosen to study some problems associated with the determination of DF by this method.

Some aspects related with precipitation of soluble constituents and retention of several substances in the fibre matrix are considered.

2 MATERIAL AND METHODS

Both orange and lemon pulp and peel were used. Raw samples were dried and ground to particles of <0.5 mm.

Analysis of starch in the samples by a UV-method[3] were performed after free sugars extraction.

A brief scheme of the methodology used is shown in Figure 1.

The fibre content was determined by the enzymatic-gravimetric A.O.A.C. method[2] omitting α-amylase and amyloglucosidase treatment, because the samples did not contain starch. Corrections of blank, protein and ash were carried out in the IDF and SDF residues.

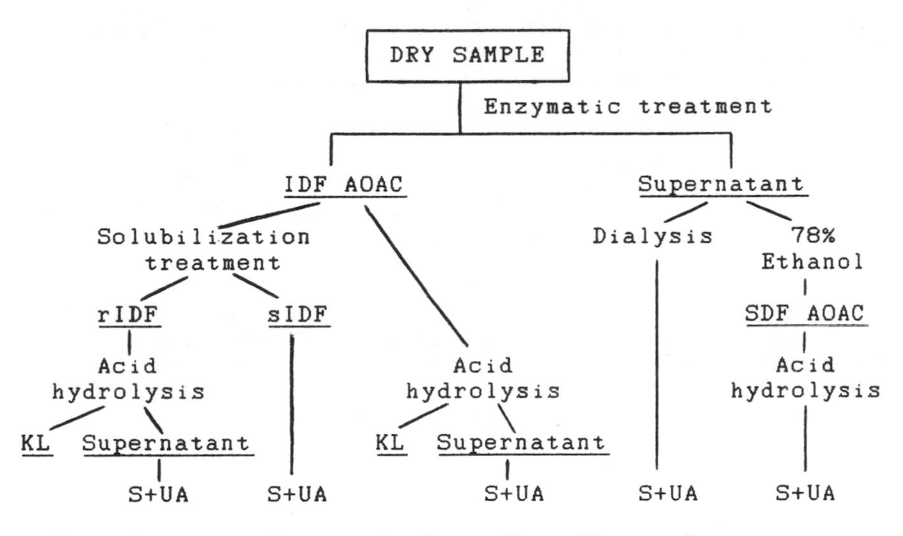

S+UA = Sugars + Uronic Acids ; KL = Klason Lignin

Figure 1 Scheme of the methodology

Some IDF residues were solubilized in water (100ºC, 30 min twice). Both supernatant (sIDF) and IDF residue (rIDF) were collected.

Additionally to the determination of SDF by 78% ethanol precipitation, some SDF values were obtained by dialysis. After the protease treatment of the raw samples, the supernatant liquid and washings were combined and transferred quantitatively into a dialysis bag (Dialysis Tubing Visking 9-32/36 mm. Medicell International, Ltd.) and dialysed against distilled water for 48 h.

Acid treatments of IDF, rIDF and SDF were carried out following conditions previously tested (IDF, rIDF: 12M H_2SO_4, 1 h, 30ºC + 1M H_2SO_4, 1.5 h, 100ºC; SDF: 1M H_2SO_4, 1.5 h, 100ºC). The residue was determinated as Klason Lignin (KL).

Neutral polysaccharides and pectines, expressed as neutral sugars (S) and uronic acids (UA) respectively, were spectrophotometrically determined[4,5] on the acid hydrolysates and solutions as it is shown in Figure 1.

3 RESULTS AND DISCUSSION

Results shown in the tables are expressed as percent of dry material of the original samples (mean values of at least three determinations ± standard deviation).

SDF AOAC values are expressed in Table 1. The major portion of the gravimetric residues correspond to inorganic components precipitated in the 78% ethanol solution (Ash= 62-82 %). The high value of the relationship Residue/SDF may be a source of error in the determination of SDF.

On the other hand, the polysaccharide content of this residue was only a part of the SDF AOAC value. That is to say, other non-fibre constituents precipitated in the 78% alcoholic solution. This could be due to the retention of non-fibre components in the polysaccharide matrix of SDF.

Table 1 Content and composition of SDF obtained by precipitation (AOAC procedure) (% dry material)

	ORANGE		LEMON
	Pulp	Peel	Pulp
Residue	37.94±1.62	46.16±2.39	56.75±0.43
Ash	30.23±1.70	37.47±1.54	36.03±0.35
Protein	0.69±0.07	0.73±0.07	0.93±0.00
SDF AOAC	7.02±1.70	7.96±2.39	19.79±0.43
Sugars	1.55±0.06	1.54±0.10	1.85±0.10
Uronic acids	0.93±0.04	0.97±0.06	7.24±0.28
*SDF	2.48±0.06	2.51±0.10	9.09±0.28

Table 2 SDF obtained by dialysis (% dry material)

	ORANGE		LEMON
	Pulp	Peel	Pulp
Sugars	2.19±0.27	2.17±0.27	3.44±0.11
Uronic acids	3.41±0.41	4.25±0.63	16.25±2.08
SDF	5.60±0.41	6.42±0.63	19.69±2.08

SDF values obtained by dialysis (Table 2) were always higher than the actual value of AOAC method (*SDF) (Table 1). This indicated that some hemicelluloses and pectins remain solubilized in the 78% ethanol solution and are lost when the AOAC method is used.

Gravimetric and spectrophotometric IDF values are listed in Table 3. Gravimetric results were about three units higher than the spectrophotometrics.

Solubilization treatment of the IDF AOAC yielded an rIDF 8-10 units lower than the first one (Table 4).

From the data on Table 5, it can be deduced that only a part of the IDF solubilized fraction (sIDF) was actually fibre (sSDF). This part could be considered as SDF while the remaining may correspond to non-fibre constituents retained in the IDF polysaccharide matrix.

In summary, both IDF and SDF values obtained by the AOAC method in citrus are higher than the actual content.

Table 3 Content and composition of IDF obtained by AOAC method (% dry material)

	ORANGE		LEMON
	Pulp	Peel	Pulp
IDF AOAC	32.08±0.30	32.64±0.69	26.04±0.19
Lignin	7.23±0.27	6.88±0.07	2.93±0.59
Sugars	13.70±0.22	14.20±0.34	15.57±0.01
Uronic acids	9.59±0.05	8.30±0.85	4.78±0.03
*IDF	30.52±0.27	29.38±0.85	23.28±0.59

Table 4 Content and composition of IDF after solubilization treatment (% dry material)

	ORANGE		LEMON
	Pulp	Peel	Pulp
rIDF	23.71±0.37	24.18±0.06	16.17±0.56
Sugars	12.27±0.29	11.62±0.45	13.07±0.34
Uronic acids	6.62±0.30	4.22±0.08	0.79±0.03

Table 5 Content and composition of solubilized fraction (% dry material)

	ORANGE		LEMON
	Pulp	Peel	Pulp
sIDF	12.33±0.37	11.51±0.69	12.72±0.56
Sugars	1.70±0.19	2.27±0.30	1.95±0.10
Uronic acids	1.51±0.04	4.73±0.24	2.73±0.30
sSDF	3.21±0.19	7.00±0.30	4.68±0.30

4 SUMMARY

Determinations of insoluble and soluble dietary fibre (IDF and SDF) were carried out in orange and lemon pulp and peel by the A.O.A.C. enzymatic-gravimetric method and by spectrophotometric procedures (as the sum of sugars and uronic acids in the acid hydrolysate of the IDF and SDF residues).

In the precipitation residue, ash weight was higher than SDF values. The spectrophotometric results were appreciably lower than the gravimetric values. Alternatively, dialysis was employed and yielded the highest SDF values.

On the other hand, an important fraction of IDF residues was solubilized in boiling water. Only a part of the solubilized fraction corresponded to fibre constituents. This indicate that other soluble compounds were retained in the insoluble polysaccharides matrix.

The results suggest that the A.O.A.C. method yields higher IDF and SDF values than the actual content. Some modifications of the A.O.A.C. method to avoid the problems associated with precipitation and retention of soluble substances in the insoluble fraction are considered. These considerations could be applied to other fruits and vegetables as well.

5 REFERENCES

1.- L. Prosky, N-G. Asp, I. Furda, J.W. DeVries, T.F. Schweizer and B.F. Harland, J. Assoc. Off. Anal. Chem., 1985, 68, 677.

2.- L. Prosky, N-G. Asp, T.F. Schweizer, J.W. DeVries and I. Furda, J. Assoc. Off. Anal. Chem., 1988, 71, 1017.
3.- Biochemical Analysis/Food Analysis. Boehringer Mannheim. Cat. No. 207748 (test-combination).
4.- D. Southgate, "Determination of Food Carbohydrates", Applied Science Publishers, England, 1976.
5.- R.W. Scoot, Anal. Chem., 1979, 51, 936.

DETERMINATION OF DIETARY FIBRE IN VEGETABLES BY HPLC

MªD. Rodriguez, MªJ. Villanueva, A. Redondo and J.Prádena

Departamento de Nutrición y Bromatología II. Facultad de Farmacia. Universidad Complutense.
28040-Madrid (España)

1 INTRODUCTION

Several works[1-3] for hydrolysis of polysaccharides from Dietary Fibre and quantification of monosaccharides have been published. These procedures differ, mainly, in hydrolysis conditions and chromatographic techniques (GLC and HPLC) used to identify and quantify the monosaccharides.

The objective of this work is to study the method of Englyst[1], but HPLC was used instead of GLC. Preparation of samples for HPLC was carried out following Slavin and Marlett[2] method.

2 MATERIALS AND METHODS

In the Englyst method, hydrolysis of polysaccharides consists of the following steps: primary hydrolysis with H_2SO_4 12M at 35ºC during 1h and secondary hydrolysis with H_2SO_4 1M at 100ºC during 2h. Preparation of samples for HPLC analysis implies a neutralization step with $Ba(OH)_2$ or $BaCO_3$.

To study primary hydrolysis microcrystalline and crystalline cellulose was used. Two conditions of the procedure were changed: temperature (35ºC and room temperature) and neutralizing agent ($Ba(OH)_2$ and $BaCO_3$). Two different kind of columns were used and resolution in both cases was compared.

3 RESULTS AND DISCUSSION

In Table 1 recoveries obtained for microcrystalline and crystalline cellulose are shown. Though Slavin and Marlett

TABLE 1 Percent Recovery of Glucose from Cellulose Using Various Chemical Hydrolysis Procedures

Cellulose	Primary Hydrolysis (Temperature)	Neutralizing Agent	%R
Microcrystalline	35ºC	$Ba(OH)_2$	44.89
"	"	$BaCO_3$	44.92
"	25ºC	"	71.79
Crystalline	35ºC	$BaCO_3$	84.01

method[2] is carried out with $Ba(OH)_2$ as neutralizing agent, neutralization step is long and tedious. It was tried to use instead $BaCO_3$ which made the procedure easier and quicker. Percentage of recovery was similar with both agents (44.89 and 44.92). Other modifications were related to temperature used during the primary hydrolysis. Results show that when microcrystalline cellulose was hydrolyzed at room temperature percentage of recovery increased (71.79). This was possibly due to a lower loss of glucose at that temperature in strong acid conditions. However, when analyzing crystalline cellulose keeping 35ºC of temperature during primary hydrolysis, it was noticed that percentage of recovery was higher (84.01). Once efficacy of hydrolysis was studied with cellulose, separation of neutral monosaccharides was studied in an amine-bonded column (Carbohydrate Analysis, Waters Associates). Resolution was not good; mannose, glucose ahd galactose elute at the same time. For this reason, this column was substituted by an aminex HPX-87P heavy metal cation exchange carbohydrate column (Bio-Rad, Richmond Laboratories,CA). Separation was excellent. Carrots were analyzed following Englyst method. Neutralization was carried out with $BaCO_3$ and analysis was done using both columns. It was confirmed that HPX-87P column gave better resolution.

REFERENCES

1. H.N. Englyst and J.H. Cumming, J. Assoc. Off. Anal. Chem., 1988, 71, 4, 808-814.

2. J.L. Slavin and J.A. Marlett, J. Agric. Food Chem., 1983, 31, 3, 467-471.
3. J.M. Ruiter and J.C. Burns, J. Agric. Food Chem., 1986, 31, 330-336.

CHARACTERISATION OF THE DIETARY FIBRE OF CASSAVA

S.H. Brough, R.J. Neale, G. Norton and J. Rickard

Department of Applied Biochemistry and Food Science
School of Agriculture
Sutton Bonington
Leics. LE12 5RD.

1 INTRODUCTION

Cassava is an important staple for some 450-500 million people in tropical regions where, in some cases it provides up to 60% of the total calorie intake. (Cock, 1982). The crop is highly productive even under adverse conditions. Approximately two thirds of the annual production is used for human consumption, around 20% for animal feed and the remainder for industrial processing.

Cassava, whilst being an excellent source of highly digestible starch (30-40% fresh wt.; 85% DM), contains little fat and protein and in some instances, has appreciable quantities of cyanogenic glycosides. (Cock, 1982). These glycosides seriously restrict the utilisation of fresh cassava for human consumption.

Thus, if cassava is to be utilised efficiently in human food and animal feed, a detailed knowledge of its composition is essential. Information on the composition of the dietary fibre of cassava is scarce. Work was initiated in this Department to characterise this component in cassava as part of a broader study to appraise the nutritional importance of all carbohydrates and the anti-nutritional factors.

2 MATERIALS AND METHODS

Cassava, cultivars HMC1 and MCol 1684 containing different levels of cyanogenic glycosides, was obtained from CIAT Columbia through ODNRI. Both cultivars underwent two drying procedures - tray and floor dried in the sun - resulting in four different samples for analysis. Dietary fibre analysis was carried out on the ground material (1mm) according to the procedure of Englyst and Cummings (1984)

3 RESULTS

TABLE 1

COMPOSITION OF THE DIETARY FIBRE OF CASSAVA.

SAMPLE	TOTAL SUGAR mg/g DM	SOL[1]	INSOL	NCP	CELL	LIGNIN[2]
1684 T	90.03	32.94	67.12	52.75	44.73	4.52
1684 F	91.85	30.93	69.87	68.21	29.82	4.64
HMC T	78.37	29.96	72.81	62.40	35.20	4.07
HMC F	90.37	27.12	74.33	60.09	33.37	4.76

1 SOL (Soluble NSP); INSOL (Insoluble NSP); NCP (Non-cellulosic polysaccharides); CELL (Cellulose). All values are expressed as total sugars (% Total NSP - Non-starch polysaccharides)

SOL NSP is calculated as TOTAL NSP - INSOL NSP
CELL is calculated as TOTAL NSP - NCP

2 LIGNIN is expressed as mg/g DM and is determined gravimetrically.

In all samples INSOL NSP and SOL NSP accounted for 70% and 30% of the TOTAL NSP respectively (Table 1). Cellulose amounted to between 30-44% of the TOTAL NSP. NCP contents varied between 53-68% TOTAL NSP. The monosaccharide composition of the TOTAL NSP of cassava, variety HMC1 Tray, is presented in Table 2 along with that for potato obtained by Englyst, Wiggins and Cummings (1982) and Graham and Aman (1988).

4 DISCUSSION

The content of soluble and insoluble NSP in cassava were similar to those obtained for raw potato (Graham and Aman, 1988) (Table 2). The monosaccharide composition of the total NSP of cassava, however, differed to that of potato with respect to rhamnose and xylose, being much higher in the former and mannose, and galactose being present to a greater extent in the latter (Table 2).

In contrast, Anderson and Bridges (1988) and Englyst and Cummings (1982) obtained much higher values of 57.2% and 53.7% respectively for the soluble NSP in potato. Also, the total NSP value for potato obtained by Englyst and Cummings (1982) was considerably lower than that reported by Graham and Aman (1988). Equally significant were the differences in the monosaccharide composition of the total NSP obtained by the two authors (Table 2).

TABLE 2

MONOSACCHARIDE COMPOSITION OF THE TOTAL NSP OF CASSAVA AND POTATO.

SAMPLE	RHA[1]	ARA	XYL	MAN	GAL	GLC	GalUA	LIGNIN[2]	TOTAL[2]
CASSAVA (HMC T)	5.07	5.68	9.15	2.64	16.71	38.19	22.54	4.07	81.82
POTATO[a]	2.3	4.9	2.03	4.06	19.8	38.9	21.54	6.46	78.90
POTATO[b]	2.5	6.6	1.95	1.17	31.12	36.19	20.43	N/A	51.40

1 Individual sugars expressed as a % of the total.
2 Total sugars and lignin expressed as mg/g DM.

a Graham and Aman, 1988
b Englyst and Cummings, 1982.

Cellulose contents of approximately 30% of the total NSP were obtained for potato by both Anderson and Bridges (1988) and Englyst and Cummings (1982) which is consistent with that obtained for cassava in this study. However, Holloway, Monro, Gurnsey, Pomare and Stace (1985) only obtained a value of 14% for cellulose.

It may be speculated therefore, that the dietary fibre of cassava will behave in a similar manner to that of potato in the gastro-intestinal tract. This aspect is currently being investigated.

REFERENCES

1. Cock J.H. , Science, 1982, 218, 755.
2. Englyst H. and Cummings J. , Analyst, 1984, 109, 937.
3. Englyst H., Wiggins H.S. and Cummings J.H, Analyst, 1982, 107,307..
4. Graham H and Aman P., J.Agric. Food Chem., 1988, 36, 494.
5. Anderson J.W. and Bridges S.R, Am. Jnl Clin. Nutr., 1988, 47, 440.
6. Holloway W.D., Monro J.A., Gurnsey J.C., Pomare E.W. and Stace N.H Jnl Food Sci., 1985, 50, 1756.

ACKNOWLEDGEMENTS

This work was supported by a research grant from ODNRI.

ANALYSIS OF THE DIETARY FIBRE FROM Olea europaea (Gordal and Manzanilla var.).

A. Heredia; R. Guillén; B. Felizón; A. Jiménez and J. Fernández-Bolaños.

Instituto de la Grasa y sus Derivados (C.S.I.C.)
Apartado 1078
41012-Seville (Spain)

1 INTRODUCTION

The olive (Olea europaea) is a foodstuff equilibrated in fibre, especially those varieties used as pickling fruits (Manzanilla and Gordal var.) which have good organoleptic characteristics and a considerable nutritive value.

The purpose of the present study has been to carry out the isolation of the different fibre fractions, using chemical and enzymatic methods, and to analyse the sugars obtained from their hydrolysis. Methodologies for fibre included neutral detergent fibre (NDF), acid detergent fibre (ADF), insoluble fibre (IF), soluble fibre (SF) and total fibre (TF).

2 MATERIALS AND METHODS

Materials.- Olives of Manzanilla and Gordal varieties were harvested in the Province of Seville (Spain). The fruits were stored at 4°C for no longer that 24 hours until used.

Fibre determination. Neutral detergent fibre were determined from the de-fatted flesh, using the procedure of Van Soest et al (1). Insoluble and soluble fibre were isolated by the method of Asp (2).

Acid hydrolysis. The polysaccharides of ADF, NDF, IF and SF were subjected to sequential hydrolysis with concentrated sulphuric acid and then more dilute.

Neutral sugars determination. The analysis of the sugars resulting from hydrolysis was carried out by GC. The alditol

The percentages of SF are close to 1% , and those of TF close to 5%. The fibre values obtained by system of Van Soest (NDF and ADF) are lower than those of IF and TF.

Acid hydrolysis of the fibre fractions. After carrying out a study of the optimun conditions, a primary hydrolysis was established with H_2SO_4 (72 %) for two hours at 40°C, and a secondary hydrolysis with H_2SO_4 (2N) for two hours at 40°C. Table III show the composition of neutral sugars (Gordal Var.) Glucose is the main major sugar in all the fractions except in SF, in which it is arabinose. The highest percentages of galactose are found in IF and SF, whereas this sugar is partially solubilized in NDF, and only traces appear in ADF. The mannose content is notably equal in IF, ADF and NDF, and much lower in SF. Part of the xylose is solubilized in the ADF treatment, only traces appear in SF, while the highest percentages are those in IF and NDF. An important fraction of arabinose is lost in the NDF treatment, practically disappears in the ADF, it is found in high percentage in IF, and in SF it is much lower. Ramnose is present in small quantities in all the fractions.

Table III. Hydrolyzates composition of the fibre fractions (%) in Gordal var.

	IF	SF	ADF Original	ADF Delignified	NDF
Glucose	26.04	6.090	54.160	39.350	21.560
Galactose	0.78	2.950	1.120	0.030	0.430
Mannose	0.085	0.400	1.490	0.890	0.680
Xylose	6.76	1.770	4.210	2.520	4.460
Arabinose	5.86	2.040	0.190	0.130	2.590
Ramnose	0.57	1.530	traces	traces	0.140

The values are the means of two determinations; coefficient of variation 10%

Cellulose. The results obtained are shown in Table IV. Significant differences between fractions, but not between varieties were observed. There aren't significant differences between the cellulose content of ADF, NDF and IF fractions, but higest values are obtained by Van Soest's method. This could be explained by the fact that sugars corresponding to hemicellulose have been detected in the de-lignified ADF fraction.

Hemicelluloses. The results showed that very low precision was got with the Van Soest's method (C.V. 40%). In relation with the hemicellulose content of ADF, NDF and IF, the higest value was found for IF. It's important to remark that an important quantity of hemicelluloses was present in ADF fractions.

acetates were prepared by the method of Englyst et al (3), with little modifications.

3 RESULTS AND DISCUSSION

Fruit composition. The fruit had been picked in the best conditions of ripeness for pickling (ripe green fruit) The results obtained for moisture, fat, proteins and free sugars are shown in Table I.

Table I Composition of Olives (Olea europaea, var. Gordal and Manzanilla) (%):

	Gordal	Manzanilla
Moisture	71.62 ± 0.06*	69.66 ± 0.04*
Dry mather	28.38	30.34
Fat	15.19 ± 0.30	16.03 ± 0.44
Protein	1.29 ± 0.14	1.20 ± 0.07
Free sugars	3.94 ± 0.55	3.75 ± 0.54
Ash	0.58 ± 0.007	0.47 ± 0.006
Free sugars		
Sucrose	0.30 ± 0,03	traces
Glucose	2.97 ± 0.35	2.33 ± 0.23
Fructose	0.37 ± 0.03	0.90 ± 0.17
Manitol	0.31 ± 0.07	0.37 ± 0.00

* Percentages refered to: fresh flesh
All values are means of four analysis ± standard desviation

Fibre determination. Table II summarizes the results of NDF, ADF, SF and IF for both varieties. The main major is IF, with very similar values for Gordal and Manzanilla.

Table II. Fibre fractions, Gordal and Manzanilla var. (%).

	Gordal	Manzanilla
ADF	2.08 ± 0.08	2.41 ± 0.08
NDF	2.82 ± 0.23	3.19 ± 0.10
IF	4.03 ± 0.09	4.26 ± 0.15
SF	1.01 ± 0.13	0.65 ± 0.32
TF	5.04	4.91

Percentages refered to fresh flesh

Table IV. Fibre fraction composition (%).

	IF	SF	ADF	NDF
(GORDAL VARIETY)				
Cellulose	26.04 ± 2.01*	6.09 ± 0.2*	54.16 ± 5.40* 59.72 ± 0.41**	21.56 ± 0.75*
Hemicelluloses	14.82 ± 1.20	18.69 ± 5.75	5.95 ± 0.20	8.30 ± 0.03 26.54 ± 14.01**
Lignin	20.59 ± 0.20	-	39.97 ± 0.38	29.32 ± 0.3
Protein	18.06 ± 1.06	3.81	3.62 ± 0.09	17.06 ± 0.8
(MANZANILLA VARIETY)				
Cellulose	18,69 ± 1.02	1.70 ± 0.11	3.4 ± 1.5 46.6 ± 3.57	17.02 ± 2.4
Hemicelluloses	8.44 ± 0.30	5.36 ± 0.20	3.12 ± 0.03	7.57 ± 1.19 24.44 ± 0.34**
Lignin	-	-	48.49 ± 0.46	-
Protein	18.66 ± 1.55	2.62	16.24 ± 1.8	14.65 ± 0.22
Ash	2.79	-	0.47 ± 0.06	-

Percentages to: * fibre; ** Determined by Van Soest method.

REFERENCES

1. Van Soest. J.P., J. Assoc. Off. Anal. Chem., 1963, 46, 825.
2. Asp. N.G.; Johansson, C.G.; Hallmen, H.; Siljestrom, M., J. Agric. Food. Chem. 1983, 31, 476.
3. Englyst, H.; Wiggins, H.J.; Cumming, J.H., Analyst 1982, 107, 307.

ISOLATION AND CHARACTERISATION OF DIETARY FIBRE IN WHITE ASPARAGUS

A. Redondo and M.J. Villanueva
Departamento de Bromatología y Técnicas Analíticas Farmaceúticas
Facultad de Farmacia.
Universidad Complutense. 28040-MADRID (Spain)
A. Heredia; R. Guillén and A. Jiménez
Instituto de la Grasa (CSIC). Apartado 1078. 41012-SEVILLA (Spain)

1. INTRODUCTION

The production of white asparagus (Asparagus officinalis) in Andalucia (Spain) has undergone a very important expansion in recent years. The commercialization of white asparagus for its consumption is very scarce in Spain, with almost all going to the conserve industry.

The objetive of the present work has been to establish the general composition of processed white asparagus, with special attention in the fibre fraction because of its nutritive, organolectic and industrial characteristics.

2. MATERIALS AND METHODS.

The asparagus are processed by the traditional system in the conserve industry and differentiated by three zones: the apex, middle, and base, which are triturated and then individually frozen.

The fibre determinations are carried out by different methods: acid detergent fibre (ADF), and neutral detergent fibre (NDF), both described by Van Soest (1), and the total fibre (TF), insoluble fibre (IF) and soluble fibre (SF), developed by Asp (2).

The different fibre fractions are subjected to the acid hydrolysis, and the resulting sugars analysed by gas chromatography as alditol acetates (3).

The pectic substances are similarly analyzed by the method of Blumenkrantz (4).

In each fibre fraction obtained, the residual content of proteins and uronic acids is determined.

Results analysis have been made by ANOVA techniques. Means comparison have been made by Duncan's method.

3. RESULTS AND DISCUSSION

The distribution of the chemical constituents of the different zones of white asparagus (apex, middle and base) is studied, as well as that of the whole product. Table I shows the percentages of the main components, expressed as fresh product. No significant differences are seen in the composition of the different zones of this vegetable.

TABLE I.- Composition of different zones or white asparagus (%).

	APEX	MIDDLE	BASE	WHOLE
Moisture	93.49±0.41	94.70±0.01	94.83±0.14	94.25±0.19
Dry Matter	6.51±0.41	5.31±0.01	5.17±0.14	5.76±0.19
Proteins	3.26±0.11	2.56±0.10	2.39±0.17	2.65±0.16
Free Sugars	22.34±1.83	22.73±0.86	24.94±1.07	24.35±1.24
Uronic Acids	0.39±0.01	0.28±0.01	0.31±0.04	0.34±0.01
Ash	1.13±0.08	1.00±0.08	1.14±0.01	1.14±0.04

The results obtained for the fibre (NDF, ADF, IF, SF, TF) are shown in Table II, expressed as dry matter. No significant differences (= 0.05) are found between the various parts of the asparagus. Significant differences between the fibre fractions have been found. The great superiority of NDF over ADF is indicative of the high hemicellulose content of asparagus. Ratio ADF: Hemicelluloses 1:1

It is interesting to remark that the NDF values are higher than IF ones.

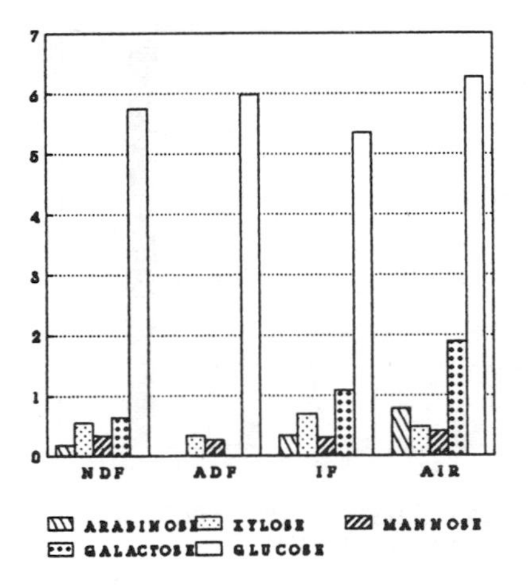

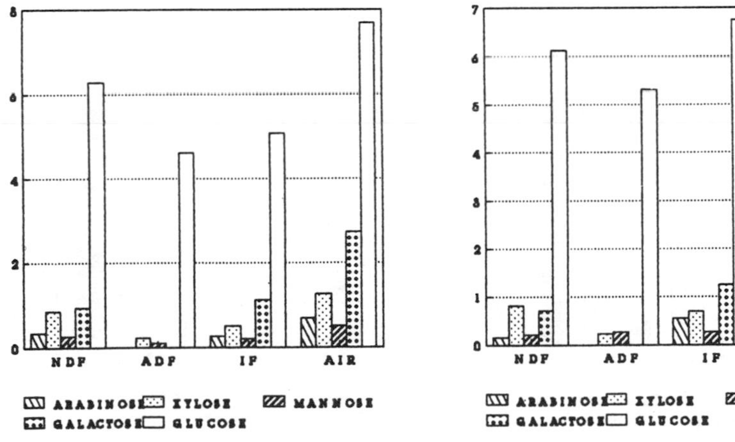

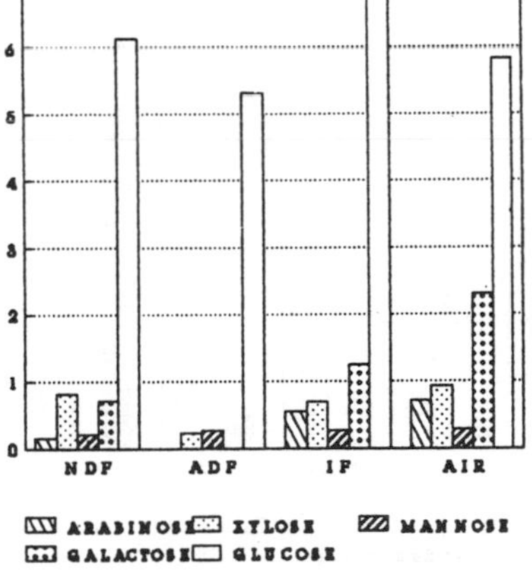

Fig 1.- Hydrolyzates composition of fibre fractions (% dry matter)

APEX ZONE

BASE ZONE

WHOLE ASPARAGUS

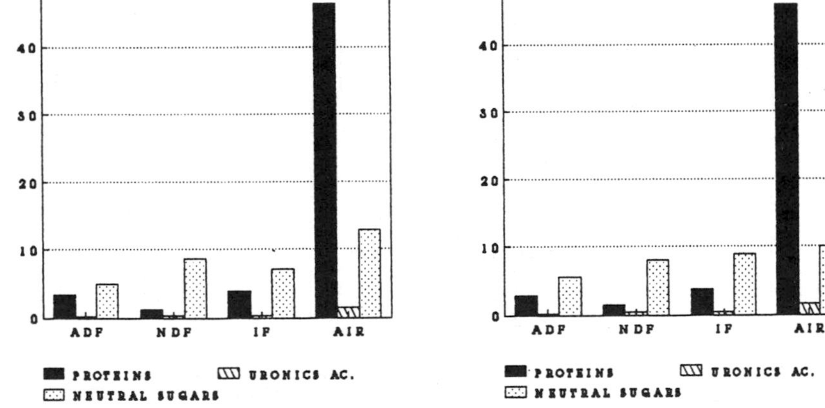

Fig. 2.- Fibre fractions composition (% dry matter)

TABLE II. Fibre fractions of white aparagus (%)*.

	APEX	MIDDLE	BASE	WHOLE
ADF	10.17±0.06	12.26±0.42	12.77±0.12	11.35±0.13
NDF	24.79±0.62	20.74±0.40	19.92±2.26	23.65±0.27
FI	20.97±0.04	18,87±1.44	21.60±0.27	19.38±0.15
FS	6.35±0.43	6.01±0.23	5.81±0.51	5.05±0.33
FT	27.31±0.47	24.88±1.21	27.41±0.24	24.43±0.47

* refered to dry matter

Figure 1 shows the composition of the hydrolyzates of the fibre fractions. In all fractions, except ADF, the main sugar in the different parts is glucose, followed by galactose and xylose, and in lower concentration, mannose and arabinose. In fraction ADF small quantities of hemicellulosic sugars are detected (mannose and xylose).

The alcohol insoluble fraction (AIR) composition is also showed as a measure of the recovery of each monosaccharide.

Figure 2 shows the composition in proteins, uronic acids and neutral sugars of the different fibre residues.

Significant differences between fibre fractions in uronic acids, but not in protein content, are found. Uronic acids in NDF and IF are statistically equal and higher than ADF. The protein solubilisation has been very good in all treatments.

REFERENCES

1 Van Soest, P.J. J. Assoc. Offic. Anal. Chemist (1963) 46, 829.

2 Asp, N.G.; Johansson, G.C.; Hallmen, H.; Siljestron, H. J. Agric. Food Chem. (1988), 31, 476.

3 Englyst, H.; Wiggins, H.J.; Cumming, J.H. Analyst (1982), 107 302.

4 Blumenkrantz, X. and Asboe-Hansen, G. Anal. Biochem. (1973) 54, 484.

WORKSHOP REPORT: ANALYSIS OF PLANT CELL WALLS IN RELATION TO FIBRE

H. Englyst

MRC Dunn Nutrition Unit
100 Tennis Court Road
Cambridge CB2 1QL

The original hypothesis was related to plant cell-wall material and it is suggested by Dr. Eastwood, Dr. Englyst and Professor Southgate that the measurement of dietary fibre (DF) should represent an index of this. Non starch polysaccharides (NSP) are a measure of cell-wall polysaccharides and therefore represent a good index of cell-wall material.

It was shown that for some types of unprocessed foods similar DF values are obtained by both the gravimetric AOAC-Asp technique and by the measurement of NSP by the Englyst procedure. However, for many types of processed foods, quite different values are obtained because the AOAC-Asp procedure includes part of the starch retrograded during processing, Maillard reaction products and other lignin-measuring substances.

Either type of method can possibly be used for some unprocessed foods, especially cereals, as an index of cell-wall material. For processed foods, the inclusion of lignin-measuring substances and retrograded starch makes the AOAC-Asp procedure unsuitable as an index of plant cell-wall material. Furthermore, the inclusion of substances formed during food processing, as in the AOAC-Asp procedure, make it impossible to calculate the DF content using the recipe and food table values for the DF content of the raw foods.

Dr. Asp suggested that starch retrograded during processing should be regarded as a legitimate component of DF. However, Dr. Cummings and Dr. Englyst have shown that the amount of starch included in the measurements by the AOAC-Asp procedure represents only a small proportion of that escaping digestion in the small intestine, and see no reason for its inclusion as DF. They propose that all the starch escaping digestion is important and that it should be measured separately.

Dr. Englyst presented a rapid colorimetric technique allowing total, soluble and insoluble dietary fibre (NSP) to be measured within a normal 8 hour working day, whilst analysis by the AOAC-Asp procedure and by the Englyst GLC procedure requires 1.5 days.

It was mentioned by Professor Asp that the gravimetric AOAC-Asp procedure could be modified so that it would not include retrograded amylose. Dr. Jeraci presented a gravimetric technique where all starch, including retrograded amylose, was dispersed by urea and it was pointed out that in Canada three methods - the gravimetric AOAC-Asp, the Englyst NSP and the gravimetric procedure developed by Dr. Mongeau, Ottawa - were accepted as official methods for the measurement of dietary fibre.

It was generally agreed that it was advantageous to have more than one type of technique for the measurement of DF provided that the various methods measure the same components of the diet.

Professor Southgate and others agreed that methods giving information about the properties of the different types of DF are required for research purposes. For routine analysis including food labelling the detailed information is not always required, but it is essential that the overall values obtained by research methods and the more rapid methods for routine analysis are in agreement.

Effect of Fibre on the Small Intestine: Implications for Digestion and Nutrient Absorption

THE BIOLOGICAL EFFECTS OF DIETARY FIBRE IN THE SMALL INTESTINE

I. T. Johnson

AFRC Institute of Food Research
Colney Lane
Norwich NR4 7UA

Most of our present knowledge of nutrient absorption has been achieved using tightly controlled experimental conditions, and relatively simple physiological media. However, during normal nutrition, the intestinal lumen provides a much more complicated environment. The dietary fibre hypothesis has done much to focus attention on the fact that nutrients are usually absorbed from complex food materials that must be physically degraded and digested before transport processes can take place. In most human diets, plant cell wall polysaccharides are an essential component of the intraluminal milieu, and it is now becoming clear that they can play an important regulatory role in many aspects of intestinal physiology.

Barrier Effects of Cell Walls

The cellular organisation of plant tissues depends upon the presence of non-starch polysaccharides which comprise the structural components of cell walls. There is now extensive evidence to show that the physical characteristics of fruits and lightly processed cereal products can influence the rate of digestion and assimilation of carbohydrate (1,2). Furthermore, the cellular structure of particularly robust plant tissues can survive processing, cooking and mastication, and hence moderate the rate of digestion of starch and other nutrients in the small intestine. Jenkins et al (3) have systematically investigated the variation in human glycaemic response to carbohydrate foods, and have shown that many legumes give particularly small blood glucose excursions following a test-meal. Some have ascribed this effect to the supposed presence of viscous "gel-forming" fibre in such foods (4). In fact viscosity appears not to be relevant in this particular context. (5,6). Wursch *et al* (7) studied the relationship between cellular integrity and the rate of starch hydrolysis in a variety of legumes, and confirmed that whereas cooking alone would ensure the gelatinisation of starch granules in white kidney beans, the starch remained encapsulated within intact cell walls. The presence of an intact diffusion barrier impeded the access of α-amylase and significantly slowed starch hydrolysis *in vitro*. Rigorous processing can disrupt cell walls however, and increase the rate of starch digestion both *in vitro* and *in vivo*.

Such studies suggest that for some diets one of the most important functions of fibre is to maintain the structural organisation of plant foods during digestion, and hence to exert a moderating influence over nutrient digestion and post-prandial metabolism (8,9). However these effects cannot be described or predicted by any currently used chemical assay for the routine determination of total fibre in foods, nor can they be reproduced by the addition of isolated polysaccharides as a supplement to the diet. Failure to allow for this important functional aspect of dietary fibre may explain why the beneficial effects of legume-rich diets in the management of non-insulin dependent diabetes (10) have not always been observed when simple fibre supplements have been used (11).

Binding Effects of Dietary Fibre

The factors which control the disintegration of plant tissues and cell walls during digestion have not been thoroughly investigated. The residual material in the lumen of the small intestine after a meal containing plant foods will include partially degraded cell walls with a range of physical and chemical properties. Most foods will yield insoluble but finely divided particles, together with some soluble polysaccharides such as pectins and β-glucans. There is a growing tendency to distinguish between soluble and insoluble fibre in physiological studies, and, in some cases even on food labels. This is a potentially useful development but the currently available techniques for the analysis of fibre do not necessarily predict the solubility of fibre under physiological conditions (12).

In many foods, the barrier effects of intact cell walls will be lost as the result of disruption by processing, cooking, and the mechanical effects of gastrointestinal motility. However the constituents of disrupted cell walls may continue to influence nutrient absorption by binding both enzymes and food components in the gut lumen. This topic has received a considerable amount of attention, both because of fears that binding of inorganic micronutrients may have adverse effects on mineral balance in persons consuming high fibre diets, and because the adsorption of bile salts and lipids has been proposed as a mechanism to account for the hypocholesterolaemic effects of some forms of fibre.

The classic studies of Widdowson and McCance (13) established that subjects consuming whole-meal bread showed reduced iron absorption in comparison to that from white-bread, and this observation has since been confirmed by other groups. Isolated components of fibre have been shown to bind iron in vitro, but the extent of such binding depends on the physical conditions of such studies and the presence of other ligands. Cereal brans are complex materials which contain high levels of phytic acid in addition to dietary fibre, and when wheat bran is dephytinised much of the iron-binding activity is lost (14). Therefore, although diets with high levels of fibre may lead to negative mineral balances, it is not necessarily the dietary fibre component of complex carbohydrate foods which is primarily responsible for the mineral binding effect.

Many components of dietary fibre have the capacity to bind bile salts and the lipid components of mixed micelles in vitro (15), and this effect has long been thought to play some role in human lipid metabolism (16). Under normal circumstances bile salts are efficiently reabsorbed in the distal ileum. If a sufficiently high proportion of bile salts are bound to some non-absorbable phase in the small intestinal lumen, this process of entero-hepatic circulation is partially interrupted and bile salt excretion is increased, leading to a compensatory increase in hepatic bile salt synthesis. This is accompanied by a fall in serum LDL cholesterol, presumably as a result of increased LDL uptake by the liver and perhaps other tissues (17). The mechanism is exploited clinically in the treatment of hereditary hypercholesterolaemia by the use of ion-exchange resins such as cholestyramine.

Both soluble and insoluble forms of dietary fibre have been shown to bind bile salts in vitro. Lignin is particularly efficient, but the binding capacity depends upon the presence of methoxyl and β-carboxyl groups, and most importantly, on the ambient pH (18). Removal of lignin substantially reduces the bile acid binding capacity of alfalfa, but residual activity appears to be associated with the hemicellulose fraction (19). Pectin (20) and other soluble polysaccharides, such as the D-mannopyrannose polymer of guar gum and the β-glucan of oat bran, also exhibit bile-acid binding capacity *in vitro* (21). Nevertheless the physiological implications of these results are uncertain. Bile salt binding is most efficient at acidic pH, and it declines markedly at neutral pH, which is characteristic of the distal ileal lumen. Furthermore, attempts to correlate the effectiveness of bile salt binding determined *in vitro* with faecal bile salt excretion and hypocholesterolaemic effects *in vivo* have not been consistently successful. It is therefore unlikely that this mechanism provides more than a partial explanation for the well established cholesterol-lowering properties of some high-fibre diets (22).

Some components of dietary fibre can bind to the endogenous enzymes of the gastrointestinal tract. *In vitro* studies have shown that wheat bran, pectin and guar gum reduce the activity of pancreatic amylase and lipase, at least partly as a result of adsorption (23,24). There are substantial reserves of digestive capacity in man and it seems unlikely that this effect is a rate-limiting factor during digestion. However there is some evidence of an adaptive increase in pancreatic output in rats fed wheat bran (25). If feedback regulation of pancreatic function occurs in man it might be of some physiological significance.

The Effects of Soluble Polysaccharides on Luminal Convection

Amongst various proposals for mechanisms whereby dietary fibre could influence the energy content of the human diet, Southgate suggested that the presence of the fibre matrix in the gut lumen might reduce the diffusion of solutes to the mucosal surface and thereby lead to malabsorption of nutrients (26). This concept was first tested experimentally by Jenkins and co-workers (27) who showed that when isolated components of fibre were added as supplements to test-meals, only soluble materials with a high viscosity reduced the post-prandial blood glucose response. The mechanism underlying this observation remained conjectural however until *in vitro* experiments carried out by Elsenhans *et al* (28), and by Johnson and

Gee (29) established that dilute solutions of the highly viscous hydrocolloid guar gum could markedly reduce the rate of absorption of monosaccharides. The use of in vitro methods enabled quantitative kinetic studies to be undertaken which showed that the reduction in transport rate was greatest at intermediate substrate concentrations, implying an increase in the kinetic parameter Km, the apparent substrate affinity of the mucosal carriers (30). This was subsequently shown to be true also for other soluble gums such as carboxymethylcellulose (31) and β-glucan (32).

The epithelial surfaces of transporting tissues are overlain by a poorly stirred boundary-layer or "microclimate" in which solute concentrations differ from those in the well-stirred aqueous "bulk phase". This phenomenon is well known to biochemists wishing to measure transport kinetics in vitro because the presence of an unstirred water leads to a reduction in the substrate concentration at the cell surface, and hence to an over-estimate of the true Km value. There is no doubt that such a layer also exists in the gut in vivo, although both the complex morphology of the mucosa and the irregular fluid movements induced by the peristaltic motion of the gut wall make its dimensions and physiological significance uncertain (33). The concept of the unstirred water layer provides a useful model to describe the effects of viscous media on nutrients transport in the small intestine. By measuring the time-course for the development of transmural electrical potentials, induced by imposing an osmotic load on isolated mucosal tissue incubated with polysaccharide-enriched media, it has been possible to demonstrate that the apparent thickness of the unstirred layer increases in a viscosity-dependent manner (34,31). Moreover this critical range of viscosities is consistent with that observed in the small intestinal contents recovered from rats after dietary supplementation with guar gum (35).

In isolated loops of intestine, a retardation of sugar and water absorption due to intraluminal guar gum has been demonstrated by Blackburn and Johnson (36), using a two-stage perfusion of rat jejunum. Similar results were obtained by Rainbird et al using a single-stage perfusion in the conscious pig (37), and by Blackburn et al using intubated human jejunum (38). In general the effect has not been observed in closed intestinal loops where both flow and peristalsis are absent (39), or in preparations where perfusion at high viscosity has led to distension of the gut, with a consequent increase in mucosal surface area (40). In an elegant paper, Anderson et al have recently shown that these various observations are consistent, and in the intact intestine, a viscosity-dependent reduction in the rate of transport only occurs when the lumen is well stirred by peristalsis (41).

The combined evidence obtained from these various approaches to the problem seems to confirm that viscous forms of dietary fibre do indeed impede access of solutes to the mucosal surface, and other mechanisms, such as a reduction in the rate of gastric emptying, are probably only of secondary importance (38). The increased resistance to transport appears to be due primarily to viscous damping of convective fluid movement in the vicinity of the mucosa, so that the path-length for solute diffusion is increased. The presence of a dispersed hydrocolloid must also impede free diffusion to some degree, but the reduction in the diffusion coefficient appears to be negligible for small solutes and low concentrations of polymer (42,43). Precise values for the viscosity of guar and other polymers during digestion in vivo are difficult to estimate because their rheological behaviour is non-Newtonian. Viscosity will vary with the intra-luminal shear-rate, and this itself is an unknown quantity during peristalsis. There is no doubt however that intraluminal viscosity can be increased in man after dietary supplementation with guar or similar polysaccharides, and perhaps after eating oats (44). In practice, the precise values achieved in vivo are probably not of great significance.

Guar gum (45) and oat gum (32) both reduce the rate of uptake of cholesterol from mixed micelles in vitro. Apart from the mechanism already discussed, there is also some evidence from in vitro experiments to suggest that these polysaccharides may directly impede the diffusion of mixed micelles which are very much larger than sugar molecules (46). Furthermore, the rate of cholesterol uptake is proportional to that of water absorption in perfused preparations of small intestine (41). Viscous polysaccharides reduce the rate of water uptake, and this effect probably also favours a reduction in the rate of lipid absorption.

Secondary Effects of Delayed Nutrient Absorption

All the various mechanisms which have been outlined so far have the effect of delaying nutrient absorption, but it is not clear to what extent they also cause losses of nutrients to the large bowel. An important distinction needs to be drawn between the barrier and binding effects of fibre, which probably prevent absorption of some nutrients entirely, and the convective effects which probably only delay absorption, but in so doing modify the normal intraluminal concentration profile (47). This latter effect is important because the small intestinal mucosa shows an adaptive response to exposure to nutrients. Both types of mechanism can lead to important metabolic consequences through a reduction in the rate of uptake of carbohydrate as glucose, and perhaps because of an increase in the quantity of carbohydrate absorbed as bacterial fermentation products. A full discussion of these effects is outside the scope of the present paper, but mention will be made here of the possible impact on gastrointestinal endocrinology.

The relatively low post-prandial glycaemic response seen after carbohydrate meals containing large quantities of legumes, or those enriched with viscous gums, are associated with a reduced insulin response in normal subjects or non insulin-dependent diabetic patients (48). Morgan et al observed a statistically significant reduction of about 70% in post-prandial gastric inhibitory peptide (GIP) levels in healthy volunteers given a guar-enriched mixed meal (49), and a smaller reduction in glucagon-like immunoreactivity (GLI). More recently Trinick et al reported that the relatively low glucose and insulin response to a guar-enriched liquid meal was associated with a statistically insignificant reduction in GIP, but observed no effect on glucagon, C-peptide and pancreatic polypeptide (50). The reduction in GIP secretion after a test-meal containing guar gum is probably a reflection of the reduced rate of glucose absorption in the proximal small bowel. The reduced GLI levels reported by Morgan et al (49) are more difficult to explain because enteroglucagon is released by specialised mucosal cells in the distal small bowel and colon. A delay in absorption would therefore be expected to increase GLI release. However, much may depend upon the duration of the experiment and the design of the sampling procedure. Paganua et al have reported raised enteroglucagon levels in diabetic children given diets supplemented with guar gum and fructose (51), but this increase may have been due to stimulation of enteroglucagon secreting cells by volatile fatty acids, following fermentation of guar. It is unclear what proportion of enteroglucagon in normal human subjects is derived from the large bowel in man, but it has been observed that ileostomists have lower enteroglucagon levels than control subjects (52).

Adaptation to Fibre in the Small Intestine

Most of our knowledge concerning the adaptive response of the small intestine to high fibre diets is derived from studies with the rat, which exhibits morphological adaptation to plant materials in the diet (53). Plant foods are highly complex in composition however, and substances such as protein and pharmacologically active plant cell constituents can influence small intestinal growth. Only recently have the effects of cell wall materials begun to be studied in isolation.

Several groups have observed an increase in small intestinal length in growing rats given diets enriched with viscous dietary fibre. This gross morphological response is accompanied by an increase in mucosal mass and DNA content (54,55,56). The intestinal mucosa is a highly proliferative tissue in a constant state of cell renewal, and it seems probable that the morphological changes are a reflection of an increased rate of cell proliferation. Such an effect was reported by Jacobs (56) and confirmed by other groups using the metaphase arrest technique to measure absolute rates of crypt cell proliferation (57,58). An enhancement of cell proliferation is seen throughout the small intestine of rats given pectin, guar gum and sodium carboxymethylcellulose (59), but it does not occur in rats fed gum arabic which is soluble in the small intestine but non-viscous (60).

Several groups have reported changes in mucosal enzyme activity in rats fed isolated sources of fibre, but the reports are not entirely consistent. For example whereas Brown et al (54) and Thomsen and Tasman-Jones (61) observed decreased mucosal disaccharidase activity in rats fed pectin or galactomannan-supplemented diets, other groups have reported unchanged or slightly elevated levels (62,63). In the present author's experience, polysaccharides which stimulate cell proliferation are associated with reduced alkaline phosphatase and lactase activity compared to rats fed a fibre-free diet, whereas maltase displays a complex change in the distribution of activity throughout the small intestine (58). It is probable that mucosal enzyme activities reflect changes in the average maturity of mucosal cells in the epithelium, complicated by the induction of specific enzymes, perhaps in response to luminal nutrient concentration.

If the reduced mucosal enzyme activity seen in response to some forms of dietary fibre is one expression of a less "mature" population of mucosal cells, then a similar reduction in carrier transport capacity might also be expected. The maximum jejunal transport rate for 3-O-methyl glucose in rats maintained for 4 weeks on a semi-synthetic diet containing guar gum was approximately 40% lower than in rats fed a similar diet containing insoluble cellulose (57). Schwartz et al (64) reported changes in enzyme activity and a reduction in glucose and leucine transport in animals fed diets containing pectin for 6 weeks. There is also evidence that lipid absorption in the rat is modified by an adaptive response to prolonged feeding with various components of dietary fibre (65).

Several mechanisms have been proposed to explain the changes in cell proliferation which seem to underlie the adaptive response to fibre in the small intestine. Short-chain fatty acids are produced in the colon by fermentation of carbohydrates, and Sakata has proposed that these are growth-promoting factors for the small intestine (66). However this does not provide an entirely satisfactory explanation for the trophic effects of dietary fibre when the responses to particularly well-defined polysaccharides are considered. Sodium carboxymethylcellulose is a highly viscous gum which stimulates cell proliferation (60) but it is almost entirely resistant to fermentation in the rat colon (67). Conversely, gum arabic, which is non-viscous but highly fermentable, has no trophic effect in the rat small intestine (60). An increase in the level of enteroglucagon in the blood usually accompanies trophic stimuli such as massive resection of the small bowel, and this peptide has long been regarded as a probable mucosal growth hormone (68). The high levels of enteroglucagon seen in rats fed guar gum and other viscous polysaccharides seemed to corroborate this hypothesis. However enteroglucagon levels are also increased in rats given gum arabic, suggesting that there is no obligatory trophic response to this peptide (60). Germ-free rats also have high basal levels of enteroglucagon with no corresponding increase in cell proliferation compared to normal controls (69).

It seems probable that the increase in cell proliferation seen in animals given certain soluble polysaccharides is an adaptive response to high intraluminal viscosity. The increased pre-mucosal diffusion resistance which occurs under these conditions may be assumed to delay nutrient clearance, thus displacing the intra-luminal concentration profile in such a way that the more distal mucosa is exposed to unusually high concentrations of nutrients, and of lipids in particular. Fatty acids have recently been shown to elicit a particularly significant trophic response (70).

Conclusions

The presence of dietary fibre in the gut lumen exerts a number of primary effects on the digestion and absorption of nutrients which have important metabolic consequences, some of which can be usefully exploited in man. The barrier effect of fibre depends upon the cellular structure of plant tissues. It is perhaps the most important small intestinal effect in relation to the dietary management of diseases such as diabetes, but, being a structural attribute, it cannot be predicted by the analysis of total fibre-content, nor can it be achieved simply by adding fibre supplements to manufactured food products. The secondary effects of fibre on the structure and function of the small intestine which are observed in laboratory animals seem to reflect an adaptive response to alterations in the distribution of nutrients in the small bowel, as a result of high luminal viscosity. There are few sources of viscous polysaccharides in conventional western diets, but the pharmaceutical use of guar gum and other polysaccharides, and perhaps high intakes of oat bran, may lead to similar effects in man. The long-term consequences of small bowel adaptation to dietary fibre are unknown.

REFERENCES

1. G.B. Haber, K.W. Heaton, D. Murphy and L.F. Burroughs, Lancet, 1977, ii, 679.
2. J.C. Brand, P.L. Nicholson, A.W. Thorburn and A.S. Truswell, Am.J.Clin.Nutr., 1985, 42, 1192.
3. M.J. Thorne, L.V. Thompson and D.J.A. Jenkins, Am.J. Clin.Nutr., 1983, 38, 481.
4. J.I. Mann, Brit.Med.J., 1984, 288, 1025.
5. S. Wong and K. O'Dea, Am.J.Clin.Nutr., 1983, 37, 66.
6. J.M. Gee and I.T. Johnson, J.Sci.Food Agric., 1985, 36, 614.
7. P. Wursch, S. Del Vedovo and B. Koellreutter, Am.J. Clin.Nutr., 1986, 43, 25.

8. D.J.A. Jenkins, T.M.S. Wolever, R.H. Taylor, C. Griffiths, K. Krzeminska, J.A. Lawrie, C.M. Bennet, D.V. Goff, D.L. Sarson and S.R. Bloom, Am.J.Clin. Nutr., 1982, 35, 1339.
9. L. Tappy, P. Wursch, J.P. Randin, J.P. Felber and E. Jequier, Am.J.Clin.Nutr., 1986, 43, 30.
10. S.E. Lousley, D.B. Jones, P. Slaughter, R.D. Carter, R. Jelfs and J.I. Mann, Diabetic Medicine, 1984, 1, 21.
11. V.A. Beattie, C.A. Edwards, J.P. Hosker, D.R. Cullen, J.D. Ward, N.W. Read, Brit.Med.J., 1988, 296, 1147.
12. J.A. Marlett, J.G. Chesters, M.J. Longacre and J.J. Bogdanske, Am.J.Clin.Nutr., 1989, 50, 479.
13. E.M. Widdowson and R.A. McCance, Lancet, 1942, 1, 588.
14. L. Hallberg, C. Rossander and A-B Skanberg, Am. J. Clin.Nutr., 45, 988.
15. G.V. Vahouny, R. Tombes, M.M. Cassidy, D. Kritchevsky and L.L. Gallo, Lipids, 1980, 15, 1012.
16. R.M. Kay, J.Lipid Res., 1982, 23, 221.
17. P.N. Hopkins and R.R. Williams, Am.J.Clin.Nutr., 1981, 34, 2560.
18. R.M. Kay, S.M. Strasberg, C.N. Petrunka and M. Wayman, "Dietary Fibres: Chemistry and Nutrition", G. Inglett and I. Falkenhag Eds, Academic Press, New York, 1979, 57-66.
19. J.A. Story, A. White and L.G. West, J.Food Sci., 47, 1276.
20. R.R. Selvendran, Chem.Ind., 1978, 428.
21. J.A. Story, "Dietary Fibre: Basic and Clinical Aspects", G.V. Vahouny and D. Kritchevsky Eds, Plenum Press, New York, 1986, 265-274.
22. J.W. Anderson, L. Story, B. Sieling, W.-J.L. Chen, M.S. Petro and J.A. Story, Am.J.Clin.Nutr., 1984, 40, 1146.
23. G. Isaksson, I. Lundquist and I. Ihse, Gastroenterology, 1982, 82, 918.
24. W.E. Hansen and G. Schulz, Hepatogastroenterol., 1982, 29, 157.
25. N.F. Sheard and B.O. Schneeman, J.Food Sci., 1980, 45, 1645.
26. D.A.T. Southgate, Proc.Nutr.Soc., 1973, 32, 131.
27. D.J.A. Jenkins, T.M.S. Wolever, A.R. Leeds, M.A. Gassul, P. Haisman, J. Dilawari, D.V. Goff, G.L. Metz and K.G.M.M. Alberti, Brit.Med.J., 1978, 1, 1392.
28. B. Elsenhans, V. Sufke, R. Blume, W.F. Caspary, Clin.Sci., 1980, 59, 373.
29. I.T. Johnson and J.M. Gee, Proc.Nutr.Soc., 1980, 39, 52A.
30. B. Elsenhans, V. Sufke, R. Blume and W.F. Caspary,

Digestion, 1981, 98.
31. I.T. Johnson and J.M. Gee, Pflug.Arch., 1982, 393, 139.
32. E.K. Lund, J.M. Gee, J.C. Brown and I.T. Johnson, Brit.J.Nutr., 1989, 62, 91.
33. F.A. Wilson and J.M. Dietschy, Biochim.Biophys.Acta , 1974, 363, 112.
34. I.T. Johnson and J.M. Gee, Gut, 1981, 22, 398.
35. N.A. Blackburn and I.T. Johnson, Br.J.Nutr., 1981, 46, 239.
36. N.A. Blackburn and I.T. Johnson, Pflug.Archiv., 1983, 397, 144.
37. A.L. Rainbird, A.G. Low and T. Zebrowska, Br.J.Nutr., 1984, 52, 489.
38. N.A. Blackburn, J.S. Redfern, H. Jarjis, A.M. Holgate, I. Hanning, J.H.B. Scarpello, I.T. Johnson and N.W. Read, Clin.Sci., 1984, 66, 329.
39. H. Forster and I. Hoos, Nutr.Metab., 1977, 21 (Suppl. 1), 262.
40. B. Elsenhans, D. Zenker and W.F. Caspary, Gastroenterology, 1984, 86, 645.
41. B.W. Anderson, J.M. Kneip, A.S. Levine and M.D. Levitt, Gastroenterology, 1989, 97, 938.
42. G. Holzheimer and D. Winne, Naun.Schmeid.Arch. Pharmacol., 1986, 334, 514.
43. C.A. Edwards, I.T. Johnson and N.W. Read, Eur.J. Clin.Nutr., 1988, 42, 307.
44. K.W. Heaton, S.N. Marcus, P.M. Emmett and C.H. Bolton, Am.J.Clin.Nutr., 1988, 47, 675.
45. J.M. Gee, N.A. Blackburn and I.T. Johnson, Br.J.Nutr., 1983, 53, 215.
46. D.R. Phillips, J.Sci.Food Agric., 1986, 37, 548.
47. N.J. Brown, J. Worldling, R.D.E. Rumsey and N.W. Read, Br.J.Nutr., 1988, 59, 223.
48. D.J.A. Jenkins, A.R. Leeds, M.A. Gassull, B. Cochet and K.G.M.M. Alberti, Ann.Intern.Med., 1977, 86, 20.
49. L.M. Morgan, T.J. Goulder, D. Tsiolakis, V. Marks and K.G.M.M. Alberti, Diabetologia, 1979, 17, 85.
50. T.R. Trinick, M.F. Laker, D.G. Johnston, M. Keir, K.D. Buchanan and K.G.M.M. Alberti, Clin.Sci., 1986, 71, 49.
51. A. Paganus, J. Maenpaa, H.K. Akerblom, V.-H. Stenman, K. Knip and O. Simell, Acta.Paediatr.Scand., 1987, 76, 76.
52. H.J. Kennedy, "The Health of Ileostomists", DM Thesis, University of London, 1981.
53. M.K. Younosjai, M. Adedoyin and J. Ranshaw, J.Nutr., 1978, 108, 341.
54. R.C. Brown, J. Kelleher and M.S. Losowsky, Br.J.Nutr., 1979, 42, 357.

55. B. Elsenhans, R. Blume and W.F. Caspary, Am.J.Clin.Nutr., 1981, 34, 1837.
56. L.R. Jacobs, Am.J.Clin.Nutr., 1983, 37, 954.
57. I.T. Johnson, J.M. Gee and R.R. Mahoney, Br.J.Nutr., 1984, 52, 477.
58. R.A. Goodlad, W. Lenton, M.A. Ghatei, S.R. Bloom and N.A. Wright, Gut, 1987, 28, 171.
59. I.T. Johnson and J.M. Gee, Br.J.Nutr., 1986, 55, 497.
60. I.T. Johnson, J.M. Gee, J.C. Brown, Am.J.Clin.Nutr., 1988, 47, 1004.
61. L.L. Thomsen and C. Tasman-Jones, Digestion, 1982, 23, 253.
62. S.E. Schwartz, C. Starr, S. Backman and P.G. Hotzapple, J.Lipid Res., 1983, 24, 746.
63. R. Calvert, B.O. Schneeman, S. Satchithanandum, M.M. Cassidy and G.V. Vahouny, Am.J.Clin.Nutr., 1985, 41, 1249.
64. S.E. Schwarz, C. Starr, S. Bachman and P.G. Holtzapple, J.Lipid Res., 1983, 24, 746.
65. G.V. Vahouny, S. Satchithanandam, I. Chen, S.A. Tepper, D. Kritchevsky, G.G. Lightfoot and M.M. Cassidy, Am.J.Clin.Nutr., 1988, 47, 201.
66. T. Sakata, Brit.J.Nutr., 1987, 58, 95.
67. G.M. Wyatt, N. Horn, J.M. Gee and I.T. Johnson, Brit.J.Nutr., 1988, 60, 197.
68. L.C. Jacobs, S.R. Bloom and R.H. Dowling, Life Sci., 1981, 29, 2003.
69. R.A. Goodlad, B. Ratcliffe, J.P. Fordham, M.A. Ghatei, J. Domin, S.R. Bloom and N.A. Wright, Quart.J.Exper. Physiol., 1989, 74, 437.
70. A.P. Jenkins and R.P.H. Thomson, Clin. Sci., 1989, 77, 555.

INFLUENCE OF WHEAT BREADS CONTAINING GUAR FLOUR SUPPLEMENTS OF HIGH AND LOW MOLECULAR WEIGHTS ON VISCOSITY OF JEJEUNAL DIGESTA IN THE PIG

F. G. Roberts[1,2], H. A. Smith[1], A. G. Low[1] and P. R. Ellis[2]

[1]AFRC Institute for Grassland and Animal Production
Church Lane, Shinfield, Reading RG2 9AQ, Berkshire

[2]Kings College, London,
Department of Food and Nutritional Sciences
Campden Hill Road, London W8 7AH

1 INTRODUCTION

Guar gum, a high molecular weight galactomannan, is known to be of therapeutic value in the treatment of diabetes largely because of its capacity to reduce post-prandial blood glucose and improve lipid metabolism[1]. The mechanism by which guar improves glycaemic control is not, as yet, fully understood but current evidence indicates that a primary factor is the way in which guar slows the rate of carbohydrate absorption in the small intestine by modifying the rheological behaviour of digesta[2]. There is however a lack of information about the levels of viscosity induced by guar *in vivo* and how this relates to changes in blood nutrients and hormones.

We have recently demonstrated in pigs that peak viscosities of guar-containing digesta (collected 2.0m distal to the pylorus) occurred within the first 60 min. of the post-prandial period corresponding with the time during which the greatest decreases in glycaemia and insulinaemia have been observed in previous studies[3]. The test meals used in our study were fully-hydrated solutions of guar gum added to a low-fat, semi-purified diet. However, guar-containing foods, which are more palatable than pure solutions, contain the polymer in a relatively unhydrated form. Hence the aim of this study was to determine whether a guar-containing wheat bread, when administered to pigs, elicits comparable rheological changes in digesta to those achieved with fully-hydrated, pure guar solutions[3].

2 METHODS

Four Large White x Landrace boar pigs of initial weight 28-30kg were fitted with re-entrant cannulae 2.0m distal to the pylorus in the mid-jejunum. The pigs were fed twice daily at a level of 4% body weight. Each meal consisted of a low fat, semi-purified (SP), wheat starch-based diet and/or white wheat bread (based on the recipe of Apling and Ellis[4]). The test bread meals contained 60g/kg of either M150 (high viscosity and high molecular weight) or M60 (low viscosity and low molecular weight) guar flour (Meyhall Chemical UK Ltd., Wirral). Each pig received a total of seven diet combinations (3 control and 4 test) as follows: 100% semi-purified (SP) diet, 100% control bread, 100% M60 bread, 100% M150 bread, 50:50 SP diet:control bread, 50:50 SP diet:M60 bread and 50:50 SP diet:M150 bread. Each diet was fed to each pig for 7 days.

Samples (2 x 30 ml) of jejeunal digesta were collected immediately following the morning feed (time 0 min.) and at 10, 20, 30, 40, 50, 60 ,90, 120, 150, 180, 210 and 240 min. Apparent viscosity was measured immediately at 39°C (Brookfield DV-II rotoviscometer, Stoughton, USA) across a range of shear rate conditions. 'Zero shear' viscosity[5] was then calculated.

3 RESULTS AND DISCUSSION

The maximum viscosities for all treatments occurred within 90 min. of feeding (Table 1) and this is consistent with our earlier studies using fully-hydrated guar solutions[3]; in both cases peak viscosities corresponded with maximal falls in glycaemia and insulinaemia in similar studies in man[6]. The viscosity of jejeunal digesta was considerably higher for all guar bread treatments than for the controls (Figures 1 and 2). The viscosities were considerably lower for the 100% guar bread treatments than for our previous study with pure guar solutions[3]. In the latter case, although the concentration of guar in the diet was much lower, it was fully hydrated and this induced a higher initial viscosity. The much more sustained high levels of viscosity when the guar bread meals are fed were of interest in relation to between-meal control of glycaemia; this effect may be the result of progressive hydration of guar within the bread matrix. It is also notable that the viscosity profile induced by the low and high viscosity guar in the 100% bread meals was very similar and may explain why these two grades of gum elicit similar

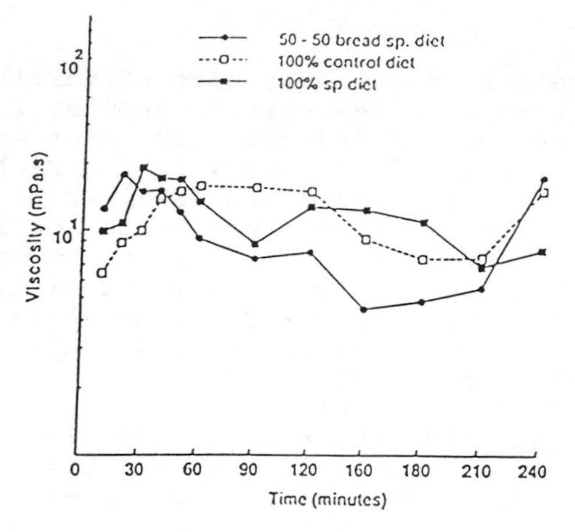

Figure 1 Zero shear viscosity of jejeunal digesta following control treatments.

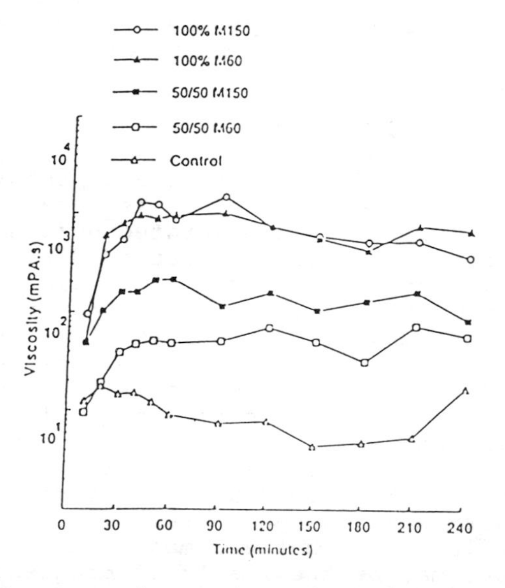

Figure 2 Zero shear viscosity of jejeunal digesta following guar bread treatments.

Table 1 Maximum 'zero shear' viscosity (mPa.s) of jejeunal digesta ($\bar{x} \pm$ se)

	Control	M60	M150
100% SP	18 (1)	---	---
50:50 SP Bread	17 (1)	70 (3)	257* (10)
100% bread	16 (1)	1005** (11)	1454** (16)

significantly different from control; * $p>0.05$; ** $p>0.01$

falls in post-prandial plasma insulin in healthy human subjects (Ellis et al., unpublished observations).

4 CONCLUSIONS

1. Bread containing guar flour induced higher jejeunal viscosities than either control bread or semi-purified diet.

2. The longer period of high viscosity seen with bread meals, compared to meals containing fully-hydrated guar solutions, may be the result of progressive hydration of the guar over several hours.

3. Similar increases in viscosity were seen for 100% bread meals whether high or low viscosity guar was incorporated, and may explain why low viscosity guar reduces post-prandial insulinaemia in man.

F.G. Roberts gratefully acknowledges receipt of an AFRC research studentship. The guar gum was kindly donated by R.M.W. Hopkins of Meyhall Chemical (UK) Ltd. We thank E.R. Morris for helpful advice.

REFERENCES

1. D.B. Peterson, In: 'Dietary fibre perspectives', (ed. A.R. Leeds), John Libbey, London, 1985, p.47.

2. C.A. Edwards and N.W. Read, In: 'Dietary fibre perspectives', (ed. A.R. Leeds), John Libbey, London, 1990, p.52.

3. F.G. Roberts, H.A. Smith, A.G. Low, P.R. Ellis, E.R. Morris and I.E. Sambrook, Proceedings of the Nutrition Society, (in press), 1990.

4. E.C. Apling and P.R. Ellis, Chem. Inds. Lond., 1982, 950-954.

5. G. Robinson, S.B. Ross-Murphy and E.R. Morris, Carbohyd. Res., 1982, 107, 17.

6. D.J.A. Jenkins, T.M.S. Wolever, A.R. Leeds, M.A. Gassull, P. Haisman, J. Dilawari, D.V. Goff, G.L. Metz and K.G.M.M. Alberti, Br. Med. J., 1978, i, 1392.

EFFECTS OF SOME SOLUBLE NON-STARCH POLYSACCHARIDES ON FOOD INTAKE, CELL TURNOVER AND GASTROINTESTINAL PEPTIDE LEVELS IN THE RAT

J. M. Gee and I. T. Johnson

AFRC Institute of Food Research
Colney Lane
Norwich NR4 7UA

1 INTRODUCTION

In the growing rat, the consumption of some soluble non-starch polysaccharides leads to an enlargement of the small intestine, increased mucosal cell proliferation, and reduced levels of blood cholesterol compared to control groups[1]. These changes, which are frequently accompanied by increased circulating levels of the small intestinal peptide enteroglucagon, may be an adaptive response to a high luminal viscosity in the proximal small intestine, associated with a reduced rate of nutrient transport[2]. However, the interpretation of the results from such experiments is frequently complicated by differences in food intake. In the present study we have compared the effects of two soluble non-starch polysaccharides on the release of gastrointestinal peptides and intestinal mucosal cell proliferation, in rats, with and without food restrictions.

2 METHODS

In the present study three groups of weanling rats were fed ad libitum on a fibre-free semisynthetic diet, (FF), or diets containing guar gum, (GG), or gum arabic, (GA), both of which are readily fermentable, but only one of which, (GG), is highly viscous in solution. Three other groups received either the fibre-free diet, a cellulose-supplemented diet, (C), or a starch-supplemented diet, (S), at levels paired to the intakes of the guar-fed rats. The composition of the diets is given in Table 1; food intakes were monitored throughout the experimental period. After 28 days the animals received injections of the anti-mitotic drug vincristine sulphate (1mg. kg^{-1}, Sigma, Poole,

UK), and were subsequently sacrificed at approximately equal intervals between 12 and 120 minutes thereafter. Each animal was deeply anaesthetised by an intraperitoneal injection of sodium barbiturate (Sagatal; M & B Dagenham, UK), and, after laparotomy, venous blood samples were collected, followed by cervical dislocation. The small intestine was removed and its unstretched length measured. Segments (1cm) of distal ileum (95% of length) were removed and placed in fixative (75% ethyl alcohol : 25% acetic acid) for subsequent microdissection and determination of crypt cell production rate by the Stathmokinetic method[3]. Serum and plasma samples were prepared from the collected blood. Glucagon-like immunoreactivity (GLI) and neurotensin in plasma were determined by radioimmunoassay (Novo Biolabs and Immunodiagnostics UK, respectively. Gastrin was measured in serum by radioimmunoassay (Immunodiagnostics, UK).

Table 1 Composition of diets ($g.kg^{-1}$)

Constituent	FF	GG	C	S	GA
Starch	330	330	330	430	330
Sucrose	330	330	330	330	330
Casein	200	200	200	200	200
Corn oil	80	80	80	80	80
Mineral mix+	40	40	40	40	40
Vitamin mix+	20	20	20	20	20
D-L-methionine	2.5	2.5	2.5	2.5	2.5
Guar gum	-	100	-	-	-
Cellulose (Solka Floc)	-	-	100	-	-
Gum arabic	-	-	-	-	100

+ Johnson, et al.[2].

3 RESULTS

Total food intake by the GG-fed rats was approximately 15% lower than that of the FF and GA groups fed ad lib (Fig 1). GG-fed rats had serum gastrin and neurotensin levels significantly lower than those fed the FF diet ad lib, but not compared to the pair-fed animals (Fig 2). In contrast, GLI activity was approximatley 3 fold higher in both the GG and GA fed groups compared to all others (Fig 2).

Cell proliferation in the distal ileum was significantly increased in the GG-fed rats compared to all other groups, and the total small intestinal length was greater (Fig 3).

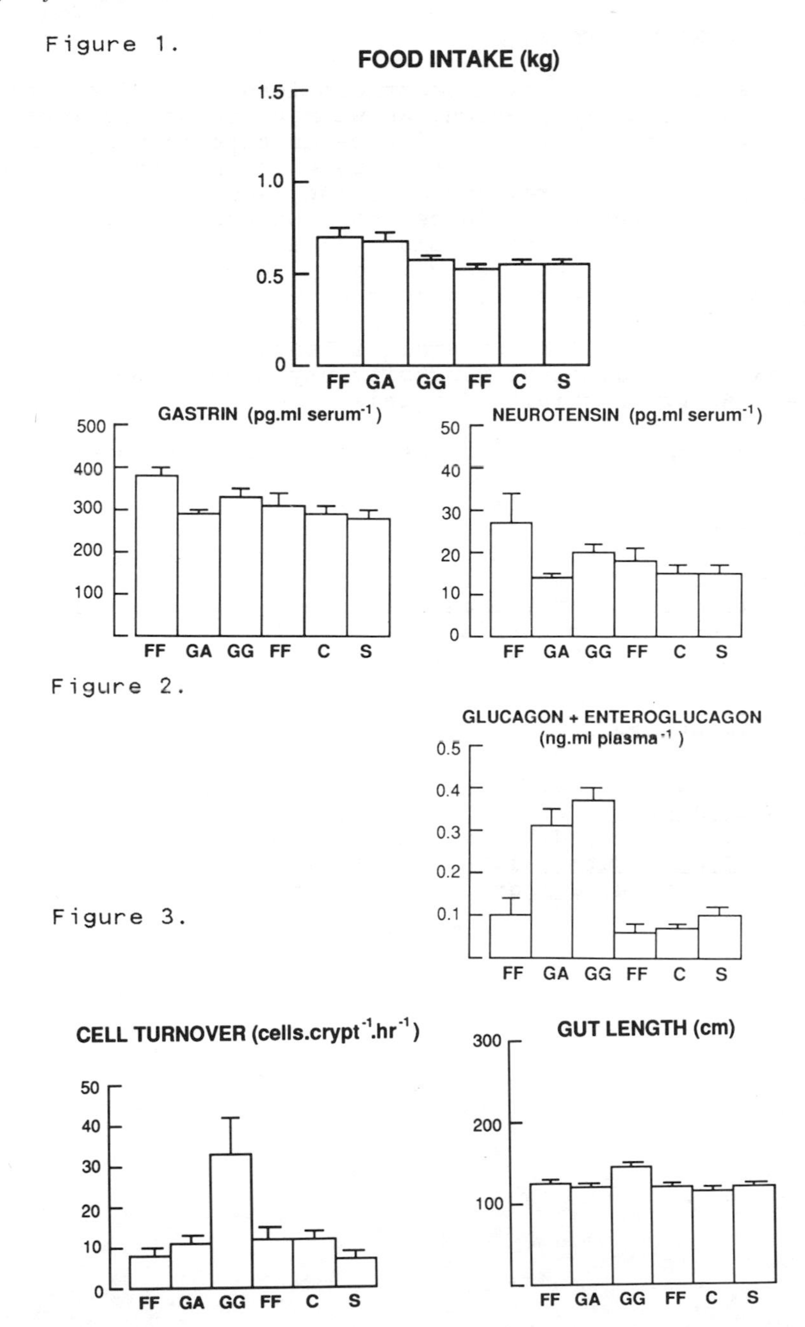
Figure 1.
FOOD INTAKE (kg)
1.5
1.0
0.5
0
FF GA GG FF C S
GASTRIN (pg.ml serum-1)
500
400
300
200
100
FF GA GG FF C S
NEUROTENSIN (pg.ml serum-1)
50
40
30
20
10
0
FF GA GG FF C S
Figure 2.
GLUCAGON + ENTEROGLUCAGON
(ng.ml plasma-1)
0.5
0.4
0.3
0.2
0.1
FF GA GG FF C S
Figure 3.
CELL TURNOVER (cells.crypt-1.hr-1)
50
40
30
20
10
0
FF GA GG FF C S
GUT LENGTH (cm)
300
200
100
FF GA GG FF C S

4 CONCLUSIONS

Soluble non-starch polysaccharides lead to an increase in plasma-GLI activity, which, as we have shown elsewhere, is accounted for primarily by increased enteroglucagon[2]. The present study illustrates that this is independent of any effect of food intake, and does not depend on increased viscosity in the small intestine. However, faster cell proliferation and greater small intestinal length were observed only in rats fed guar gum, which suggests that neither the provision of a fermentable substrate, nor high levels of enteroglucagon in the plasma provide sufficient stimuli for this form of small intestinal adaptation. Increased intraluminal viscosity may however be an essential pre-requisite for this effect.

Serum levels of both gastrin and neurotensin were reduced by the modest food restriction imposed in this study, but there was little evidence of any effect of guar gum. However the GA-fed rats, which showed no significant decline in food consumption, had lower levels of both peptides compared to FF-fed controls. The reason for this is unclear but it may indicate an effect of soluble polysaccharides on nutrient flow or luminal concentration of nutrients in the small bowel which is independent of viscosity.

REFERENCES

1. I.T. Johnson, J.M. Gee and R. R. Mahoney, Brit.J.Nutr., 1984, 52, 477-487.
2. I.T. Johnson, J.M. Gee and J.C. Brown, Am.J.Clin.Nutr., 1988, 47, 1004.
3. R.M. Clarke, J.Anat., 1970 107 519-529.

FIBRE AND INTESTINAL EPITHELIAL CELL PROLIFERATION

R.A. Goodlad, B. Ratcliffe*, J.P. Fordham ** C.Y. Lee and N.A. Wright.

Imperial Cancer Research Fund, Histopathology Unit, 35-43 Lincoln's Inn Fields, London WC2A 3PN.
* Polytechnic of North London, Holloway Road, London N7 8DB.
** AFRC Institute of Food Research, Reading RG2 9AT.

INTRODUCTION

The gastrointestinal tract has one of the most rapid rates of cell renewal in the body,[1] intestinal epithelial cell proliferation and intestinal absorption[2] are profoundly influenced by food intake, and a considerable atrophy of the gut is seen when it is deprived of food or 'luminal nutrition',[3] this is perhaps best seen when the animal is fed intravenously.[4] A similar atrophy is seen in the colon of animals fed a fibre-free 'elemental' diet[5].

The Effects of Dietary Fibre

The colonic atrophy associated with fibre-free diets can be reversed by the addition of dietary fibre.[6] This phenomena was further investigated using the starvation-refeeding model, as this is one of the best ways of potentiating a proliferative response.[7] Refeeding starved rats with an elemental diet supplemented with a variety of dietary fibres provoked a marked proliferative response in the colon, and to a lesser extent in the small intestine, the magnitude of the response being greater in rats fed more fermentable fibres, with no response to inert bulk observed[8].

The effects of the intestinal microflora

Starvation Refeeding Experiment The starvation-refeeding model, (in which the rats were starved for three days and then refed for two days), was also utilised to investigate the contribution of the intestinal microflora to the proliferative response to dietary fibre. Crypt cell production was assayed using vincristine to arrest cells as they entered metaphase, and measuring the rate of accumulation of arrested cells in microdissected crypts.[9] Refeeding conventional rats with an elemental diet or with the elemental diet supplemented with inert bulk had little effect on intestinal proliferation. Crypt cell production in the colon increased sixfold when the diet was supplemented with a fermentable fibre mixture.[10] A smaller response was also observed throughout the small intestine. This response was not observed in germ-free rats. See Figure 1.

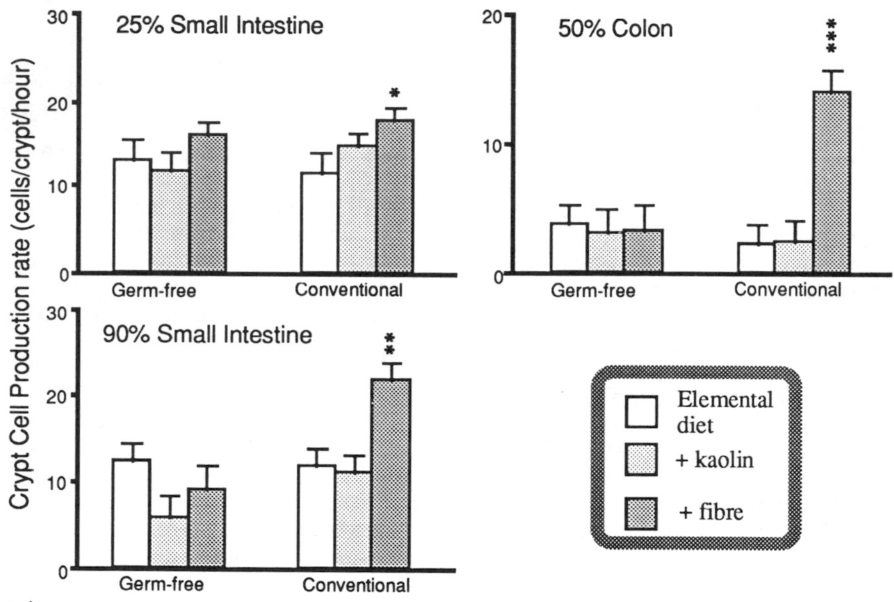

Figure 1. The effects of refeeding starved rats with a fibre-free elemental diet supplemented with inert bulk or a dietary fibre mixture.

Steady State Experiment The marked response to refeeding indicated that the response to fibre may also be seen in rats fed continuously. The rats were fed the elemental diet plus the fibre mixture (1 part ispaghula mucilage (Reckitt & Coleman Ltd) to 9 parts wheat grain fibrous extract (Labaz Sanofi Ltd). Crypt cell production rate was more than doubled in the colon of conventional rats, a small response was noted in the distal small intestine, but not in the proximal small intestine (Figure

2). Administration of the fibre-supplemented diet to the germ-free rats again had no proliferative effect.

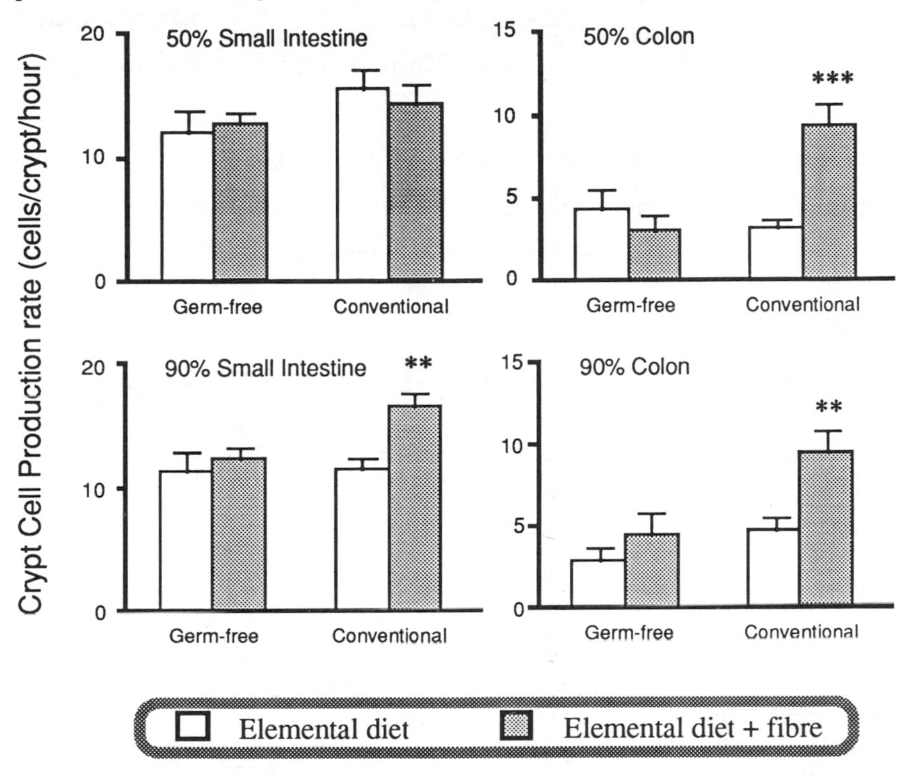

Figure 2. The effects of 2 weeks feeding with a fibre-free elemental diet or with the same diet supplemented with a dietary fibre mixture.

The wet weights of the various sections of the gastrointestinal tract are shown in Table 1. Two-way analysis of variance showed that diet had a significant effect ($P<0.001$) on the stomach, small intestine, caecum and colon weights. The microflora had significant ($P<0.001$) effects on the caecum and colon weights only.

Table 1 Tissue weight (as percentage of total body weight) mean ± standard error

	Germ-Free		Conventional	
	Flexical	Flexical + Fibre	Flexical	Flexical + Fibre
Stomach	0.366 ± 0.004	0.473 ± 0.011***	0.390 ± 0.006	0.459 ± 0.011***
Small intestine	2.056 ± 0.037	2.140 ± 0.033	1.944 ± 0.034	2.120 ± 0.036**
Caecum	0.537 ± 0.023	0.630 ± 0.014**	0.276 ± 0.006	0.353 ± 0.013***
Colon	0.360 ± 0.014	0.561 ± 0.015***	0.336 ± 0.008	0.498 ± 0.015***

Significantly greater than respective elemental diet** $P<0.01$, *** $P<0.001$.

DISCUSSION

We have shown that dietary fibre can stimulate epithelial cell proliferation in the colon, and (to a lesser extent) in the small intestine. This effect is dependent on the presence of the colonic microflora, indicating that it is the products of hind-gut microbial breakdown of fibre that has the trophic effect. The effect of diet on the colonic and caecal weights of the germ-free rats suggests that there is also an effect on the muscle layers, as was observed in the mouse.[6]

Fibre is broken down to short chain fatty acids, which can stimulate cell proliferation in ruminants[11] and in rodents.[12] This could either be a direct effect or moderated by some systemic influence.

Our own preliminary results also indicate a trophic effect of the direct infusion of short chain fatty acids. Sakata[12,13] has shown that this effect is systemic, which has been confirmed by Rombeau et al.[14]

The implications of a proliferative response to fibre are wide ranging, and sound a note of warning. Stimulation of intestinal proliferation can act as a promoter of carcinogenesis, as demonstrated by Jacobs.[15] Most animal carcinogenesis experiments use very powerful agents to initiate the process, thus these studies may concentrate excessively on the end stages of what is a complex and multistage process.

Dietary fibre has a multitude of effects on the gut, and proliferation is just one of these. Its other effects may well outweigh the proliferative risk, but this must be established, especially in the light of the rather dubious nature of most of the much publicised data used to advocate the health benefits of a greatly increased intake of dietary fibre.

REFERENCES

1 N.A. Wright and M.R. Alison. The biology of epithelial cell populations. Oxford University Press, Oxford, 1984, Vol.2.

2 Goodlad RA, Plumb JA & Wright NA. Epithelial cell proliferation and intestinal absorptive function during starvation and refeeding in the rat. Clin. Sci., 1988; 74: 301-306.

3 R.A Goodlad. Gastrointestinal cell proliferation. Dig. Dis., 1989; 7: 169-177.

4 R.A. Goodlad, T.G. Wilson, W. Lenton, N.A. Wright, H. Gregory and K.G. McCullagh. Intravenous but not intragastric urogastrone-EGF is trophic to the intestinal epithelium of parenterally fed rats. Gut 1987; 28: 573-582.

5 P. Janne, Y. Carpenter and G. Willems. Colonic mucosal atrophy induced by a liquid elemental diet in rats. Am. J. Dig. Dis., 1977; 22: 808-12.

6 R.A. Goodlad and N.A. Wright. The effects of addition of cellulose or kaolin to an elemental diet on intestinal cell proliferation in the mouse. Br. J. Nutr., 1983; 50: 91-98.

7 R.A. Goodlad, M.Y.T. Al-Mukhtar, M.A. Ghatei, S.B. Bloom & N.A.Wright. Cell proliferation, plasma enteroglucagon and plasma gastrin in starved and refed rats. Virch. Arch. (B), 1983; 43: 55-62.

8 R.A. Goodlad, W. Lenton, M.A. Ghatei, S.R. Bloom and N.A. Wright. Effects of an elemental diet, inert bulk and different types of dietary fibre on the response of the intestinal epithelium to refeeding in the rat and relationship to plasma gastrin, enteroglucagon and PYY levels. Gut 1987, 28: 171-180.

9 R.A. Goodlad and N.A. Wright. Quantitative studies on epithelial replacement in the gut. In techniques in the life sciences. Techniques in digestive physiology. Vol P2, Ed Titchen TA. Elsevier Biomedical Press Ireland, 1982; p212/1-212/23.

10 R.A. Goodlad, B. Ratcliffe, J.P. Fordham and N.A.Wright. Does dietary fibre stimulate intestinal epithelial cell proliferation in germ-free rats?. Gut 1989; 30: 820-825

11 R.A. Goodlad. Some effects of diet on the mitotic index and the cell cycle of ruminal epithelium of sheep. Q. J. exp. Physiol., 1981; 66: 487-499

12 T. Sakata and T. Yajima. Influence of short chain fatty acids on the epithelial cell division of digestive tracts. Q. J. exp. Physiol, 1984-; 69: 639-48.

13 T Sakata. Stimulatory effects of short-chain fatty acids on epithelial cell proliferation of isolated and denervated jejunal segment of the rat. Scand. J. Gastroenterol., 1989; 24: 886-890.

14 M.J. Koruda, R.H. Rolandelli, R.G. Settle, D.M. Zimmaro and J.L. Romneau. Effects of parenteral nutrition supplemented with short-chain fatty acids on adaptation to massive small bowel resection. Gastroenterology 1988; 95: 715-20.

15 L.R.Jacobs. Effect of dietary fibre on colonic cell proliferation and its relationship to colon carcinogenesis. Prev. Med., 1987; 16: 566-71.

THE INFLUENCE OF DIFFERENT DIETARY FIBRE SOURCES ON DIGESTIBILITIES OF PROTEIN AND DRY MATTER IN RAT

Santosh Khokhar and A.C. Kapoor

Department of Foods & Nutrition
Haryana Agricultural University
Hisar, Haryana - 125004, India

1 INTRODUCTION

With increase in awareness of the importance of consuming more fibres in the human diet due to its protective role against a number of diseases of modern societies, its nutritional importance has become increasingly recognised. It is now evident that not only the amount of fibre in diet but also the type and source of dietary fibre is important. Inclusion of guar gum in an otherwise complete diet causes a significant decrease in apparent protein digestibility and a significant rise in faecal dry matter[1]. Less common fruits found in semi-arid zones of India have been reported to affect the biological utilization of the diet[2]. This paper presents the result on comparative affects of different sources of dietary fibre on digestibilities of protein and dry matter, and their relationship.

2 MATERIALS AND METHODS

Fibre Sources

Cellulose was procured from Patel Chest Institute, New Delhi, Isabgol (*Plantago ovato*) and cabbage (*Brassica oleracea*) from local market and fresh guave fruit (*Psidium guajava*) from Horticulture Farm of Haryana Agricultural University, Hisar, India.

Cabbage leaves were cleaned, cut into small pieces, spread under fan at room temperature (40-42°C) to make it dry and then ground to fine powder using 0.5 mm sieve in a cyclotec mill. After cleaning guava fruit, the pulp (free from seeds) was prepared by using an electric pulper, dried at 50°C in dehydrating chambers and ground to a fine powder (0.5 mm sieve). Both plant foods, cabbage and guava were analysed for netural detergent fibre (NDF), acid detergent fibre (ADF), cellulose and lignin by the method of Van Soest and Wine[3]. Hemicellulose was determined as the difference between NDF and ADF. Pectin was determined as calcium pectate[4]. Total dietary fibre was calculated as a sum of hemicellulose, cellulose, lignin and pectin.

Cellulose, pectin, cabbage and guava were incorporated in the diets so as to provide 5 and 10 per cent dietary fibre. As Isabgol has very high water holding capacity (about 40 times) and it is used in very small amounts for medical purposes it was fed at low levels (1 and 2 per cent). The fibre free diet was employed as control diet. The protein content of diets was adjusted to 10 per cent with egg albumin after taking into account the true protein content of plant foods. Other dietary ingredients were added as per standards for nutritional studies[5].

Feeding Experiment

Eighty eight male weanling rats, weighing 27-35 g were randomly divided into eleven groups and housed in individual cages in an air-conditioned room maintained at 22±2°C with 12 hours dark and light period.Food and water were provided *ad libitum*. Feeding lasted for 37 days. Nine days before the expiry of the feeding, the rats were transferred to metabolic cages for estimations recorded in Table 2. Diet residue, faecal matter and urine were collected daily. The rats were weighed twice a week. The faeces and urine were collected separately from each rat and pooled at the end of experiment. Pooled samples of faeces and diet residues were dried in hot air oven at 80°C for 12 hr, cooled in a desiccator and weighed. To the urine a few drops of H_2SO_4 were added and kept under frozen condition. Nitrogen in urine and faeces was determined by the micro-kjeldahl method[6]. Endogenous N in the faeces and urine was determined by feeding a group of rats on protein free diet containing cellulose (5 per cent) as dietary fibre for five days.

Apparent protein digestibility (APD) and true protein digestibility (TPD) were determined by the method of Chick *et al*[7]. Dry matter digestibility (DMD) was worked out as the proportion of dry matter intake appearing in the faeces i.e. the indigestible dry matter. *In vitro* protein digestibility was determined by incubating sodium caseinate (Sigma) as reference protein with different fibres at protein weight to fibre weight basis (Casein:fibre, 1:1 and 1:2) following the modified method of Singh and Jambunathan[8].

Statistical Analysis

The data were subjected to analysis of variance and correlation coefficients between APD, TPD, DMD and *in vitro* protein digestibility were worked out[9].

3 RESULTS

Apparent protein digestibility and true protein digestibility receiving different dietary fibres had significantly ($P < 0.05$) lower intake of nitrogen compared to control (fibre free group) (Table 2).

The group fed on fibre diet showed significantly ($P < 0.05$) higher nitrogen absorption as compared to that of those fed on different dietary fibres. Further, increase in amount of dietary fibre from different sources significantly ($P < 0.05$) lowered the nitrogen absorption except Isabgol fed groups. This decrease in nitrogen absorption is due to increased nitrogen loss in faeces.

Apparent and true digestibilities of protein of diets containing different dietary fibres varied from 64.4 to 88.2 per cent and 68.4 to 93.7 per cent, respectively being the lowest in diet containing 10 per cent cabbage and the highest in fibre free diet. Increasing dietary fibre (either from cabbage or guava or pectin) in the diets caused a significant decrease in APD whereas TPD decreased significantly only in case of plant foods. Apparent as well as true digestibilities of all the diets containing different fibres varied significantly ($P < 0.05$) among themselves except cellulose and pectin diets.

Dry Matter Digestibility

Inclusion of dietary fibre in the diet resulted in a significant ($P < 0.05$) decrease in dry matter digestibility of diets (Table 2). DMD of different diets varied from 82.6 (cabbage at 10 per cent level) to 96.9 per cent (fibre free diet).

With the increase of dietary fibre in the diets there was a decrease in dry matter digestibility. However, significant ($P < 0.05$) reduction was found only in case of diets containing cabbage and guava. DMD of diets containing cellulose, pectin or Isabgol (at both levels) did not differ significantly ($P < 0.05$) among themselves but had significantly ($P < 0.05$) higher DMD than that of diets containing cabbage and guava.

In vitro Protein Digestibility

The *in vitro* protein digestibility of casein (fibre free) was 87.7 per cent but the addition of various dietary fibres caused significant reduction. The maximum decrease was observed when guava fibre and casein were incubated at the ratio of 2:1. When pectin and Isabgol were incubated with casein there was a slight decrease in protein digestibility at ratio 2:1 but in case of cellulose, cabbage and guava, protein digestibility decreased significantly ($P < 0.05$).

4 DISCUSSION

Low APD observed on inclusion of dietary fibre in the diets seems to be because of increased nitrogen losses in faeces. Total faecal N tended to increase with increased levels of fibre (cellulose and pectin) in the diets resulting in decreased apparent digestibility of protein[10]. The decreased absorption of N following ingestion of diets containing different fibres may lead to increased faecal N excretion which is probably because of interference of these polysacchrides with protein digestion. Apparent as well as true digestibility of protein decreased with increase in level of plant foods as source of dietary fibre[11-12]. Increasing the level of cellulose to influence dietary protein use by decreasing its apparent digestibility has also been observed earlier[13]. This decrease in APD and TPD by causing increased excretion of faecal as well as urinary N may be due to a number of factors. Increased secretion of digestive enzymes has been shown to occur when pectin was added to the diet[14]. Reabsorption of endogenous amino acids secreted into the gut has been found to be low on feeding a fibre-supplemented diet[15].

Table 1 Effect of different dietary fibres on nitrogen absorption, apparent protein digestibility, true protein digestibility and dry matter digestibility

Dietary fibre (per cent)		Nitrogen consumed (mg)	Nitrogen absorbed (mg)	Apparent protein digestibility (per cent)	True protein digesti-bility (per cent)	Dry matter digestibility (per cent)
Cellulose	5	498	425	85.3	91.9	93.2
	10	462	379	82.0	88.0	92.5
Pectin	5	485	420	86.6	90.9	95.8
	10	433	336	77.6	88.0	93.2
Isabgol	1	502	438	87.3	93.7	95.5
	2	497	424	85.3	90.0	93.4
Cabbage	5	453	338	74.6	80.5	86.9
	10	450	290	64.4	68.4	82.6
Guava	5	457	369	80.7	86.3	92.0
	10	450	340	75.5	79.6	84.5
Fibre free		527	465	88.2	92.2	96.9
SE(m)		±8.0	±10.1	± 1.7	±1.3	±1.2
CD(P < 0.05)		22.6	28.6	4.7	3.7	3.5

Table 2 Effect of different dietary fibres on in vitro protein digestibility

Fibre: Protein		Protein digestibility (in vitro) (per cent)
Cellulose	1:1	85.4(2.6)
	2:1	82.7(4.6)
Pectin	1:1	83.6(4.8)
	2:1	82.9(5.5)
Isabgol	1:1	83.4(4.9)
	2:1	83.2(5.1)
Cabbage	1:1	83.1(5.3)
	2:1	81.6(7.0)
Guava	1:1	82.3(6.2)
	2:1	81.2(7.4)
Casein (no fibre)		87.7
SE(m)		±0.32
CD (P < 0.05)		0.9

Table 3 Correlation coefficient (r) among apparent protein digestibility, true protein digestibility, *in vitro* protein digestibility and dry matter digestibility of different diets

	Per cent digestibility			
	APD (*In vivo*)	TPD (*In vivo*)	Protein digestibility (*In vitro*)	DMD (*In vivo*)
Apparent protein digestibility (APD) (*in vivo*)	-	0.982**	0.719*	0.714*
True protein digestibility (TPD) (*in vivo*)	0.982**	-	0.672*	0.708*
Protein digestibility (*in vitro*)	0.719*	0.672*	-	0.519
Dry matter digestibility(DMD)	0.714*	0.708*	0.519	-

Based on 9 degrees of freedom

* Significant at 5 per cent level

** Significant at 1 per cent level

Low APD and TPD observed by feeding composite dietary fibre may be attributed to their relatively high lignin content in addition to hemicellulose, cellulose and pectin[16-17]. Any reduction in intestinal transit time associated with fibre-containing diets could also leave less time for digestion and absorption of dietary protein[18]. More pronounced effects of plant foods in lowering either APD or TPD in this study and that cabbage too which also contributed some of its N to diets was most effective, it can be inferred that it is rather the protein-N associated with the fibre source that is poorly digested and excreted in the faeces. Variations in DMD of different food groups may be explained on the basis that different constituents of dietary fibre are fermented by microbial flora to varying extent[19]. Higher dry matter digestibility of cellulose and pectin containing diets may be due to higher fermentability rates of these fractions in the large intestine of rats. Cellulose can be fermented upto 50 per cent whereas water soluble polysacchrides like pectin is completely fermented, thus resulting in decreased faecal bulk excretion[20]. Analysis of faeces reveal that lignin in all diets is indigestible but hemicellulose or cellulose is partially digested or fermented in certain diets[21].

Lower dry matter digestibility of cabbage and guava may be due to presence of lignin, the undigestible constituent of dietary fibre of these foods. As lignin is known to inhibit bacterial activity, it may be responsible for lowering the DMD of food. In general, the more lignified a cell was on plant structure, the less liable it is to complete fermentation in the gut.

In vitro protein digestibility of casein was markedly reduced when incubated with guava followed by cabbage and gums (Pectin and Isabgol) and the decrease was even greater when ratio of fibre to protein was increased. Higher levels of fibre especially pectin, xylan and Karaya gum were more active in interfering with casein digestibility as compared to lignin and hemicellulose and cellulose generally considered to be non-reactive[22]. Low protein digestibility in case of guava and cabbage may be due to presence of pectin and lignin which together exert greater effect than when pectin alone is present. Lignin may exert an effect on casein hydrolysis through physical/or hydrogen bonding since it contains numerous internal ether and ester linkages within a phenylpropane polymer structure.

Protein digestibility in vitro was positively and significantly related with APD and TPD. DMD was positively and significantly correlated with APD and TPD but non-significantly with protein digestibility (In vitro). This shows that even in the presence of various dietary fibres, the in vitro and in vivo protein digestibilities are positively and significantly associated. Thus in vitro protein digestibility can be used to assess the protein quality of various foods which are even rich in fibres without going into tedious and time consuming in vivo methods. A positive correlation between in vitro and in vivo methods reflects that the animal assays (rats) and commercial proteolytic enzymes used in vitro digestibility are identical in their abilities to digest food protein which will allow eventually to develop a rapid assay that would predict protein nutritional quality as measured by man. Four-enzyme *in vitro* assay for protein digestibility was able to predict in man and rat the *in vivo* protein digestibility[23].

Therefore, it can be concluded from this study that with increase in amount of dietary fibre especially composite dietary fibre from 5 per cent to 10 per cent the different parameters of protein digestibility were adversely affected to varying extent indicating thereby that higher levels of dietary fibre lowers the absorption of protein, which is usually low in the diets of low socio-economic groups[1]. *In vitro* and *in vivo* protein digestibility were positively and significantly correlated which reveals that *in vitro* methods can be used to assess the protein quality.

REFERENCES

1. A.E.Harmuth-Hoene and E.Schwerdtfeger. Nutr.Metab.,1979, 23, 399.
2. V.Agarwal and B.M.Chauhan. M.Sc. Thesis, Haryana Agricultural University,Hisar.
3. P.J.Van Soest and R.H.Wine. J.Assoc.Offi.Anal.Chem.,1967, 50, 50.
4. S.Ranganna. Mannual analysis of fruit and vegetables products. 1977, Tata McGraw Hill Publishing Co. Ltd.
5. AIN Adhoc Committee on Standards for Nutritional Studies. J.Nutr. 1977, 107, 1340.
6. AOAC. 1980. Official Methods of Analysis (13th ed.). Association of Official Analytical Chemists. Washington, D.C.
7. H.Chick,J.C.D. Hutchinson and H.M.Jackson, 1935. J.Biochem. 29, 1702.
8. U.Singh and R.Jambunathan. J.Food Sci. 1981, 46, 1364.
9. Y.G.Panse and P.V.Sukhatme. Statistical Methods of Agricultural Workers, 1962, 2nd ed. IARI, New Delhi.
10. K.Kiem and C.Kies. Cereal Chem. 1979, 56, 73.
11. Z.Stankiewicz, L.Bilczuk, L.Szponar and B.Majewska. Nutr.Abstr. Rev., 1974, 45, Abst.No. 6163, 1975.
12. H.Meier and S.Poppe. Zum Einslulichkeit der Aminosayreb Int. Symp. Amino Acids, 1977. P.C. 5, Budapest.
13. J.Wojcik and C.B.Delorme. Nutr.Rep.Intr. 1982, 25, 709.
14. N.F.Sheard and B.O.Schneeman. J.Food Sci., 1980, 45, 1645.
15. H.Bergner, O.Siman and M.Zinnea. Arch.Tilrermachr., 1975, 25, 95.
16. B.N.Mitaru and R.Blair. J.Sci.Food Agric., 1984, 35, 625.
17. M.A.Eastwood, J.R.Kirkpatrick, W.D.Mitchell, A.Bome and T.Hanitton. Br.Med.J., 1973, 4, 392.
18. J.H.Cummings, D.A.T.Southgate, W.J.Branch, H.S.Wiggins, H.Houstan, D.J.A.Jenkins, T.Jivray and M.J.Hill. Br.J.Nutr., 1979, 41, 477.
19. F.W.Sosulski and A.M.Cadden. J.F ood Sci., 1982, 47, 1472.
20. C.M.Donangelo and B.O.Eggum. Br.J.Nutr., 1985, 54, 741.
21. C.M.Gagne and J.C.Acton. J.Food Sci., 1983, 48, 734.
22. N.Rich, L.D. Satterlee and J.L.Smith. Nutr. Rep. Intr. 1980, 21, 285.
23. C.E.Bodwell,L.D.Satterlee and L.R.Hackler, Am.J.Clin.Nutr. 1981, 34, 678.

EFFECTS OF TWO INCORPORATION RATES OF GUAR GUM ON DIGESTIBILITY, PLASMA, INSULIN AND METABOLITES IN RESTING DOGS.

A. Delaunois, K. Neirinck, A. Clinquart, L. Istasse and J. M. Bienfait.

Service de Nutrition - Faculte de Médecine Vétérinaire 45, rue des Vétérinaires 1070 Bruxelles, Belgium.

1. INTRODUCTION

In a previous work (De Haan et al., 1989) three different fibres (cellulose, pectin and guar gum) were incorporated at 3.5% on a dry matter (DM) basis in diets for resting dogs. Guar gum, the most viscous fibre, induced a decrease of gastric emptying, a reduction of digestibility mainly of proteins and lipids, a reduction of the postprandial peak of glucose, α-amino-N, urea, triglycerides, cholesterol and non-esterified fatty acids (NEFA).

The present experiment was designed to compare two incorporation rates of guar gum on digestibility and blood metabolites in the resting dog.

2. MATERIALS AND METHODS

Animals and diets.

Six adult Beagle dogs were used in 2 (3 x 3) latin square design. They were offered a control diet or a diet supplemented with fibre. The control diet was based on steam-treated rice (475 g/kg DM), minced meat (419 g/kg DM) maize oil (84 g/kg DM) and minerals + vitamins (22 g/kg DM). The diet was calculated to supply 550 kj of metabolizable energy/kg 0.75 body-weight.

Measurements.

Water and DM intakes were measured. The apparent digestibility was determined by total collection of faeces on a one-week period. During the following week, blood samples were collected before feeding and during 6 hours after the meal.

3. RESULTS AND DISCUSSION

Composition of the faeces.

The DM content of the faeces decreased with the level of guar gum: 46.61, 34.47 and 24.82% (P<0.001) with 0, 3.5 and 7% of guar gum, respectively (Table 1). These results are in agreement with previous observations.[1] The concentrations in ash and insoluble ash decreased in the faeces with the incorporation rate of fibre. By contrast, the proteins, ether extract (EE), and nitrogen free extract (NFE) contents of the faeces increased. Such findings were previously reported in humans with whole bread, fruits and vegetables [2], and with cellulose [3]. The increase in proteins could partly be associated with a microbial fermentation in the large bowel, as indicated by an increase of volatile fatty acids in the caecum of rats offered guar gum and arabic gum [4]. The content of Na and K in the faeces increased with the addition of guar gum, while that of P, Ca, Cu, Zn and Mn decreased. It was with the inclusion of 7% that the largest changes were observed for all the chemical components of the faeces.

Table 1. Composition of the faeces.

Guar gum		0%	3.5%	7%	SED
DM(%)		46.61	34.47	24.82	1.76
Ash	(g/kg DM)	467.88	374.33	295.55	14.55
Insoluble ash	(")	26.96	21.63	17.55	2.64
Proteins	(")	277.95	274.27	317.46	11.41
EE	(")	101.20	117.87	201.50	16.44
NFE	(")	206.46	221.47	232.62	12.84

Digestibility coefficients.

The digestibility of DM was 9%, 31% with the control diet (Table 2). This value, along with the high digestibility of the organic components, indicated the high quality of the diet offered to the dog (steam-treated rice and minced meat). The incorporation of fibre reduced the digestibility of all organic components, the difference being significant for proteins (P<0.001 with 3.5% and P<0.001 with 7%). The incorporation rate of 7% induced the largest reductions mainly for EE (91.74 vs 96.71%). The digestibility of Na and K was high (95.33 and 96.24% respectively) and was reduced by the fibre. By contrast, the digestibility of P was relatively low at 41.04% and was increased with the two levels of guar gum (P<0.05 or 0.001). The digestibility of Ca was very low.

Table 2. Digestibility coefficients (%)

Guar gum	0%	3.5%	7%	SED
DM	92.31	91.21	90.73	0.52
Crude proteins	94.33	91.86	89.68	0.01
EE	96.71	95.51	91.74	0.01
Ca	4.42	1.43	10.63	0.05
P	41.04	47.00	55.58	0.02
Na	95.33	91.77	82.33	0.01
K	96.24	92.20	87.53	0.01

Plasma insulin and metabolites.

The pattern of plasma insulin was quite similar between the 3 diets (Figure 1a). However, the extent of the post prandial rise was smaller with 7% of guar gum. No significant differences were observed in plasma glucose before and after feeding (Figure 1b).

The concentration in α-amino-N (Figure 1c) was lower than the control ($P<0.01$ from 2 till 5 hours after feeding with 3.5%, and $P<0.1$ from 1 h 30 till 6 h with 7% of fibre). Amino acids are bound on the digesta in the small intestine [5], and therefore, the utilization of dietary proteins is reduced, as indicated by the reduction of protein digestibility and the lower plasma α-amino-N concentrations. Plasma urea (Figure 1d) was reduced by addition of guar gum at the 2 incorporation rates: such effect could be associated with the delay in the absorption of amino acids and their catabolism by the liver.

The incorporation of guar gum did not change to a large extent the post prandial pattern of plasma triglycerides (Figure 1e), although there was a trend for lower concentrations with 7% of fibre. Guar gum also significantly reduced plasma concentrations of cholesterol (Figure 1f). These reductions have to be related to different mechanisms such as a reduction in transit time, a lower absorption or a higher faecal excretion [6].

4. CONCLUSION

From this experiment, it appeared that proteins and lipids metabolisms mainly were influenced by increasing doses of guar gum. Various mechanisms were involved: reduction of digestibility and absorption of the nutrients, decrease of gastric emptying, gelification of the digesta [7], binding of amino acids [5] or bile acids [8], changes in the activity of enterocytes or in the mucosal thickness [9].

REFERENCES

1. V. De Haan, L. Istasse, S. Jakovjevic, I. Dufrasne, J. M. Bienfait. Proc. Nutr. Soc., Aberdeen 26-28 Sept. 1989.
2. D. A. T. Southgate and J. V. Durnin, Brit. J. Nutr., 1970, 24, 517.
3. J. L. S. Slavin and J. A. Marlett, J. Nutr., 1980, 110, 2020.
4. B. Tulung, C. Remesy, C. Demigne, J. Nutr., 1987, 117, 1556.
5. P. Howard, R. B. Mahoney and T. Wilder, Nutr. Report Int., 1986, 34, 135.
6. J. Rautureau, Th. Coste and P. Karsenti, Cah. Nutr. Diet. 1983, 28, 85.
7. N. A. Blackburn and I. T. Johnson, Brit. J. Nutr., 1981, 46. 239.
8. D. Gallaher and B. O. Schneeman, Am. J. Physiol., 1986, 250, G420.
9. A. G. Low and A. L. Rainbird, Brit. J. Nutr., 1984, 52, 499.

WORKSHOP REPORT: RELATION BETWEEN PHYSICAL PROPERTIES/PHYSIOLOGICAL EFFECTS OF UNAVAILABLE POLYSACCHARIDES

N. Read

Floor K
Royal Hallamshire Hospital
Glossop Road
Sheffield S10 2JR

1. How does structure relate to physical properties?

Soluble polysaccharides are often charged. Viscosity is related to the entangling of long chains as ribbons, rather than densely packed clumps. Formation of a gel involves chemical bonding of adjacent chains.

Pectin can change from a viscous sol to a gel at acid pH. This can alter physiological effects. When a mixture of pectin and glucose gels in the stomach, postprandial blood glucose levels are not reduced.

Gums with densely packed side chains resist bacterial attack, but bacteria are crafty and can adapt to break down even densely packed molecules like xanthan gum.

Application of heat in cooking can markedly alter physical properties; different effects can be seen with different gums.

2. How can viscosity alter absorption?

Viscous solutions alter the convection of luminal contents by inhibiting digestion and by decreasing the access of products of digestion to the cell surface. The latter can be regarded as having two components, but both are probably different manifestations of the same physiological effect. They are: a decrease in the mixing in the bulk phase and an increase in the dimensions of the poorly stirred layer immediately adjacent to the epithelium. Some polysaccharides, particularly the substituted celluloses, may stick to the epithelium.

3. <u>Is it possible to predict the effects of viscous polysaccharides in vivo from their physical properties in vitro?</u>

The viscosity of some gums will change when they are exposed to solutions of differing pH and ionic composition. It is not possible to predict the exact ionic environment of different regions of the gut. Viscosity of a solution is related to the shear characteristics; these will also vary in different parts of the gut and are impossible to predict. Dr Morris has shown how to define the viscosity of a polysaccharide over the whole range of shear rates by two values; maximum viscosity at theoretical zero shear rate and the shear rate that gives half the maximum viscosity.

Dr. Edwards suggested that in theory, shear characteristics could be determined by studying the relative behaviour of a range of polysaccharides that do not alter their viscosity with changes in ionic strength, and reading off the point at which the gums rank as they do <u>in vivo</u>..

4. <u>Are effects of polysaccharides on absorption in the small bowel related to their viscous properties?</u>

They probably are. Dr Heaton showed that the plasma insulin response of wheat and rice but not oats was related to the size of the particles fed. Presumably the reason that oats decreases insulin response independent of particle size relates to the fact that only oats produce viscous solutions. In another study, Dr Schneeman treated oats with β-glucanase and abolished the reduction of plasma cholesterol.

Although we eat few viscous polysaccharides, cooking can produce viscous mixtures and gut contents can be quite viscous.

5. <u>What is the physiolgical basis of laxative action</u>

There are several factors. Bacterial cell mass is probably less important than some of the others. Dr Eastwood felt that totally fermented fibre gave no increase in stool weight even though it stimulated bacterial proliferation.

Water holding capacity after fermentation is a much more important factor and Dr Tomlin's experiments indicate that retention of polysaccharide structure is essential for stool bulking. Stool bulking is however not necessarily associated with an acceleration of transit time. Some bacterial degradation of the polysaccharide structure appeared to be necessary to induce an acceleration of transit time. It is not yet clear what fermentation products do to transit time and colonic motility.

Another recent study showed that inert particles of polyvinyl tubing or polystyrene have a similar laxative action as bran, reintroducing the concept of roughage. Particles may stimulate multimodal receptors to induce secretion and propulsion, but there are other possibilities. Dr Berry suggested that it would be interesting to investigate the effect of coating bran with silicon.

MINERAL ABSORPTION OF DIETARY FIBRE FROM PROCESSED VEGETABLES

Y. Saito and Y. Ohiwa

Department of Food Science and Nutrition
Koriyama Women's College
Koriyama, Fukushima, 963 Japan

1 INTRODUCTION

Some studies have shown that vegetables usually contain soluble and insoluble dietary fibre (DF)s, and the amount of both DFs vary separately depending on cooking method.[1,2] The variation in the amount of DFs by cooking probably is a result of the diversity of chemical composition. Many studies have indicated that the physiological effects depend on the kinds of DF, such as pectin, hemicellulose or cellulose. Thus it was shown by our studies that the effect of vegetable DF on the availability of Ca and Mg in rats was changed by cooking.[3)] These results may be partly due to mineral absorption, since McConnel *et al*[4] reported that cooking had a varied effect on the cation capacity. Therefore, the present study was undertaken to determine the effect of cooking on the Ca and Na absorption of DF separated from carrot, spinach and the root of edible burdock (burdock) *in vitro*.

2 METHODS

Preparation of dietary fibre. Samples of sliced carrot and burdock, and cut spinach were either raw, boiled for 10, 15 and, one and a half minutes, or fried for 3, 3 and 2 min. respectively. Experiment 1 : The raw, boiled and fried vegetables were homogenized in acetone, and dehydrated and defatted with acetone and ether(fibre A). The dehydrated vegetables were boiled in 85% methanol for 30 min., and filtered. The residues were washed with boiling water, acetone and ether (fibre B). Experiment 2: The dehydrated vegetables were extracted by boiling in a solution of 85% methanol three times.

The residues were washed with acetone and ether (fibre C). A part of the residue was washed with boiling water 5 times and then washed with acetone and ether (fiber D). All air-dried DFs were got through a 32-mesh sieve.
Determination of Ca and Na absorption. Two vasking tubings containing 5 ml water were hung in a 300 ml tall beaker and 200 ml 0.1% $CaSO_4$ or 0.05% NaCl solution (pH 5.4 or 8.0), and 0.5g DF were entered in the beaker and stirred at once. 0.1 ml of solution was taken out from the water in the tubings at 30, 60, 120 and 180 min. after stirring. The solution was diluted with 0.1N HCl. The amount of Ca or Na in the solution was determined by atomic absorption spectrophotometry. The concentration of Ca and Na in the tubing became maximum at 120 and 90 min. after stirring respectively, and were in a state of equilibrium to the concentration of Ca and Na in the beaker. The Ca and Na absorption of DFs were estimated from the concentration of Ca and Na in the tubing in the state of equilibrium.

3 RESULTS AND DISCUSION

Experiment 1: Table 1 shows the Ca and Na absorption of fibre A and B at pH 5.4 and pH 8.0. The Ca absorption of carrot fibre A at pH 5.4 was significantly increased over that of the fibre A of raw carrot by boiling. It was further increased by frying, but the difference between the values of boiled and fried fibre A was not significant. The Ca absorption of carrot fibre B was the most in the boiled DF, but that was not significantly chnged by cooking. In spinach fibre A and B the Ca absorption was more than that of carrot and burdock. But that of fibre A was increased by cooking, whereas that of fibre B was decreased by cooking. That part of fibre B which probably did not contain soluble fraction was less than that of fibre A, with no relation to cooking. This result suggests that a part of the mineral absorption in spinach DF may be ascribed to the soluble fraction. Fried fibre A and fibre B absorbed significantly more or less Ca than the raw fibre. In burdock fibre A absorbed almost no Ca, whereas fibre B absorbed some Ca. Cooking had almost no effect on the Ca absorption of both fibre A and B. DF of burdock was mostly inulin so the difference in Ca absorption of fibre A and B might be ascribed to inulin content. In spinach DF the amount of Na absorption was much less than the amount of Ca absorption. The Na absorption was changed only by frying significantly. The Ca absorption at pH 8.0 tended to be less than that at pH 5. From this result it seemed that the effect of vegetable DFs on Ca availability was probably lower in small intestine than in stomach. The fibre A of raw carrot absorbed more Ca than the cooked fibre. Ca absorption of carrot fibre A

Table 1 Ca and Na absorption of fibre A and B separated from raw and cooked carrot, spinach and the root of edible burdock. (mg/g of dry matter)

Cooking	Fibre A or B	Carrot	Spinach	The root of edible burdock
		Ca absorption at pH 5.4		
Raw	A	3.37±0.02* [a b]	20.92±2.66	-1.65±1.89 [a]
	B	8.42±5.99	17.81±0.68 [a]	8.36±2.21 [a]
Boiling	A	7.04±1.65 [a c]	23.04±5.66	0.73±2.81 [b]
	B	11.87±0.02 [c]	14.27±4.42	8.47±2.83 [b]
Frying	A	11.68±4.25 [b]	25.83±2.17 [b]	-1.13±1.40 [c]
	B	9.93±2.55	11.19±2.15 [a b]	8.69±4.55 [c]
		Ca absorption at pH 8.0		
Raw	A	9.45±0.35 [a b c]	17.91±1.51 [a b]	4.34±0.35
	B	2.97±1.89 [a]	12.52±0.71 [a d e]	3.94±1.98
Boiling	A	1.69±0.07 [b]	15.87±1.05 [b]	2.89±0.00
	B	1.27±0.60 [d]	6.02±2.88 [b d]	2.91±1.02
Frying	A	2.93±1.27 [c]	22.38±1.01 [c b]	3.13±1.23 [a]
	B	3.81±0.00 [d]	7.29±3.13 [c e]	1.46±0.69 [a]
Spinach		Na absorption		
		pH 5.4	pH 8.0	
Raw	A	5.41±2.45	2.50±0.00	
	B	5.41±1.79 [b]	2.40±1.12 [b]	
Boiling	A	5.75±2.57	1.10±1.57	
	B	6.87±2.37	2.60±1.36 [c]	
Frying	A	3.34±0.98 [a]	0.42±1.56 [a]	
	B	10.18±3.64 [a b]	6.33±2.40 [a b c]	

* Mean ± SD (n=3 or 5) Same superscript are significantly different from each other.

and B was significantly decreased by boiling, and the change in Ca absorption at pH 8.0 by cooking was different from that at pH 5.4. In spinach DF the difference of pH did not have much effect on the Ca absorption. The fried fibre A absorbed more Ca than the raw fibre A, whereas the fried fibre B absorbed less Ca than the raw fibre did. The Ca absorption of burdock fibre A and B was low at pH 8.0, and was mostly uneffected by cooking.

Experiment 2. Fibre C was purified more than fibre A,

since acetone dehydrated vegetables were treated with boling 85% methanol, whereas fibre D was similar to fibre B, since the preparation methods of both fibres were almost the same. Table 2 shows Ca and Na absorption of fibre C and D.

Table 2 Ca and Na absorption of fibre C and D separated from raw and cooked carrot, spinach and the root of edible hurdock. (mg/g of dry matter)

Cooking	Fibre C or D	Carrot	Spinach	The root of edible burdock
		Ca absorption		
Raw	C	$9.33 \pm 0.44^{*b}$	30.66 ± 1.87^{adf}	9.07 ± 1.28
	D	9.94 ± 1.13^{c}	11.27 ± 1.07^{aeg}	9.18 ± 1.42^{b}
Boiling	C	10.83 ± 0.76^{b}	17.53 ± 1.06^{bd}	6.72 ± 1.64
	D	11.03 ± 1.09^{cd}	7.78 ± 0.53^{be}	6.11 ± 1.70
Frying	C	10.14 ± 0.82^{a}	25.86 ± 1.51^{cf}	5.39 ± 0.21^{ab}
	D	7.87 ± 1.32^{ad}	5.86 ± 2.81^{cg}	8.23 ± 0.70^{a}
		Na absorption		
Raw	C	10.91 ± 0.84^{a}	1.49 ± 0.33^{ab}	7.62 ± 0.00^{ac}
	D	9.77 ± 1.69^{b}	1.51 ± 0.00^{c}	4.33 ± 1.42^{ad}
Boiling	C	5.62 ± 0.85^{ac}	3.35 ± 0.80^{a}	5.02 ± 0.00^{bc}
	D	6.83 ± 0.84^{d}	1.49 ± 0.00	2.18 ± 0.84^{bd}
Frying	C	18.71 ± 3.36^{c}	6.09 ± 1.50^{b}	7.18 ± 0.55^{e}
	D	20.09 ± 3.33^{bd}	4.64 ± 1.17^{c}	4.48 ± 1.04

* Mean ± SD (n=3 or 5) Same superscript are significantly different from each other.

In carrot, the boiled fibre C and D absorbed more Ca than raw fibres, while the Ca absorption of fried fibre D was less than that of raw and boiled fibre D. In spinach fibre C, raw and fried fibre especially absorbed significant amounts of Ca. THe Ca absorption of fibre D was less than that of fibre B, but the change of fibre D by cooking was similar to that of fibre B.
In burdock, fibre C and D absorbed Ca, but the effect of cooking on Ca absorption was slight. The Na absorption of fibre C and D was more than the Ca absorption , and the change in the Na absorption of fibre C by cooking was similar to that of fibre D. Boiled fibre C and D absorbed significantly less Na than raw fibres, whereas fried fibres significantly more Na than the raw fibres. In spinach, the Na absorption of fibre C and D was less than the Ca absorption. In burdock, the Na absorption of fibre D was less than that of fibre C,

but the changes resultant from cooking method on both fibres tended to be the same.

The above results suggested that the effects of cooking on the Ca and Na absorption of the vegetable fibre differed in the cooking methods, the kinds of vegetable involved and the methods of fibre separation.

Refrences

1. K. Yoshita and Y. Saito, J. Home Economics Japan, 1985, 36, 721
2. M. Nyman, N. G. Asp and K. E. Plsson, Lebensm. Wiss. Technol., 1987, 20, 29
3. Y. Saito, S. Sato and Y. Ooiwa, J. Home Economics Japan, 1989, 40, 1039
4. A. A. McConnell, M. A. Eastwood and W. D. Michell, J. Sci. Food Agric., 1974, 25, 1457

BINDING OF MINERALS BY NEW SOURCES OF HIGH DIETARY FIBRE CONTENT

M. Torre, A. R. Rodríguez and F. Saura-Calixto

Department of Analytical Chemistry
Alcalá Hres. University
28871 Alcalá de Henares
Madrid

1 INTRODUCTION

A major negative effect to increased consumption of dietary fibre may be a reduction in the bioavailability of trace minerals.[1] There are many 'in vivo' and 'in vitro' studies concerning this subject.[1,2] In vitro studies of mineral bioavailability in the presence of dietary fibre are useful because they are faster and less expensive than 'in vivo' studies and because they offer better control over experimental variables, such as pH, type(s) of fibre, mineral concentration, etc.[3]

Certain by-products from Spanish food and agricultural processing plants have been shown to have high dietary fibre content.[4,5] The interactive effect of some of these subproducts with Fe^{3+} and Ca^{2+} have been investigated 'in vitro', in order to know the percentage of these minerals bound in an insoluble form and to understand the possible mechanisms of these interactions.

2 MATERIALS AND METHODS

Materials

Citrus wastes (lemon peel, LP), cider wastes (CW) and olive pomace (OP), were obtained from Spanish food processing plants. Alcohol-insoluble solids (AIS) of these ground materials ($<$0.5 mm) were prepared and dried. The standard fibre components were: lignin (Lg) (Therapharm Ltd., Dowhan Market, Norfolk), and cellulose (C) (Aldrich).

Procedure

Solutions containing the metal ions Fe^{3+} and Ca^{2+} at gastric pH (pH 2-4) and pH of the small intestine (pH 5-8), respectively, were added to centrifuge tubes which contained ca. 100 mg of sample. The suspensions were shaken for 3 h (time of an 'in vivo' digestion) at room temperature. Supernatants were discharged after centrifugation at 3000 g for 15-30 min. The equilibrium metal ion concentrations were determined by A.A.S. The amount of mineral elements bound in an insoluble form (BC, $\mu g\ ml^{-1}\ g^{-1}$ of sample) was calculated from the residual equilibrium concentration in the supernatant, corrected by the endogenous metal ion concentration, (EC), the initial metal ion concentration (IC) and the mass of AIS or standard fibre components.

Analyses were performed at least twice. Results of bound concentrations are expressed as the difference between mean values of two determinations.

3 RESULTS AND DISCUSSION

Dietary fibre composition and other constituents of the food by-products, determined in earlier investigations,[4,5] is described in Table 1.

Mineral binding properties of AIS and standard fibre components were studied as a function of sample mass, the initial metal ion concentration and the pH of the medium.

Effect of Sample Mass on Binding

Figure 1 shows the relationship between sample mass and the extent of mineral binding by the AIS of two food by-products and standard components of fibre, when other variables are held constant. In all cases, bound mineral

Table 1 Dietary Fibre Content and Main Constituents of food by-products (% dry matter), determined by the AOAC method

Sample	TDF	IDF	SDF	TP	KL	I-UA
Cider waste	62.5	48.3	14.2	5.4	18.2	5.75
Olive pomace	69.4	65.7	3.7	5.3	37.2	2.30
Lemon peel	47.7	28.2	19.5	1.3	5.5	5.33

TDF: Total dietary fibre; IDF: insoluble dietary fibre; SDF: soluble dietary fibre; TP: total protein; KL: Klason lignin; I-UA: insoluble uronic acids.

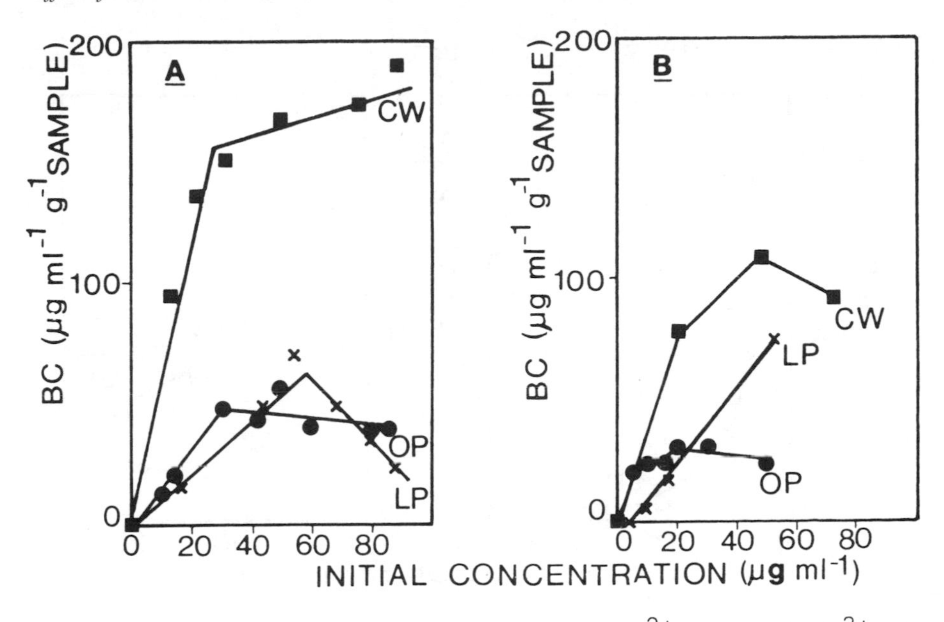

Figure 1 Influence of sample mass on Ca^{2+} (A) and Fe^{3+}(B) binding
IC ($\mu g\ ml^{-1}$) and pH values were: (A) LP: 32.3, 5.9; Lg: 45.25, 6.3; C: 44.26, 5.6; and (B) CW: 48.4, 2.5; Lg: 45.67, 2.5

concentration decreased as the sample mass increased, up to a mass greater than 200 mg, when almost a constant bound concentration value is reached. This behaviour may indicate that most of the metal ions were bound at the surface, as was previously reported.[6]

Effect of Metal Ion Concentration on Binding

The effect of initial solution concentration, at a constant pH value, of both Ca^{2+} and Fe^{3+} on calcium and iron binding is shown in Figure 2. From this figure, it is obvious that binding of both ions increased with increasing solution concentrations, up to a certain value of maximum binding, which varies with the nature of sample and the mineral considered.

The binding affinity to the by-products investigated and the maximum amount of metal ion bound per 100 g of sample ($\mu mol\ g^{-1}$ of sample) can be ranked as

Ca^{2+} : CW (>118.2) $\gg$ OP (31.2) > LP (14.1) and

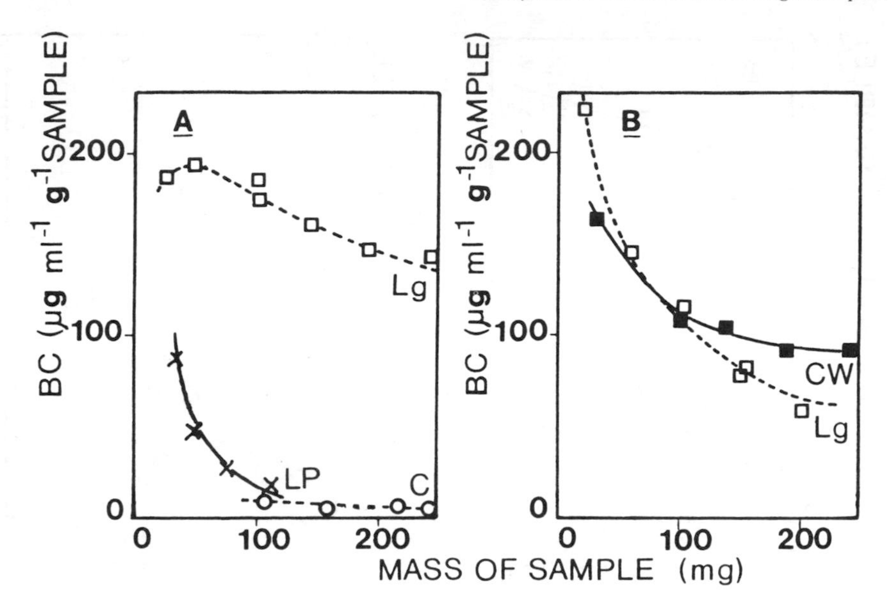

Figure 2 Influence of initial Ca^{2+} (A) and Fe^{3+} (B) concentration on binding
Sample mass (mg) and pH values were: (A) CW: 100, 5.5; OP: 100, 5.8; LP: 50, 5.7; and (B) CW: 100, 2.0; OP: 100, 1.8; LP: 100, 1.9

Fe^{3+} : CW (48.3) > LP (>33.6) > OP (13.9)

The contribution of lignin and cellulose to Fe^{3+} and Ca^{2+} binding was also studied. The order of affinity was:

Ca^{2+} : Lg (>114.7) $\gg$ C (9.4) and
Fe^{3+} Lg (145.5) $\gg$ C (18.0)

As shown, cellulose has essentially no ability to bind either ferric or calcium ions, while lignin interactions with both elements are very strong. On the basis of these results, the presence of both carboxyl and hydroxyl groups in the phenylpropanoid units of lignin, together with the great content of this fibre component (Klason lignin) in CW an OP may indicate the contribution of these functional groups to cation binding by fibre. Moreover, the significant content of pectic substances, with carboxyl groups in their structure, in a sample with a low content in Klason lignin, LP, corroborates this conclusion. On the other hand, contribution to mineral binding of soluble uronic acids was not considered since recent 'in vitro' studies showed the release of cations bound to soluble fibre in a pH range of 6-8.[7]

Scatchard analyses of Ca^{2+} and Fe^{3+} binding by the AIS of food by-products, lignin, and cellulose, showed the existence of one specific binding site for Ca^{2+} in CW (log $K_{eff.}$ = 4.13) and one specific binding site for this ion in Lg (log $K_{eff.}$ = 4.34). No specific binding sites were observed either in the OP-Ca, LP-Ca, and C-Ca systems or in the binding of ferric ions by the by-products, lignin and cellulose. This fact helps us to explain the great interaction of lignin and cider waste with calcium.

Effect of pH on Mineral Binding

The effect of pH on binding, at a constant metal ion concentration (Table 2) showed that mineral binding increased with increasing pH; the increase was much more marked in the case of Fe^{3+} (which tends to hydrolyse in aqueous systems, with the formation of insoluble polynuclear species at pH ⇒2.0). The small increase of Ca^{2+} bound to Lg and CW in the pH range 4-7.5 may suggest involvement of groups with high pK_a, probably phenolic groups, while the greater increase of Ca^{2+} bound to LP and OP in this pH range may indicate involvement of carboxyl groups (of lignin and/or uronic acids).

Table 2 Mineral binding ($\mu g\ ml^{-1}\ g^{-1}$) as a function of pH

Ion	CW		OP		LP		Lg		C	
	pH	BC	pH	BC	pH	BC	pH	BC	pH	BC
Ca	5.5	154.8	5.2	35.3	5.9	50.2	3.9	152.0	3.3	7.7
Ca	6.7	172.4	6.6	55.5	6.8	57.9	6.3	185.5	5.1	11.2
Ca	7.4	188.4	7.4	93.0	7.4	81.6	7.1	196.8	7.0	17.0
Fe	2.1	56.6	1.9	24.5			2.0	8.9	2.6	34.7
Fe	2.5	128.6	2.9	62.6			2.7	298.6	3.1	156.7
Fe	2.9	198.1	3.9	28.5			3.0	445.3		

IC ~50 $\mu g\ ml^{-1}$ in all cases, except for OP-Fe (~10 $\mu g\ ml^{-1}$)

REFERENCES

1. D.A.T. Southgate, Am.J.Clin.Nutr., 1987, 45, 1256.
2. J.L. Kelsay, 'Dietary Fiber in Health and Disease', G.V. Vahouny and D. Kritchevsky, N.Y. , 1982, 91.
3. P.E. Johnson, Biol. Trace Elem. Res., 1989, 19, 3.
4. I. Goñi, M. Torre and F. Saura-Calixto, Food Chem., 1989, 33, 151.
5. F. Saura-Calixto, I. Goñi, E. Mañas and R. Abia, Grasas y Aceites, 1989, 40, 241.

6. K. Lee and J.S. García-López, J.Food Sci., 1985, 50, 651.
7. U. Schlemmer, Food Chem., 1989, 32, 223.

EFFECTS OF DIETARY FIBER ON MINERAL BALANCES IN HUMANS

E. Wisker[1] , T. F. Schweizer[2] and W. Feldheim[1]

[1] Institute of Human Nutrition, University of Kiel, D-2300 Kiel (Federal Republic of Germany)

[2] Nestlé Research Centre, Nestec Ltd., Vers-chez-les-Blanc, CH-1000 Lausanne 26 (Switzerland)

1. INTRODUCTION

Although there were reports that fibers from cereals and also from fruits and vegetables can reduce the bio-availability of minerals and trace elements[1,2,3], the subject is controversial. It has been shown that cations will bind to dietary fiber in vitro[4,5]. In vivo, the binding will be influenced by several factors in foods like phytic and oxalic acid, vitamin C , protein and other minerals and the pH of the gastrointestinal tract. Those factors may enhance or inhibit binding. The effects of fiber and/or other agents on cation availability may be studied by test meals. However, test meals do not give a picture of the utilization of minerals for a longer period of time. It was the objective of our studies to measure by balance experiments the influence of various kinds and levels of dietary fiber on mineral availability. The fiber containing foods under study were constituents of mixed diets as were consumed in the Federal Republic of Germany.

2. MATERIALS AND METHODS

Subjects and experimental design. Four studies were performed with healthy female students (21-27y). Each study was divided in a low fiber control and one or more high fiber experimental periods. All food eaten during the experiments was prepared in the institute kitchen and weighed exactly. As beverages various kinds of tea, coffee and water were allowed. The beverage consumption was recorded. Each experimental period lasted 21 d, between the periods there was a break of at least 3 wk.

Table 1 Fiber sources and increase in fiber intake on the high fiber diets

Study		Fiber source	Increase in DF intake (g/d)	Phytic acid (mg/d)
1	**a.**	Bread enriched with barley fiber (High protein intake)	15	low
	b.	Bread enriched with barley fiber (Medium protein intake)	15	low
2	**a.**	Whole grain bread and rolls	29	1200
3	**a.**	Raw carrots	15	n.d.
	b.	Blanched carrots	15	n.d.
	c.	Canned carrots	15	n.d.
4	**a.**	Various fruits and vegetables	33	n.d.

Diets. In each study, the basal diet (15 to 19 g DF/d) was the same during the control and the experimental periods. The basal diet consisted of two 1 d menus, which were fed in rotation. During the high fiber experimental periods, high fiber breads were substituted for low fiber breads (studies 1 and 2) or vegetables and fruits were added to the basal diet (studies 3 and 4) (Table 1).

Balance technique and analytical methods. From d 15 to 20 duplicates of all foods and beverages, from d 16 to 21 all feces and urine were collected. Dietary fiber was analyzed by the AOAC procedure[6]. Ca, Mg, Zn and Fe were determined by AAS.

3. RESULTS AND DISCUSSION

When the intake of minerals and trace elements was elevated due to the consumption of the fiber sources, the absolute excretion of those cations by the feces was increased. This could be measured with whole grain bread, carrots and fruits/vegetables. These observations are in agreement with other reports[3,7] and may be the result of the increase in fiber and/or phytic acid. Balances were not decreased by the elevated fecal excretion. Concerning Zn and Fe, balances were improved due to the higher intake with the whole grain bread and the fruit/vegetable diets.

Table 2 Intake and balance of calcium (mean values)

Study	Period	Intake (mg/d)	Balance (mg/d)	n
1	Control	979	+ 11	12
	a.	1020	+ 86	12
	b.	916	- 32	12
2	Control	1190	- 7	6
	a.	1259	+ 20	6
3	Control	1098	+ 84	12
	a.	1273	+ 87	12
	b.	1311	+ 41	12
	c.	1284	+ 25	12
4	Control	1094	+ 27	8
	a.	1009	- 18	8

Table 3 Intake and balance of magnesium (mean values)

Study	Period	Intake (mg/d)	Balance (mg/d)	n
1	Control	252	+ 8	12
	a.	245	- 5	12
	b.	243	- 11	12
2	Control	221	- 15	6
	a.	434	- 12	6
3	Control	261	+ 3	12
	a.	316	+ 10	12
	b.	305	+ 1	12
	c.	303	- 6	12
4	Control	232	- 9	8
	a.	368	- 8	8

Table 4 Intake and balance of zinc (mean values)

Study	Period	Intake (mg/d)	Balance (mg/d)	n
1	Control	10.8	+ 0.2	12
	a.	10.9	- 0.3	12
	b.	7.8	- 1.2	12
2	Control	10.1	0	6
	a.	15.8	+ 1.1	6
3	Control	10.8	- 0.6	12
	a.	12.1	- 0.7	12
	b.	11.9	- 0.7	12
	c.	11.8	- 0.8	12
4	Control	9.6	- 0.3	8
	a.	14.1	+ 0.8	8

Table 5 Intake and apparent absorption of iron (mean values)

Study	Period	Intake (mg/d)	Apparent absorption (mg/d)	n
1	Control	8.6	+ 0.9	12
	a.	10.4	+ 0.3	12
	b.	8.6	- 1.3	12
2	Control	7.3	+ 1.0	6
	a.	17.0	+ 3.6	6
3	Control	7.5	+ 0.8	12
	a.	8.8	+ 0.8	12
	b.	9.0	+ 0.8	12
	c.	9.7	+ 0.2	12
4	Control	6.8	+ 0.3	8
	a.	14.8	+ 2.7	8

In this study, the processing of carrots did not influence the balances of the elements studied. Compared to the control, no differences were observed. In an investigation with rats, raw carrots decreased the balances of Ca and Mg compared to sauted carrots[8].

When the intake of minerals was not changed due to the increase in fiber intake, fecal excretion and also the balances were not significantly affected. However, a lower consumption of animal protein decreased Zn and Fe balances.

4. CONCLUSIONS

Concerning mixed diets with a relative high amount of animal foods, our studies indicate that dietary fiber intakes up to 50 g/d have no adverse effects on mineral balances. However, a decrease of animal products in the diet may influence trace elements balances.

REFERENCES

1. J.G. Reinhold, B. Faradji, P. Abadi and F. Ismail-Beigi, J.Nutr., 1976, 106, 493.
2. J.L. Kelsay, K.M. Behall and E.S. Prather, Am.J. Clin.Nutr., 1979, 32, 1876.
3. J.L. Kelsay and E.S. Prather, Am.J.Clin.Nutr., 1983, 38, 12.
4. W.P.T. James, W.J. Branch and D.A.T. Southgate, Lancet, 1978, 1, 638.
5. L. Camire and F.M. Clydesdall, J.Food Sci., 1981, 46, 548.
6. L. Prosky, N.G. Asp, I. Furda, J.W. DeVries, T.F. Schweizer and B. Harland, J.Assoc.Off.Anal.Chem., 1985, 68, 677.
7. W. Van Dokkum, A. Wesstra and F.A. Schippers, Br.J. Nutr., 1982, 47, 451.
8. Y. Saito and S. Sato. In: D.A.T. Southgate, I.T. Johnson and G.R. Fenwick, Eds. Nutrient Availability: Chemical and Biological Aspects. Royal Society of Chemistry, Cambridge. 1989, p. 262.

Dietary Fibre in the Large Intestine: Implications for Colorectal Function and Energy Metabolism

Function of Dietary Fibre in the Large Intestine

Dr Martin Eastwood
Gastro Intestinal Unit
Western General Hospital
Edinburgh EH4 2XU

1 INTRODUCTION

The colon is one of the least understood organs in man and animals. Problems of nomenclature arise as knowledge accumulates.

The colon is tubular, capacious and contractile, and a reservoir which conserves water, electrolytes, anions and cations. One end is in continuity with, though functionally separated from, the ileum. The left side controls the orderly evacuation of faeces. The right side, or caecum, contains a large bacterial population which is metabolically active within the constraints of a nutrient starved environ. The sources of nutrition to the caecal bacterial flora are dietary, exfoliated cells, intestinal and biliary secretions[1]. The nutrients include polymeric carbohydrates (eg fibre, resistant starch)[2], proteins[3], lipids[4] and small molecular weight molecules secreted in bile which have been conjugated with glucuronic acid, amino acids or sulphate moieties[5].

An important element in colonic physiology is dietary fibre. Hugh Trowell defined dietary fibre as plant cell wall material resistant to the intestinal secretions of the host[6]. A problem of nomenclature has arisen with dietary fibre[7]. This problem is, however, indicative of a more basic dilemma concerning fibre on the one hand and caecal bacterial metabolism on the other. Implicit in Trowell's definition is the bacterial breakdown of fibre in the colon. In the caecum there is a large number of bacteria, some 500 or more separate species. These caecal bacteria can however be regarded as a functional single metabolic entity, a bacterial mass or even an organ.

Chemical and Physical Properties of Dietary Fibre

There is a poor correlation between the chemistry (total amount and the constituent sugars) and the functions of various fibres[8,9]. Different fibre types may initiate a different response from bacteria within the colon. There may be a generalised increase in bacteria leading to an increase in bacterial mass and metabolism. It is not known whether increases in bacterial populations occur uniformly in all species or are restricted to certain bacterial types that benefit from and are implicated in the metabolism of the fibre that has stimulated the increase.

Physical properties that influence function along the gastrointestinal tract are a combination of the colligative properties of the water soluble fibre components and the surface properties of the water insoluble fibre components. These properties include water holding, cation exchange, organic absorption and gel filtration. These may be influenced by particle size distribution. These physical properties are interrelated. It has been suggested that dietary fibre can act as an ion exchanger and molecular sieve chromatography system throughout the gastrointestinal tract. The plant cell wall provides a water soluble or insoluble matrix or phase through which the soluble content of the gut can variably penetrate and partition with[8].

The biological potency of dietary fibre depends upon its relationship with water along the gastrointestinal tract. Holding water as a solvent affects the biological availability of solutes. In the colon water is distributed in three ways[10]

1. free water that can be absorbed from the colon
2. water that is incorporated into bacterial mass
3. water that is bound by a fibre.

The measurement of the amount of water in fibre can be followed in a number of ways by centrifugation or by suction pressure methods [10,11]. A quick useful comparative method is using flow characteristics of insoluble fibre and this enables the effect of cooking to be identified[12].

Caecal Fermentation. The process whereby a compound is bacterially dissimilated in the caecum under anaerobic conditions is complex and varied, leading to partial or complete decomposition with the end products being either absorbed from the colon to be utilised as nutrients, or absorbed and re-excreted in the enterohepatic circulation or to be excreted in stool. Many compounds are variably and simultaneously decomposed in the mixture and apparently unrelated compounds may well influence the metabolism of each other.

Recent work has demonstrated that dietary fibre can be extensively degraded to volatile fatty acids and other metabolic products[13] including methane, hydrogen and carbon dioxide. The degree to which this occurs depends on the type of fibre and also the amount of fibre[14]. The duration of fibre feeding is important. Experiments feeding gum arabic[15], or carrot[16], show that a period of ingestion is required to develop evidence of fermentation of these fibres. This is in contrast to acute exposure when there may be no evidence of fermentation[17]. The proportion of the major fatty acids, acetate, butyrate and propionate varies with diet and other conditions. Gum arabic results in an increase of production of short chain fatty acids in the rat colon[14]. The relative amounts of acetate and butyrate resulting from the fermentation of gum arabic in the rat, is dependent upon the amount of gum arabic fed and also whether the gum arabic is fed with an elemental diet or standard rat pellet diet[14]. The metabolic products maybe absorbed from the colon and utilised by the colonic mucosa or more remotely in the body. It has been suggested in man that on a control diet, only 20 per cent of the overall ingested fibre is recovered in the stools. It has been suggested that approximately 20g of cell wall polysaccharides and other carbohydrates are fermented in the human colon each day, so that approximately 200mmol of short chain fatty acids are produced[13]. Whilst a considerable amount of short chain fatty acids are produced, only 7 - 20mmols per day are excreted in the stool, therefore there must be substantial absorption and metabolism of short chain fatty acids in the colon. The residual fibre, some of the metabolic products and bacteria are excreted in the faeces[8].

Hydrogen, methane and carbon dioxide constitute flatus. The volume of flatus passed per rectum has been estimated to range from 200 to 2,400 mls per day[18]. Certain foods produced flatulence, particularly beans. Hydrogen production in the colon is dependent upon the delivery of ingested non-absorbable fermentable substrates to a high concentration of bacteria[19]. Normally these substrates consist of non-absorbable oligosaccharides such as stachyose, raffinose and possibly more complex polysaccharides such as starch[20]. The evolution of breath hydrogen has a regular pattern through the day, with a comparatively high volume in the morning, falling until midday and rising during the early afternoon. The mean hydrogen excretion at any time in the day is $\leq$ 0.5μmol/L and individual breath samples rarely contain more than 0.9μmol/L[21]. The average daily breath hydrogen excretion by an individual is of the order of 50-200mls per day. In acute studies few of the different components of dietary fibre, cellulose, pectin and lignin resulted in an increased production of hydrogen[22]. The single dose administration of unprocessed bran results in a modest increase in breath hydrogen[17], though it may be that there is contamination of this wheat bran by starch. If a fibre preparation is taken for 3 weeks, then there is an increase in breath hydrogen[15]. Therefore the breath hydrogen is a useful indicator of caecal fermentation activity.

There are wide variations reported in the proportion of individuals who exhale methane in their breath. This ranges from 33 - 55 per cent in North America, 40 - 61 per cent in Britain and 75 - 80 per cent in Nigeria[18]. The methane excreting status of an individual has been shown to remain stable over periods of up to 3 years. The fermentation of faecal material from both methane and non methane breath excretors leads to methane production. This suggests that methane is produced by all subjects but in varying concentrations. That only when the production reaches a threshold level does methane appear in the breath[23]. In vitro fermentation studies have shown methane production to be stimulated by added amino acids and glycoproteins[24].

It has now been shown that a proportion of dietary starch may be resistant to pancreatic enzymes and passes to be fermented in the caecum[2,20]. The question now arises, is this resistant starch, which is dietary plant material, a fibre? The term fibre implies structural plant cell wall composed of polysaccharides and lignins. By narrowing the concept of fibre to caecally degraded polymeric substances the potential for understanding the role of fibre and other substances along the gastrointestinal tract is reduced. Mucoprotein and biliary excretion compounds are also metabolised in the caecum[1,5,24].

Caecal metabolism and enterohepatic circulation.

Water soluble conjugates arriving in the caecum usually biliary excretion products which have not been absorbed in the intestine, e.g. bile acids and bilirubin, are modified by bacteria to less water soluble compounds. Such modified metabolites may be adsorbed to bacteria or fibre, or be absorbed into the circulation through the colonic mucosa. This enterohepatic circulation contributes to the conservation of compounds in the body, e.g. bile acids and possibly other compounds of intermediate metabolism. This process is of physiological importance. It has been shown that bile acids can be adsorbed to dietary fibre[25]. However, those fibres which are most important in increasing faecal bile acids excretion are those which are extensively fermented in the colon, e.g. pectin and gum arabic[9].

Bile acids are 7α-dehydroxylated in the colon yielding the secondary bile acids, lithocholic and deoxycholic. It has been suggested that the amount of faecal bile acids excreted may be of importance in the aetiology of colonic neoplasm[26]. There is an in verse relationship between stool weight and the lithocholate-deoxycholic acid ration in faeces. It is possible that primary bile acids reaching the colon are more efficiently converted to secondary bile acids when there is a longer transit time and a smaller stool weight. Populations with an increased lithocholic to deoxycholic ratio and low stool weight may be more likely to develop neoplastic change. Alternatively however it may be that low stool weight is characterised by a high lithocholic acid to deoxycholic acid ratio and is a feature of a normal population and not primarily a predictor of colon cancer[27].

Colonic Fermentation

Dietary fibre can usefully be regarded as plant

cell wall material, consisting of polysaccharides and lignin passing as a hydrated sponge, modulating intestinal function along the foregut to the hindgut where the fibre may be degraded by bacteria[8].

The factors that affect the fermentation of polysaccharides in the colon include the inherent characteristics of fibre; water solubility, lignification, particle size and molecular structure[8]. The interplay of water, bacteria and fibre must mean that no single factor stands on its own, so that bacteria which get ready access to the interior of a fibre, such as pectin, by virtue of water solubility will also be influenced by particle size and molecular structure. Similarly, the resistance to fermentation of wheat bran is influenced by particle size which is in part mediated through chemistry, the availability of bacteria and also water holding capacity.

There is also a range of chemicals which pass to the caecum providing bacteria with an energy source. Such compounds of exogenous and endogenous origin[5] hormones, mucoproteins, metabolic breakdown products will be completely, partially or minimally metabolised by bacteria.

Faeces

There is no simple logic available based on the reaction of complex carbohydrates which predicts biological activity in the colon. The disparate actions with each dietary fibre source can only be defined by complex and time consuming experiments[9].

Faeces are complex and consist of 75% water, the dry weight being composed of bacteria, the residue being fibre, and excreted compounds[28]. There is a wide range of individual and mean stool weights[29]. In a study in Edinburgh the variation in stool weight was between 19 to 280g for 24 hours. Faecal constituents, bile acids, sterols, fat, electrolytes, correlated strongly with faecal mass. There was no relationship between age and faecal weight and transit time. Of the dietary constituents, only dietary fibre influenced stool weight[30]. The most important mechanism whereby stool weight increases is through the water holding capacity of unfermented dietary fibre, e.g. wheat bran. The greater the water holding capacity of the bran, the greater the effect on stool weight[8]. Fibre may influence faecal output by another mechanism. The stimulation of colonic microbial growth is the result of ingestion of such fermentable fibre sources as apple, guar or pectin. This is an uncertain method as there is not always an increase in stool weight as a result of eating these fibres[28].

Action of Dietary Fibre on Human Colon and Faeces

Human studies suggest that the physiological action of dietary fibre in the colon may be in one of three ways[9] -

An ingested fibre may

1. Increase stool weight with no effect on serum cholesterol, faecal bile acids or caecal fermentation as demonstrated by hydrogen excretion (e.g. wheat bran, gum tragacanth).

2. No effect on stool weight, but with a reduction in serum cholesterol, a change in faecal bile acid excretion and an increase in breath hydrogen excretion (e.g. gum arabic, pectin as raw carrot).

3. No effect on stool weight, serum cholesterol, faecal bile acids or breath hydrogen (gum karaya).

Table 1 Summary of Results from experiments wherein a specific fibre source is added to the diet[9]

Fibre Source and daily intake production g	Stool weight initial/after 21 days g/day mean ± s.d.	
16g Wheat bran	120 ± 18*	183 ± 22*
10g Gum tragacanth	125 ± 40*	188 ± 48*
25g Gum arabic	147 ± 40	161 ± 43
200g Raw carrot (pectin)	$142 \pm 83^{+}$	$177 \pm 74^{+}$
10g Gum karaya	134 ± 25	139 ± 39

$*p \leq 0.01$; $^{+}p \leq 0.05$

Table 2 Summary of results from experiments wherein a specific fibre source was added to the diet[9]

Fibre source and daily intake production g	Breath-hydrogen mean increase ppm mean ± s.d.
16g Wheat bran	no change
10g Gum tragacanth	no change
25g Gum arabic	6 to 17
200g Raw carrot(pectin)	17 to 34
10g Gum karaya	no change

Table 3 Summary of results from experiments wherein a specific fibre source was added to the diet

Fibre source and and daily intake g	Total faecal bile acids initial mmol/d	after 21 days
16g Wheat bran	0.55 ± 0.33	0.53 ± 0.25
10g Gum tragacanth	0.92 ± 0.09*	1.88 ± 0.28*
25g Gum arabic	1.08 ± 0.55	0.99 ± 0.41
200g Raw carrot(pectin)	0.81 ± 0.10	0.95 ± 0.10
10g Gum karaya	1.01 ± 0.41	0.79 ± 0.22

Summary of Action of Fibre on Colon and Faeces

In an attempt to symbolise the action of any dietary fibre or in the colon, the following suggestions are made.

1. Colonic metabolism is dependent upon nutrient availability to the caecal bacterial mass, therefore colonic metabolism will be inversely proportional to the efficiency of upper gut digestion and absorption (ingestion).

   Protein and fat are 95% absorbed in the upper intestine, 5 - 10% starch evades the digestion as retrograde starch[2, 20] and fibre passes virtually intact to the caecum[6].

2. Stool weight is affected by the extent of fibre fermentation and hence is influenced by the water distribution as a result of

fermentation
bacterial mass
water holding capacity of the residual fibre
transit time
chemical or osmotic water from fibre
fermentation

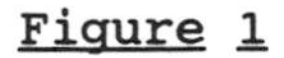

Figure 1

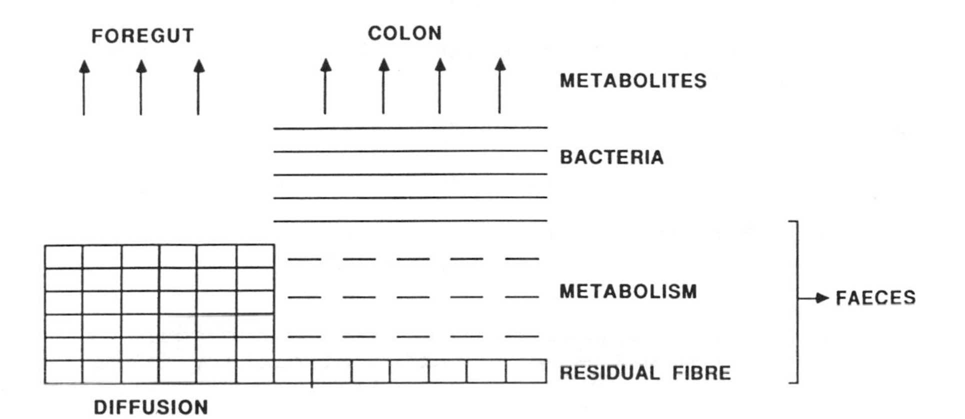

Legend

Action of dietary fibre along the intestine
(1) A sol modulating upper intestinal physiology and absorption
(2) A nutrient for bacteria in the caecum
(3) Key contributor to stool weight and constituents

3. It is possible to symbolize the biological effect of a fibre in the caecum, colon or faeces as

$$\alpha \quad \frac{F_R}{F_O} \qquad \frac{WHC_F}{WHC_O} \qquad \frac{B_F}{B_O} \qquad \frac{M_F}{M_O}$$

F Fibre (g)

WHC Water holding capacity
B Bacterial mass
M Metabolic product

O original
F after fermentation
R residue

A more direct symbolizing of the potency of a fibre on stool weight might be -

Effectiveness of 1g ingested fibre to influence stool weight

$$\alpha \quad (F_R\ WHC) + (B_F\ WHC) + M_F$$

Water holding capacity (WHC)
Bound water to
F_R fibre residue after fermentation
B_F bacterial mass after fermentation
M_F

This chemically or osmotically active water could be created following the fermentation of dietary fibre. For example, short chain fatty acids.

REFERENCES

1. R.Freter, 'Human Intestinal Microflora in Health and Disease', Academic Press, New York, 1983.
2. A.M.Stephen, A.C.Haddad and S.F.Philips, Gastroenterology, 1983, 85, 589.
3. T.Fushiki, N. Yamamoto and K. Iwai. Agric.Biol. Chem, 1985, 49, 1335.
4. M.A.Eastwood and D. Hamilton, Scand.J. Gastroent., 1970, 5, 143.
5. R.L. Smith, 'The Excretory Function of Bile', Chapman and Hall, London, 1973.
6. H. Trowell, D. Burkitt and K. Heaton. 'Dietary Fibre, Fibre-depleted Foods and Disease, London, Academic Press, 1985.
7. M.A. Eastwood and R. Passmore, The Lancet, 1983, 202.
8. M.A.Eastwood and W.G.Brydon, 'Dietary Fibre, Fibre-depleted Foods and Disease, London, Academic

Press, 1985.
9. M.A.Eastwood, W.G.Brydon and D.M.W.Anderson, Am.J.Clin.Nutr, 1986, 44, 51.
10. J.A.Robertson and M.A.Eastwood, Brit.J.Nutr. 1981, 46, 247.
11. J.A.Robertson and M.A.Eastwood, Brit.J.Nutr., 1981, 45 83.
12. A.Anderson and M.A.Eastwood, J.Sci.Fd.Agric., 1987, 31, 185.
13. J.H.Cummings, Gut, 1975, 16, 323.
14. A.H. McLean Ross , M.A.Eastwood, W.G.Brydon, A. Busuttil, L.F. McKay and D.M.W. Anderson Brit.J.Nutr., 1984, 51, 47.
15. A.H.McL. Ross, M.A.Eastwood, W.G.Brydon, J.R.Anderson, and D.M.W.Anderson, Am.J.Clin. Nutr., 1983, 37, 368.
16. J.A.Robertson, W.G.Brydon, K.Tadesse, P.Wenham, A. Walls, and M.A.Eastwood, Am.J.Clin.Nutr., 1979, 32, 1889.
17. K.Tadesse and M.A.Eastwood, Brit.J.Nutr. 40, 393.
18. W.G.Brydon, L.F.McKay and M.A.Eastwood, Dig.Dis., 1986, 4, 1.
19. J.H.Bond and M.D.Levitt, Am.J.Clin.Nutr, 1978, 31, S169.
20. H.N.Englyst, V.Anderson and J.H.Cummings, J.Sci. Fd.Agric., 1983, 34, 1343.
21. K.Tadesse, D.Smith, and M.A.Eastwood, Q.J.Exp. Physiol., 1980, 65, 85.
22. K.Tadesse and M.A.Eastwood, Brit.J.Nutr., 1978, 40, 393.
23. L.F.McKay, M.A.Eastwood and W.G.Brydon, Gut, 1981, 26, 69.
24. J.A.Perman and S.Modler, Gastroent., 1982, 83.
25. M.A.Eastwood and D. Hamilton, Biochim.Biophys.Acta, 1968, 52, 165.
26. M.J.Hill, B.S.Drasar, V.Aries, J.S.Crowther, G.Hawksworth, G. and G. Williams, Lancet, 1971, 1, 95.
27. W.G. Brydon, M.A.Eastwood and R.A.Elton, Brit.J. Cancer, 1988, 57, 635.
28. A.M.Stephen and J.H.Cummings, Nature, 1980, 284, 283.
29. J.B.Wyman, K.W.Heaton, A.P. Manning, A.C.J.Wicks, Gut, 1978, 19, 146.
30. M.A.Eastwood, W.G.Brydon, J.D.Baird, R.A.Elton, S.Helliwell, J.H.Smith and J.L.Pritchard, Am.J. Clin.Nutr., 1984, 40, 628.

THE EFFECT OF 9.5 g/d RESISTANT STARCH ON COLON FUNCTION

J. Tomlin and N. W. Read

Sub-department of Human Gastrointestinal Physiology and Nutrition,
K Floor, Royal Hallamshire Hospital,
Glossop Road, Sheffield S10 2JF.

1 INTRODUCTION

Retrograded amylose is a type of starch which resists hydrolysis with pancreatic amylase. It forms in foods which have been baked or boiled and subsequently cooled (RS_3). Starch that resists breakdown in the small intestine will pass into the large intestine where it may act in a similar manner to the unabsorbed non-starch polysaccharides of dietary fibre. One of the proposed mechanisms for their effects on the colon is that the carbohydates act as substrate to the bacteria, stimulate bacterial cell growth and increase bacterial cell mass[2] and generate fermentation end-products such as short-chain fatty-acids (SCFA) and gases which may affect colonic motility and secretion.[3] As human colonic bacteria are capable of fermenting RS in in-vitro incubations, increasing bacterial cell mass and producing SCFA,[4] RS may be involved in maintaining normal colon function. The relative importance of RS is difficult to tell as there are no published figures for the amount normally present in the diet, and the effect of ingestion of RS on stool output and gas production in humans has not been studied before.

The aims of this experiment were to investigate whether RS can influence colon function in humans by monitoring whole-gut transit-time, stool mass, frequency, consistency, flatus production and the ease of defaecation, and to test whether RS affects breath levels of one of the fermentation end-products, hydrogen, as this may indicate the amount of substrate being fermented in the colon.[5]

2 METHODS

Eight healthy male volunteers were asked to supplement their normal diets with commercial breakfast cereals containing different amounts of RS. The dietary supplements were administered in a random order during two 7 d study periods which were separated by at least a week. One of the supplements contained 350 g/d Kellogg's Cornflakes and the other contained 380 g/d Kellogg's Rice Krispies (providing 10.33 and 0.86 g/d RS respectively).[6] The major nutritional components of the two supplements were balanced to within 1% of each other by including small amounts of other substances (Table 1) using food table data.[7] The volunteers were encouraged to maintain their normal diets as far as possible and were encouraged to eat similar foods during the two study weeks, although it was accepted that the cereal supplements and accompanying milk and sugar (equivalent to six large bowls /d) would displace some of their normal food intake. They kept diaries of food intake using approximate portion sizes and alcohol consumption was restricted to no more than 15 units per week and no more than 4 units per day (1 alcohol unit = $^1/_2$ pint beer, = 1 glass wine, = 1 measure spirits).

During the study the volunteers took 15 small plastic radio-opaque markers at around the same time each day and noted the exact time in a diary. During the last 5 d of

Table 1 Composition of the two dietary supplements.

	RS (g)	Starch (g)	NSP (g)	Energy (kJ)	Pro (g)	Fat (g)	CHO (g)
DIET A							
350g Cornflakes	10.33	262.1	2.3	5.39	30.1	5.6	298
33g Boiled sweets	0	0.1	0	0.45	0	0	28
2.4g Wheat-bran	0	0.3	0.9	0.02	0.3	0.1	6
2.3g Butter	0	0	0	0.07	0	1.9	0
TOTAL	**10.33**	**262.5**	**3.2**	**5.93**	**30.4**	**7.6**	**332**
DIET B							
380g Rice Krispies	0.86	262.1	3.2	5.79	22.0	7.5	327
9g Casilan	0	0	0	0.14	8.1	0.1	0
TOTAL	**0.86**	**262.1**	**3.2**	**5.93**	**30.1**	**7.6**	**327**
DIFFERENCE (A-B)	**9.47**	**0.4**	**0**	**0**	**0.3**	**0**	**5**

Abbreviations used in table; RS = resistant starch, NSP= non-starch polysaccharides, Pro = protein, CHO = carbohydrate.

each study-week the volunteers collected all stools passed into individual plastic bags and labelled them with the time and date of defaecation. The stools were then weighed to yield the stool mass and X-rayed to visualise the radio-opaque markers. The whole-gut transit-time was calculated using the continuous method,[8] and the median value of the 5d period was used in the calculations.

Breath hydrogen measurements were made at 15 min intervals for 8h on the first day of each supplement. End-expiratory samples were obtained using a modified Haldane-Priestley tube, and the hydrogen concentration was measured on a breath hydrogen monitor (GMI, Renfrew, Scotland). The area under the hydrogen curve was calculated (with no base-line correction), and this was used to give an indication of the amount of substrate fermented. The amounts and timings of the cereals and other food and drink was identical during the two weeks.

The volunteers noted in the diary the ease of each defaecation on a visual analogue scale with the 0 and 100 points labelled 'no effort' and 'much straining'. They assessed the form of the resultant faeces on a scale from 1 to 8 by comparison with a set of standard descriptions and photographs, based on a linear scale of stool consistency.[9] They also noted all episodes of flatus passage per rectum. The volunteers assessed the effects of the two supplements compared with their usual bowel habits in questionnaires completed at the end of the study weeks.

Paired results from the two weeks were compared using Wilcoxon's matched pairs rank sum test.

3. RESULTS

There was no significant difference between the total mass of faeces collected during the last 5 d of the two study weeks ($p>0.05$; Table 2). Whole-gut transit-times were not significantly different between the two periods ($p>0.05$; Table 2). There were also no significant differences in the stool frequency, the mean stool form or in the ease of defaecations recorded by the volunteers ($p>0.05$; Table 2).

Breath hydrogen excretion, measured as the area under the 8h breath hydrogen curve, was significantly higher on Cornflakes than on Rice Krispies ($p<0.05$; Table 2). There were no significant differences in the number of flatulent episodes recorded ($p>0.05$; Table 2), although

Table 2 Results of bowel function measurements (mean values, n=8).

	CORNFLAKES	RICE KRISPIES
Faecal Output (g/5d)	893	982
Stool frequency (/5d)	5.0	5.8
Mean consistency*	5.0	5.0
Mean ease of defaecation£	28	30
Whole-gut transit time (h)	43.4	39.5
Total flatulent episodes (/wk)	35	27
Total breath hydrogen (/8h)	12.1	7.5

* - rated on a scale from 1 (liquid) to 8 (hard pellets)

£ - rated on a scale from 0 (no effort) to 100 (much straining)

Cornflakes produced 63 more episodes in total than Rice Krispies.

Volunteers reported that both supplements reduced flatus frequency compared to their normal diets. The questionnaires also revealed that both supplements seemed to reduce the amount of faeces produced, to make the stools firmer, reduce the frequency of defaecation and to make defaecation more difficult compared to normal bowel habits. There were no significant differences in the subjective assessments of the two supplements ($p>0.05$).

The order in which the supplements were fed had no effect on the results.

4. DISCUSSION

The normal intake of RS in the UK was calculated to be 2.76 g/d, using data on the RS contents of a number of foods and food ingredients,[6,10,12] and the average weekly consumption of these foods (and foods prepared from the food ingredients) by the UK population.[13] The Cornflake supplement should therefore have provided over three times the amount of RS in the UK diet to the colonic bacteria.

The difference in RS calculated to be supplied by the two supplements (9.47 g/d) was not sufficient to cause any change in faecal output, stool frequency, consistency or whole-gut transit-time. This may have been because the background diets of some volunteers contained relatively high amounts of dietary fibre (between 4.4 and 27.1 g/d), so that any effect of the RS was swamped, however the significant difference in breath hydrogen excretion on the first day of the supplements suggests that more substrate **was** reaching the bacteria from the Cornflake supplement.

It is possible that the RS in Cornflakes was primarily converted to gas rather than being used for bacterial cell growth. Another explanation is that the bacteria adapted to the steady input of RS, so that any initial impact on colon function could not be detected after measurements began on day 3.

In conclusion our data suggest that whilst on a normal Western diet, a tripling of the normal intake of RS would not cause dramatic changes in colon function.

REFERENCES

1. H.N. Englyst, H. Trowell, D.A.T Southgate, J.H. Cummings, Am J Clin Nutr 1986 44, 42.
2. A.M. Stephens, J.H. Cummings, J Med Micro 1980 13, 45.
3. S.E. Fleming, D. Marthinsen, H. Kuhnlein, J Nutr 1983 113, 2535.
4. H.N. Englyst, G.T. Macfarlane, J Sci Food Agric 1986 37, 699.
5. B. Flourie, C. Florent, J-P. Jouany, P. Thivend, F. Etanchaud, J.C.Rambaud, Gastro 1986 90, 111.
6. H.N. Englyst, V. Anderson, J.H. Cummings, J Sci Food Agric 1983 34, 1434.
7. A.A Payul, D.A.T. Southgate, 'McCance & Widdowson's The Composition of Foods', HMSO, London, 1987.
8. J.H. Cummings, D.J.A. Jenkins, H.S. Wiggins, Gut 1976 17, 210.
9. G.J. Davies, M.Crowder, B. Reid, J.W.T. Dickerson, Gut 1986 27, 164.
10. H.N. Englyst, J.H. Cummings, Analyst 1984 109, 937.
11. H.N. Englyst, J.H. Cummings, Am J Clin Nutr 1986 44, 42.
12. H.N. Englyst, J.H. Cummings, Am J Clin Nutr 1987 45, 423.
13. Annual Report of the Foosd Survey Committee, 'Household Food Consumption and Expenditure', HMSO, London, 1985.

USE OF THE WISTAR RAT AS A MODEL FOR INVESTIGATING THE FIBRE-DEPLETION THEORY OF COLONIC DIVERTICULOSIS, 1949-1988.

C.S. Berry

Flour Milling and Baking Research Association
Chorleywood
Rickmansworth
Herts., U.K.

1 INTRODUCTION

Diverticular disease is one of the degenerative conditions that have been blamed upon chronic consumption of a fibre-depleted diet. It is characterised by outpouchings of the mucosal lining through the muscle wall of the colon, and is common in the UK and other Western countries, particularly in older people. A substantial body of epidemiological evidence offers strong support to the view that dietary fibre is protective against diverticular disease[1], though other factors, notably age-related changes in connective tissue have been implicated[2]. There is also a straightforward conceptual link between fibre and diverticulosis, based on the idea that the lumen of the colon diminishes when deprived of bulking agents, as a consequence of adaptive changes in colonic muscle. In this contracted state it becomes more prone to complete constriction during normal or increased activity of colonic circular muscle. Raised pressure in these segments allows the mucosa to herniate through weak points in the muscle coat.[3]

This paper summarises attempts to obtain experimental proof of a causal link between fibre deficiency and diverticular disease using the Wistar rat as an animal model.

2 CARLSON AND HOELZEL EXPERIMENTS

Carlson and Hoelzel reported in 1949 the findings of two lifespan trials using Wistar rats.[4] In the first experiment, results of which are summarised in Table 1,

Table 1 Incidence of diverticulosis Carlson and Hoelzel (1949).

	Diet	Male Rats	Female Rats
(a)	Control	8/19 (42%)	14/26 (54%)
(b)	10% alfalfa	4/9 (44%)	3/14 (21%)
(c)	As (b) plus 5% kapoc and 5% psyllium husk	3/11 (27%)	4/12 (33%)
(d)	"Vegetarian diet" inc. 50% wholewheat flour	1/21 (5%)	3/25 (12%)

137 rats were fed diets with a variety of fibre supplements for their entire lifespan.

Colonic diverticula began to be seen after the 20th month of the trial in animals dying of natural causes. The overall incidence of diverticulosis was abut 30%, with a preponderance in animals fed a control diet described as "low residue, omnivorous, and containing 35% meat protein and 28% fat". However, the fibre content of the diets was not determined (even as "crude fibre"). Moreover, animals were allowed free access to "trimmings of head lettuce". (Fortunately lettuce leaves consist mainly of water (96%) and contain only about 1.5% dietary fibre).

The diverticula described by Carlson and Hoelzel were either solitary of multiple (as in human clinical presentation), and were generally situated close to the caecum. In the latter respect they differed from typical Western diverticular disease, which affects the more distal ("left-sided") sigmoid region of the colon. However, right-sided diverticular disease is not unknown in human populations, and is the favoured site of presentation in Oriental populations[5]. The reason for the different distributions is unknown.

Carlson and Hoelzen also described in their 1949 paper a remarkable experiment involving 115 rats divided between what appears to be 15 (or maybe 17) dietary regimen, 10 of which involved crossovers between 2 diets, and a further 2 involving crossovers between 3 diets. In some instances the group size was reduced to 1 (or zero), and there were many groups with just 2 or 3 animals. This prolific scattering of animals between treatment groups clearly precluded any kind of statistical analysis. Nevertheless, the data appeared to suggest that a rat that was raised initially on a high-fibre diet and then subsequently switched to a low-fibre diet was especially prone to diverticulosis.

3 FMBRA LIFESPAN TRIAL (WHEAT FIBRE)

Our first and largest trial at Chorleywood used a total of 1800 Wistar rats. A range of levels of fibre, supplied as white or wholemeal bread, or as unprocessed wheat bran, were incorporated into isocaloric low-fat semisynthetic diets. In addition, we tested a non-isocaloric high fat/low fibre diet and a commercial high-fibre stock diet. The first diverticulum was seen at around 18 months. The results, which we published[6] in 1985, showed a statistically highly significant inverse relationship ($p<0.001$) between total DF, measured by the Southgate procedure, and incidence of histopathologically-confirmed colonic diverticulosis (Fig. 1).

Only 27.8% of rats fed a diet containing 34% dried wholemeal bread (Diet E) developed diverticulosis compared

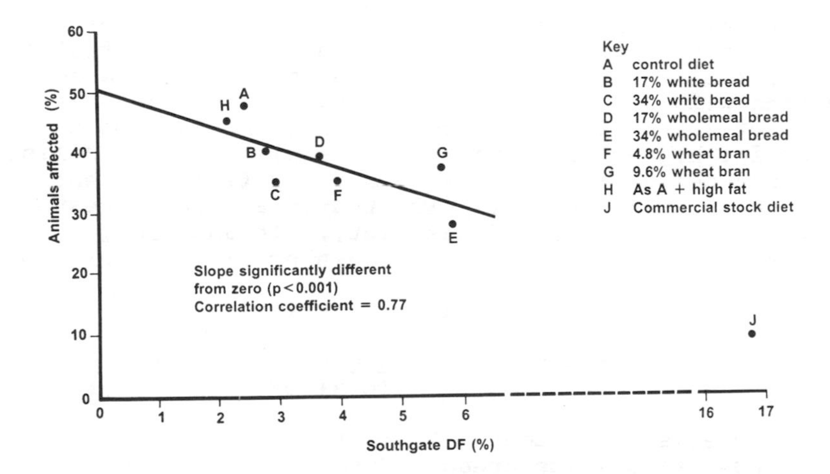

Figure 1 Effect of cereal DF on incidence of colonic diverticulosis (900 male, 900 female rats).

with 47.5% on Control diet A in which pregelatinised wheat starch was the main component. Diet G, with the higher of two levels of unprocessed wheat bran, produced an anomalously high incidence (37.2%) of diverticulosis, despite having about the same level of wheat fibre as the wholemeal bread diet E. These data give no grounds for thinking that heat processing reduces the protective effect of wheat bran; in fact, the opposite appears to be true. The lifespan trial also showed that fibre associated with starchy endosperm in white bread (Diets B and C) was at least as effective in protecting against diverticulosis as that associated with the branny outer layers (pericarp, testa and aleurone) of wheat, when compared on a 1:1 basis, despite its greater fermentability.

Further evidence of the efficacy of cereal fibre was the exceptionally low incidence of diverticulosis (9.4%) in the group of animals fed a commercial stock diet containing 47.5% whole wheat and 40% Sussex ground oats (run primarily as a check on animal husbandry).

4 SECOND FMBRA TRIAL (WHEAT VERSUS NON-WHEAT FIBRE AND RESISTANT STARCH)

The first FMBRA trial, summarised above, ran for the entire lifespan of the rat, and did not end until almost all the animals had become moribund, or had died of natural causes (almost 3 years). In the second trial a shorter feeding time (approx. 15 months) was achieved by starting with 306 ex-breeder female rats purchased at 8-9 months of age having previously been maintained on a commercial high-fibre breeding stock diet. Table 2 lists the diets tested in the trial and their fibre contents. Dried white bread was this time tested at its highest practicable level (80% of the diet). In addition, autoclaved, amylomaize starch, shown previously[7] to be a

Table 2 Diets fed in second FMBRA trial.

Diet	Source of fibre	A.O.A.C. Total DF %	Enzyme-resistant Starch
A	None (sucrose control)	0	0
B	80% dried white bread	3.4	1.4
C	80% rolled oats	6.6	0.5
D	37% extruded peas	5.5	0.7
E	15% amylomaize starch	3.4	4.0
F	15% wheat bran	7.5	0

rich source of enzyme-resistant starch (ERS), was tested at a level which gave an identical analytical value for "apparent" dietary fibre in a AOAC gravimetric procedure. In addition, three other sources of fibre, two of cereal origin (wheat bran and rolled oats) and one non-cereal (legume) were tested at the same estimated level of Southgate total fibre (6.3%). Actual levels were subsequently determined retrospectively on pooled batches using FMBRA's currently available AOAC and Englyst methodology.

Once again, a highly significant linear inverse relationship was seen between total apparent fibre and incidence of diverticulosis (Fig. 2). The rolled oats diet was as effective as the wheat bran diet, despite having a slightly lower level of AOAC total fibre, and a greater proportion of soluble fibre. In addition, the amylomaize diet, supplying apparent fibre primarily as ERS, was as effective as the "isofibrous" white bread diet in which both NSP and ERS were present in a ratio of approximately 1.8:1. The legume diet was associated with a lower incidence of diverticulosis relative to zero fibre (sucrose) control, but the difference failed to reach significance at the 5% level.

These data confirm the findings of the previous trial, namely that the protective effect of fibre in the rat model is not confined to insoluble, poorly fermentable

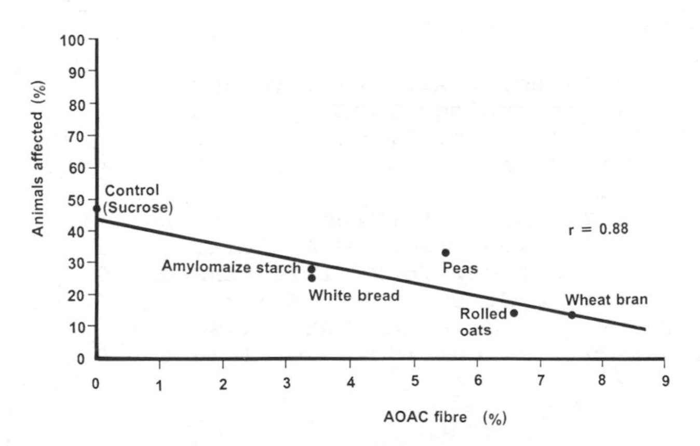

Figure 2 Effect of sources of cereal and non-cereal fibre and enzyme-resistant starch on incidence of diverticulosis (306 female rats).

wheat bran fibre, despite epidemiological evidence that the latter is more protective in Western populations. However, the caecal site of presentation of diverticula in the rat makes it likely that most if not all bulk-forming agents in the diet may be protective in this model, since bacterial degradation will be less complete than at more distal sites. It is possible that the low incidence of the predominantly right-sided diverticular disease in Oriental populations reflects an ability of all forms of NSP and ERS to protect the proximal colon. Wheat bran, on the other hand, may have a special role to play in protecting the susceptible sigmoid region of the Western bowel against this condition, by virtue of its exceptional resistance to anaerobic degradation.

ACKNOWLEDGEMENTS

Many colleagues have contributed to the FMBRA studies, notably Dr N. Fisher, Dr Joan Hardy, Dr Alison Owens, Mrs Joyce Lucas, Mr J. Gregory, Mr R. Plimsaul, Mr B. Burke, Mrs Ruth East and staff of the Animal Unit. This research was sponsored by the Ministry of Agriculture, Fisheries and Food and the results are Crown Copyright.

REFERENCES

1. N. Painter in "Dietary Fibre, Fibre-Depleted Foods and Disease" (eds. H. Trowell, D. Burkitt and K. Heaton), Academic Press, London, 1985, Chapter 8, p.145.
2. M.A. Eastwood, D.A.K. Watters and A.N. Smith, Clinics in Gastroenterology 1982, 11, 545.
3. N.S. Painter, Ann. R. Coll. Surg. Eng. 1964, 34, 98.
4. A.J. Carlson and F. Hoelzel, Gastroenterology, 1949, 12, 108.
5. M.A. Eastwood, J. Eastwood and M. Ward in "Fiber in Human Nutrition" (eds. G.A. Spiller and R.J. Amen), Plenum Press, New York, 1976, Chapter 9, p.207.
6. N. Fisher, C.S. Berry, T. Fearn, J.A. Gregory and J. Hardy, Am. J. Clin. Nutr., 1985, 42, 788-804.
7. C.S. Berry, J. Cer. Science, 1986 4, 301.

ENHANCEMENT OF COLONIC CRYPT PROLIFERATION IN MAN BY SHORT-CHAIN FATTY ACIDS (SCFA)

W. Scheppach, P. Bartram, A. Richter, F. Richter, H. Liepold, H. Kasper

Department of Medicine
University of Würzburg
D-8700 Würzburg, F.R.G.

INTRODUCTION

The major constituents of dietary fibre(DF) are non-starch polysaccharides(NSP)[1]. Together with certain fractions of starch[2] they resist enzymatic digestion in the small bowel and reach the caecum. Depending on chemical composition and physical properties most NSP are rapidly degraded under anaerobic conditions by the colonic microflora[3]. End products of this fermentative process are short-chain fatty acids(SCFA). The predominant SCFA are acetic(AC), propionic(PROP) and n-butyric(BUT) acids which are generated in an approximate molar ratio of 60 : 25 : 15[4]. They are rapidly absorbed from the colonic lumen[5]. BUT serves as a fuel for energy metabolism of colonic epithelia. Moreover, isolated colonocytes of man and rat prefer BUT to other substrates like glucose, glutamine, lactate and ketone bodies[6,7]. The metabolism of the other SCFA by colonocytes seems to be less extensive which allows AC and PROP to enter the portal circulation. At the hepatic level PROP is taken up by the hepatocytes and may affect cholesterol metabolism[8]. AC alone is recovered in peripheral venous blood[9] to be oxidised to CO_2 by many tissues including skeletal and heart muscle[10].

While the extraintestinal effects of SCFA[11] are not well understood there is evidence that these acids play a major role at their production site in the caecum. The interactions between bacterial metabolites and the colonic epithelium seems to be of special interest. As BUT is a preferred energy yielding substrate to colonocytes[6,7] the hypothesis was put forward that SCFA may also affect cell replication of the colonic mucosa in man.

METHODS

During routine colonoscopy 6 biopsies were taken in 31 patients from normal caecal mucosa. Pathologic findings elsewhere in the colon included diverticula or hemorrhoids but not inflammatory bowel disease or tumors. The samples were immersed in Eagle's medium and incubated for 3 h in the presence of SCFA (3 biopsies) or equimolar NaCl (3 biopsies, control). Then pulse labelling of proliferating cells was achieved by addition of ^{3}H-thymidine for 1 h[12]. The samples were fixed in formalin, embedded in paraplast, section-cut in 4 um slices and stained with acid Schiff's reagent (Feulgen reaction). Sections were dipped in Ilford K2 emulsion and exposed for 15 days following standard techniques of autoradiography[12].

The incubation medium contained sodium salts of SCFA at the following concentrations:

(a) AC = acetate, 60 mmol/l, n = 7;
(b) PROP = propionate, 25 mmol/l, n = 8;
(c) BUT = n-butyrate, 10 mmol/l, n = 8;
(d) SCFA = AC(60 mmol/l) + PROP(25 mmol/l) + BUT (10 mmol/l, n = 8)

The labelling frequency of colonocytes was estimated in 15-20 longitudinally sectioned crypts of each run by light microscopy.The number of labelled(l) and unlabelled (u) cells per crypt column was determined and the whole crypt labelling index (LI, l/l+u) calculated. Crypts were divided into 5 equal compartments, compartment 1 representing the crypt base and compartment 5 the crypt surface; in this way a LI for every crypt compartment was calculated. Mean LI values of 15-20 crypts per run were compared by the Wilcoxon rank sum test for paired data or by the Mann-Whitney U-test for unpaired data where appropriate. Differences were considered significant at $p < 0.05$. All values were expressed as mean ± SEM.

RESULTS

Under control conditions (NaCl 10 - 95 mmol/l) the whole crypt LI values (Figure 1) ranged from 0.062±0.008 to 0.074±0.012 without a significant difference between the groups. AC raised crypt labelling by 30% (LI: 0.084±0.005, $p<0.05$), PROP by 70% (LI: 0.124±0.014, $p<0.05$), BUT by 89% (LI: 0.140±0.016, $p<0.05$) and SCFA (AC+PROP+BUT) by 103% (LI: 0.126±0.012, $p<0.05$). Due to interindividual variation there was no difference between PROP, BUT, and SCFA, but labelling in these 3 groups was higher than in

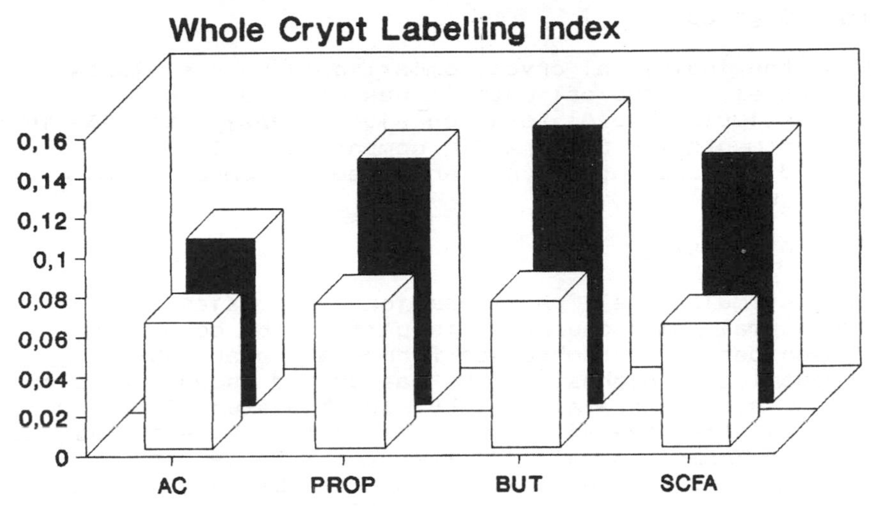

Figure 1 Whole crypt labelling index for incubation experiments with acetate(AC), propionate(PROP), butyrate (BUT) and the combination of SCFA(AC+PROP+BUT). ■SCFA, □NaCl(control).

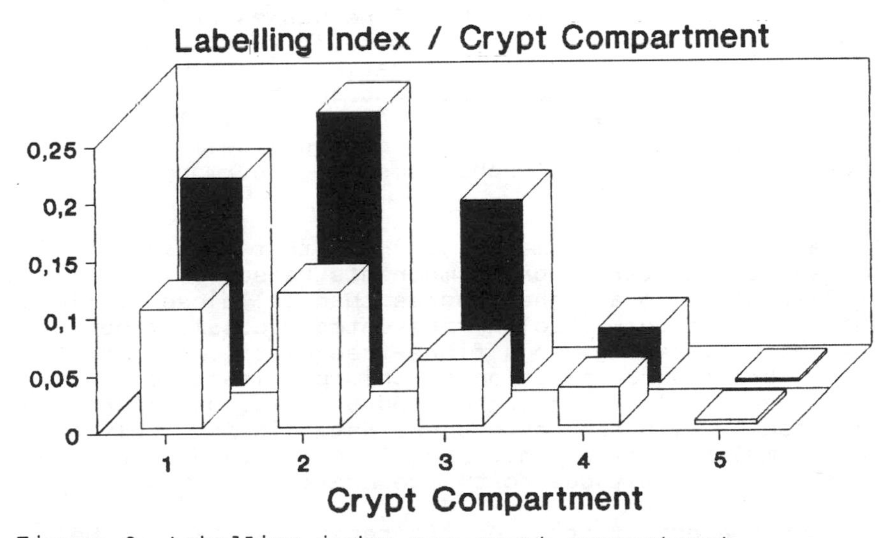

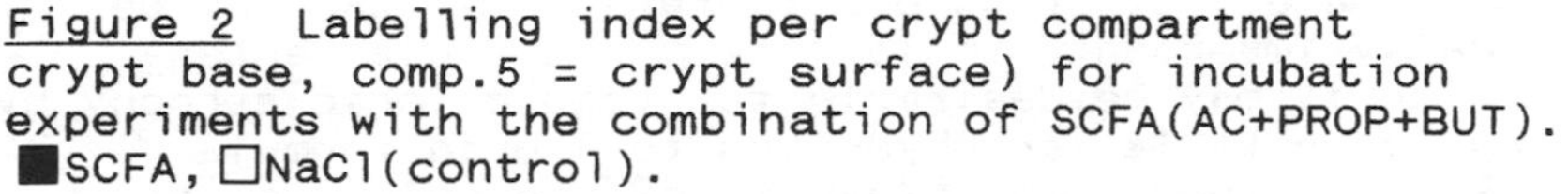
Figure 2 Labelling index per crypt compartment crypt base, comp.5 = crypt surface) for incubation experiments with the combination of SCFA(AC+PROP+BUT). ■SCFA, □NaCl(control).

the AC group.

When the individual crypt compartments were evaluated separately, a typical pattern was observed: PROP, BUT and SCFA stimulated proliferation significantly ($p<0.05$) in compartments 1 - 3 and AC in compartments 1 - 2. Thymidine labeling was not enhanced in compartments 4 and 5 (Figure 2).

DISCUSSION

SCFA stimulate cell proliferation in biopsies taken from the normal human caecum. This part of the colon was chosen because carbohydrate fermentation and SCFA production is highest in the caecum and ascending colon due to substrate availability. In this study PROP and BUT were as effective as a combination of SCFA, while AC had a minor impact on cell proliferation. These data are in good agreement with Sakata's results obtained in rat experiments[13,14]. He reported a 52 - 99% increase of cell proliferation by SCFA; butyrate exerted a stronger effect than propionate or acetate. Kripke et al.[15,16] perfused the rat colon with SCFA (AC 70 mmol/l, PROP 35 mmol/l, BUT 20 mmol/l) and found a rise of mucosal weight, protein and DNA contents; BUT alone was as effective as the SCFA combination. Thus, in man (in vitro) and in rat (in vivo) colonic crypt proliferation is stimulated by SCFA of which BUT seems to be most important. BUT is a preferred fuel for colonocytes[6,7] which may also explain a higher proliferative activity of these cells in the presence of BUT. Other mechanisms, e.g. hormonal, may play a role as well.

It was shown in experimental animals that feeding of fermentable dietary fibre components raised colonic crypt proliferation to a higher degree than cellulose[17]. Fibre but not inert bulk (kaolin) prevented mucosal atrophy in the colon of rats given a fibre-free diet[18]. In germ-free rats fibre had no effect on colonic proliferation whereas in conventional rats fibre caused a marked proliferative response[19]. With regard to these data, it is likely that the stimulation of colonic proliferation by dietary fibre is at least partly due to the generation of SCFA.

The enhancement of crypt proliferation by SCFA may be clinically important. In long-term artificial enteral nutrition it may be desirable to avoid colonic atrophy by adding fermentable carbohydrates to liquid formula diets[20]. The functional capacity of the colon may be improved by SCFA in the short-bowel syndrome; in the rat model intestinal adaptation to massive small bowel resection was enhanced by addition of pectin[21] which is fermented in the colon to give rise to SCFA. - The stimulation of caecal proliferation by SCFA suggests close interactions between the microflora and the host organism which merit further scientific attention.

REFERENCES

1. J.H. Cummings, Lancet, 1983, I, 1206.
2. H.N. Englyst, H. Trowell, D.A.T. Southgate and J.H. Cummings, Am.J.Clin.Nutr., 1987, 46, 873.
3. A.A. Salyers, A.P. Kuritza and R.E. McCarthy, Proc. Soc. Exp. Biol. Med., 1985, 180, 415.
4. J.H. Cummings, Gut, 1981, 22, 763.
5. J.H. Cummings, Scand.J.Gastroenterol., 1984, 19, (Suppl.93),89.
6. W.E.W. Roediger, Gut, 1980, 21, 793.
7. W.E.W. Roediger, Gastroenterology, 1982, 83, 424.
8. W.L. Chen, J.W. Anderson and D. Jennings, Proc. Soc. Exp. Biol. Med., 1984, 175, 215.
9. E.W. Pomare, W.J. Branch and J.H. Cummings, J.Clin. Invest., 1985, 75, 1448.
10. F. Lundquist, L. Sestoft, S.E. Damgaard, J.P. Clausen and J. Trap-Jensen, J.Clin.Invest., 1973, 52, 3231.
11. W. Scheppach, H.S. Wiggins, D. Halliday, R. Self, J. Howard, W.J. Branch, J. Schrezenmeir and J.H. Cummings, Clin.Sci., 1988, 75, 363.
12. M. Usugane, M. Fujita, M. Lipkin, R. Palmer, E. Friedman and L.Augenlicht, Digestion, 1982, 24, 225.
13. T. Sakata, Can.J.Anim.Sci., 1984, 64(Suppl.), 189.
14. T. Sakata, Br.J.Nutr., 1987, 58, 95.
15. S.A. Kripke, A.D. Fox, J.M. Berman, R.G. Settle and J.L. Rombeau, J.P.E.N., 1988, 12(Suppl.), 8S.
16. S.A. Kripke, A.D. Fox, J.M. Berman, J. DePaula, R.G. Settle and J.L. Rombeau, Clin.Nutr., 1987, 6(Suppl.), 38.
17. J.R. Lupton, D.M. Coder and L.R. Jacobs, J.Nutr., 1988, 118, 840.
18. R.A. Goodlad and N.A. Wright, Br.J.Nutr., 1983, 50, 91.
19. R.A. Goodlad, B. Ratcliffe, J.P. Fordham and N.A. Wright, Gut, 1989, 30, 820.
20. R.G. Settle, J.P.E.N., 1988, 12(Suppl.), 104S.
21. M.J. Koruda, R.H. Rolandelli, R.G. Settle, S.H. Saul and J.L. Rombeau, J.P.E.N., 1986, 10, 343.

LABORATORY ASSESSMENT AND EFFECTS OF INCREASED COLONIC FERMENTATION INDUCED BY UNABSORBABLE CARBOHYDRATES.

F. Brighenti, G. Testolin, D.J.A. Jenkins.
Dipartimento di Scienze e Tecnologie Alimentari e Microbiologiche Sezione Nutrizione - Universita' di Milano, Italy
Department of Nutritional Sciences, Faculty of Medicine, University of Toronto, Canada.

1 INTRODUCTION

Dietary fiber has been suggested to exert metabolic effects either by altering small intestinal absorption or possibly by increasing colonic fermentation[1,2] However, discriminating between the two mechanisms is difficult since the fibers that delay absorption are generally easily fermented. Moreover, data on fiber fermentability are mainly derived from in vitro studies, and only indirect evidence of the fate of the fermentation products exist for fibers mixed in actual diets. We wish to present the sequential changes in breath H_2 and circulating Short Chain Fatty Acids (SCFA) as a possible marker of colonic microclimate adaptation to fermentable substrates. The carbohydrate source used was lactulose, a non-absorbable readily fermentable sugar widely used as a laxative and for therapy of hepatic encephalopathy.

2 METHODS

SCFA analysis

SCFA were determined by HPLC in serum after sample preparation involving acid deproteinization or ultrafiltration, followed by vacuum distillation.

Sample preparation. One-hundred ul of serum or whole blood were added to 150 ul of an ice-cold standard solution containing 2 mmol/l butyric acid (internal standard) and 5% perchloric acid, into a 500 ul eppendorf centrifuge tube. After careful mixing, the tubes were centrifuged 10 minutes at 3000 x g, and 150 µl of clear supernatant were transferred to a 5x75 mm glass tube. Alternatively, 500 µl of serum were deproteinized by ultrafiltration at 1500 x g for 70 minutes at 4 °C using an MPS-1 system with a 30,000 MW cutoff (Amicon, Danvers, MA USA). To 400 µl of ultrafiltrate, 100 µl of 0.4M perchloric acid containing 2 mmol/l butyric acid were added, and an aliquot of 150 µl transferred to a glass tube. The solution was wall-frozen in liquid nitrogen on one side of the tube, and the sample was distilled at low (<5 mm Hg) pressure using an Y-shaped device similar to that described by Tollinger et al.[3].

Chromatographic conditions. The apparatus was a Hewlett-Packard Chemstation (Hewlett-Packard, Palo Alto, CA, USA) including an isocratic 1050 series pump (Hewlett-Packard), a diode-array 1040A detector (Hewlett-Packard), a Column Heater (Bio-Rad, Richmond, CA, USA), a Wisp 710B autoinjector (Waters Ass., Milford, MA, USA). The column used was a Bio-Rad HPX-87H Organic Acid Analysis Column (Bio-Rad) on line with a Cation-H Micro-Guard Column (Bio-Rad). Mobile phase was 0.005 N H_2SO_4 with a flow rate of 0.6 ml/min at a column temperature of 60 °C.
For routine analysis detection was made at 210 nm, while for purity analysis the whole spectra from 190 to 280 nm were acquired.

Breath Hydrogen analysis

Breath Hydrogen was measured on expired alveolar air samples collected by mean of a modified Haldane-Priesly tube[4] into 30 ml plastic syringes with a 3-ways luer lock tap. The subjects were not allowed to smoke during the test, and kept their normal activity during the day of the test. The samples were analysed by gas chromatography using a dedicated gas analyzer (model 12i MicroLyzer, Quintron instruments Co Inc, Milwaukee, Wi, USA).

Study design

Eight healty volunteers, mean age 37.5y±3.0y were randomly placed on two 2-week metabolic diets of identical composition, one of which was supplemented with lactulose (1g per 100 kcal, maximum 25 g/day), following a crossover design. Diets provided 30%, 30% and 40% of the calories at breakfast, lunch and dinner respectively. Diet composition was 15% protein, 33% fat and 52% carbohydrate, and apported 15g fiber per 1000 kcal, according to food composition tables. Breath H_2 was measured hourly on days 1, 7 and 13 of the lactulose period, and on day thirteen of the control period. Serum SCFA was measured every two hours on day thirteen of each phase, and, in a supplementary study, on the first day of lactulose supplementation.

3 RESULTS

The only SCFA detectable in periferal blood was acetate, identified by its retention time and UV spectrum. Mean±SD recoveries of formate, acetate, propionate and butyrate, from 100 ul of pooled serum spiked with 100 µl of standard solution, were 105±5, 100±3, and 101±1% respectively. When the perchloric acid procedure was used on pooled fasting serum, the calculated coefficent of variation for acetate analysis was 25% . To this value, distillation contributed 8% CV (10 replicates), and chromatography 2% CV (10 replicates). Thus, the coefficent of variation for deproteinization was about 15% , a quite high value possibly deriving from acetate esters hydrolysis operated by the concentrated acid at various extent. However, when samples were deproteinized by ultrafiltration, the coefficent of variation for duplicates dropped to 5% ,and for the whole procedure to 12%, due to a better reproducibility of both the deproteinization and the distillation steps. Moreover, the SD for replicates remained constantly low (about 5 to 10 umol/l) independently from the acetate concentration in blood.

During the day profile, serum acetate levels increased, though not significantly, in the latter part of the day on lactulose diet by about 22% on day 1, and 24% on day 13, (Figure 1).

A much higher difference was noted for breath H_2. The mean day-long difference in breath hydrogen levels on the lactulose diet on day 13 was 8.0±1.1 ppm ($p<0.001$, Figure 2).

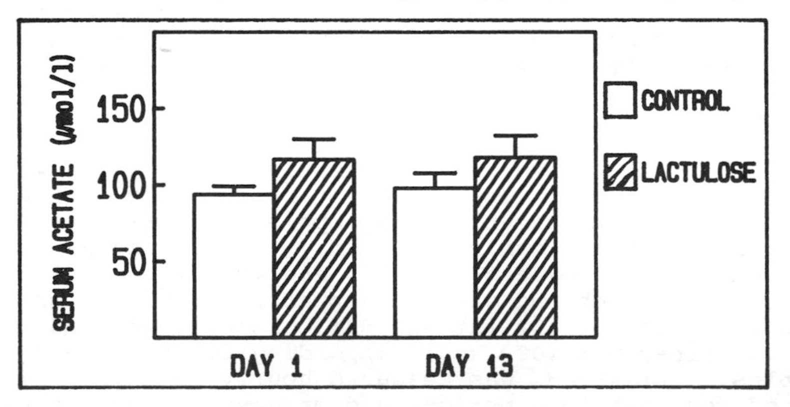

Figure 1 Mean acetate levels on the second half of the day profile (14.00 to 20.00)

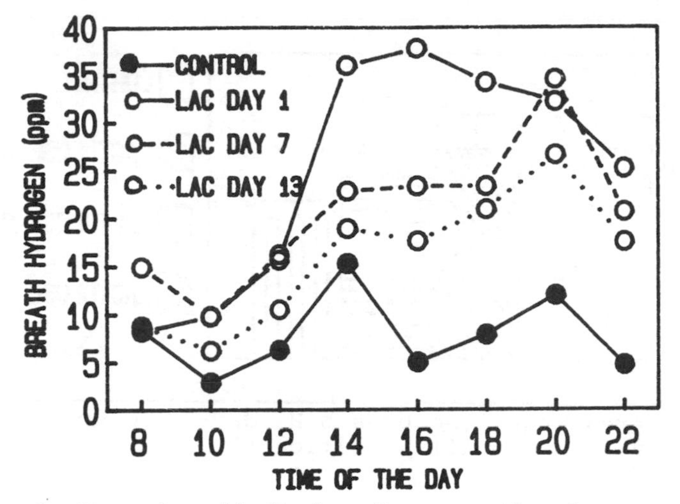

Figure 2 Mean breath H_2 levels over the day.

Considering the areas under the curve, values for lactulose on day 1, 7 and 13 were 310%, 268% and 206% of the control respectively (Figure 3).

4 DISCUSSION

The most common markers of colonic fermentation are, together with colonic pH, breath hydrogen and circulating SCFA, usually determined by Gas Liquid Chromatography. High Performance Liquid Chromatography of SCFA has been successfully used in acute studies showing colonic absorption of infused acetate and propionate[5]. This method permits good sensitivity even at the low circulating acetate levels found in fasting blood.
In this study we observed no difference between the one day and the two week acetate levels. Thus, if the microflora adaptation toward higher production of SCFA into the lumen[6] exist, it is not reflected by higher circulating levels in vivo. However, the progressive reduction of the amount of H_2 expired may reflect a modification of bacteria metabolism over the time, possibly due to a decrease in colonic pH, with consequent inhibition of hydrogen production[7,8]. As a consequence of increased colonic fermentation, due to addition of lactulose to otherwise identical diets, we observed an effect on FFA levels over the day and in fasting serum total and LDL cholesterol and apolipoprotein B. The results of this part of the study will be published in detail elsewhere[9].

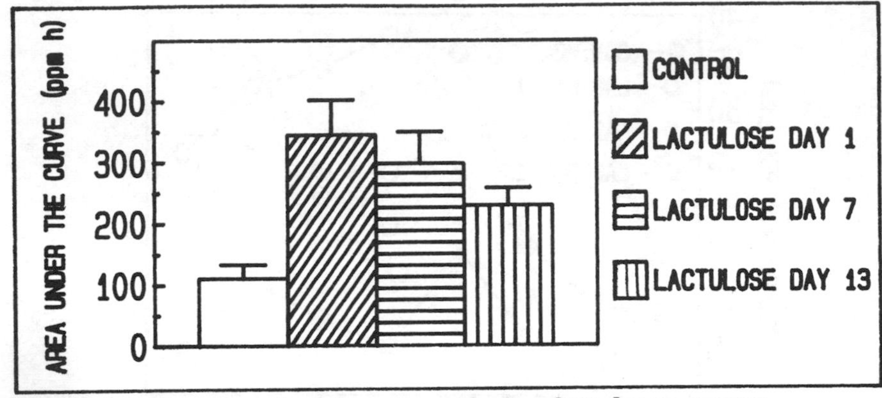

Figure 3 Area under the breath H_2 day-long curve. Lactulose values are all significantly different from control ($p<0.001$).

It can be concluded that, in assessing colonic fermentation using only breath H_2 as fermentation marker, the lenght of the study is a major factor of variability that should be taken into consideration.

ACKNOWLEDGEMENTS

We thank INALCO, Milan, for supplies of pure lactulose, and Steve Mitchell and Mark Jenkins for the help in developing the method for SCFA analysis. F.B. was partially supported by a Grant from C.N.R.

REFERENCES

1. D.J.A. Jenkins, T.M.S. Wolever, V. Vuksan et al. N.Eng. J.Med., 1989, 321, 929.
2. J.W. Anderson, S.R. Bridges. Diabetes, 1981, 30 (suppl 1), 133A.
3. C.D. Tollinger, H.J. Vreman, M.W. Weiner. Clin.Chem., 1979, 25, 1787.
4. G. Metz, M.A. Gassul, A.R. Leeds et al. Clin.Sci.Mol. Med., 1976, 50, 237.
5. T.M.S. Wolever, F. Brighenti, D. Royall et al. Am.J. Gastroenterol., 1989, 9, 1027.
6. C. Florent, B. Flourie, A. Leblond et al. J.Clin.Invest.,1984, 75, 608.
7. J.A. Perman, S. Modler, A.C. Olson. J.Clin.Invest., 1981, 67, 643.
8. H. Vogelsang, P. Ferenci, S. Frotz et al. Gut, 1988, 29, 21.
9. D.J.A. Jenkins, T.M.S. Wolever, A. Jenkins et al. Submitted to New Eng. J. Med.

DIETARY FIBRE AND COLONIC MUCIN CHANGE IN GOLDEN HAMSTER

A Morphometrical and Histochemical Study

C. B. Huang, E. Lundin, J. X. Zhang, R. Stenling,
G. Hallmans and C. O. Reuterving

Departments of Pathology and Nutritional Research,
University of Umeå, S-901 87 Umeå, Sweden

1. INTRODUCTION

Dietary fibre has been shown to have a number of impacts on physiology and pathology.[1-8] However, the influence of dietary fibre on intestinal mucins has been paid less attention to.[1-4] Gastrointestinal mucins are complex glycoproteins that are secreted from mucin-producing or goblet cells, and that form a viscoelastic gel covering the mucosal surface of the gastrointestinal tract. They have a range of known or potential functions,[2,10] such as lubrication, protection of mucosa, waterproofing of mucosa, antibacterial (viral, parasitic) action, adsorption of cholesterol, cooperation with sIgA, interaction with enzymes, and action as an enzymatic substrate. In physiological conditions, the mucin secreting cells of the gastrointestinal canal consistently produce their own specific mucin type(s) both in humans and in animals.[9-13] Qualitative changes appear in pathological conditions such as inflammation and neoplastic transformation.[10-13]

2. MATERIALS AND METHODS

Twenty-five male Syrian golden hamsters (age of 4-6 weeks) were randomly divided into a control group fed fiber free diet, a rye bran and a soybean fiber supplemented group. The mean body weight at the beginning of the study was 71 ± 2 g. All animals were individually housed with a 12 hr light/dark cycle, and had free access to their respective diets (Table 1) and to tap water. In each diet, 25% of the energy came from protein, 35% from fat and 40% from carbohydrate.

Table 1. NUTRITIONAL COMPOSITION OF THE DIETS (g/100g)

	DIET		
	Control	Rye Bran	Soybean Fibre
Milk protein	37.6	23.9	29.0
Corn oil	15.9	11.9	13.1
Wheat starch	38.2	26.2	32.6
Vitamins	1.2	1.0	1.0
Minerals	6.1	4.9	5.1
Gelatine	1.0	1.0	1.0
Rye bran	0	31.1*	0
Soybean fibre	0	0	18.2*

* Fibre source; fibre composition is shown in Table 2

Table 2. FIBRE COMPOSITION* (g/100 g diet)

	Control	Rye Bran	Soybean Fibre
Soluble Fibre	0	3.0 (20.5%)	1.6 (10.8%)
Insoluble Fibre	0	11.9 (79.5%)	13.2 (89.2%)
Total Fibre	0	14.9 (100 %)	14.8 (100 %)

* Analysed according to Asp[14];

After six weeks of feeding, all hamsters were sacrificed by heart puncture under anesthesia. From each animal, transverse circular gut specimens were taken from the proximal colon (at 1 cm distance from the caecum) and the distal colon (at 1 cm from the rectum). All samples were fixed in 10% neutral formalin, dehydrated and embedded in paraffin. Five serial sections (5 μm) were cut from each sample. One section was stained with hematoxylin and eosin (H & E); the other sections were treated by the alcian blue pH 2.5/periodic acid-Schiff (AB/PAS),[11] high-iron diamine/alcian blue pH2.5 (HID/AB),[13] and periodic acid-borohydride/potassium hydroxide/ periodic acid-Schiff/hematoxylin (PB/KOH/PAS/H)[11] methods. The AB/PAS method discriminates neutral mucin (red) from acidic mucins (involving N-acetyl sialomucin, O-acetyl sialomucin, and some sulphomucin as well)(blue). The HID/AB method stains sulphomucin (brown-black) and sialomucin (blue). By the PB/KOH/PAS/H procedure, only O-acetyl sialomucin is stained red.

For quantitatively morphological evaluation, stereological techniques[15,16] were applied. For this purpose, all histochemically stained sections were light microscopically projected on a paper with a printed quadratic lattice with 1 cm of line space. Total magnification (M) was calibrated by a 10^{-2} mm scale. Volumes of goblet cell mucin (**Vg**), mucosa (**Vmc**) and muscular layer (**Vms**) were estimated by a formula:

$$V = Z^2 * L * P$$

V = volume; **Z** = distance between lines of the lattice (10 mm in the study) divided by the total magnification (M) (131 in the study); **L** = given length of a gut tube specimen (10 mm in this current study); **P** = counted points of the lattice cross falling on target. From these figures obtained, volume density of goblet cell mucin (**Vg/Vmc**) was also calculated. A Student t test was used for statistical data.

3. RESULTS

Most of the mucin-producing goblet cells were located in the crypts, while only a small number were seen in the surface epithelia towards the intestinal lumen. The distribution of the histochemically defined mucin types differed between the proximal and the distal colon. Proximally, goblet cells of the mucosa produced preponderantly N-acetyl sialomucin, sulphomucin and O-acetyl sialomucin, and a minor amount of neutral mucin. Distally, a medium amount of N-acetyl sialo- and sulphomucin, small amounts of neutral mucin as well as a trace of O-acetyl sialomucin were observed. However, rye bran or soybean fiber

Table 3. VOLUME ALTERATIONS OF GOBLET CELL MUCINS OF COLONIC MUCOSA IN HAMSTERS AFTER SIX WEEKS OF FEEDING

	No.	**Vg** (mm^3/cm)	**Vmc** (mm^3/cm)	**Vg/Vmc** (mm^3/mm^3)
Proximal Colon:				
Control	(9)	3.41±0.45	21.62±3.26	0.17±0.02
Rye bran	(8)	3.98±0.61	15.01±1.60	0.27±0.02*
Soybean f.	(8)	4.77±0.91	17.85±1.87	0.27±0.03#
Distal Colon:				
Control	(9)	1.06±0.17	13.02±1.45	0.08±0.01
Rye bran	(8)	2.03±0.29*	10.57±1.05	0.19±0.03*
Soybean f.	(8)	1.57±0.21	10.69±1.21	0.15±0.01*

Mean value ± SE; *$P < 0.01$, #$P < 0.05$; AB/PAS staining

supplementation did not induce any qualitative changes of goblet cell mucin either in the proximal or the distal colon.

Vg of the distal colon was significantly larger in the rye bran group compared with the fibre-free group. Vg/Vmc was significantly increased both in the proximal and the distal colon and both in the rye bran and the soybean fibre group as compared to the fibre free control group (Table 3). The muscular and mucosal volumes per cm colon did not differ between the fiber free and the two fibre groups.

4. DISCUSSION

Previous studies on the effect of dietary fibre on the intestinal mucins have mainly concerned the small intestinal mucosa.[1,3,4] Using scanning electron microscopy and autoradiographic analysis, an increased goblet cell secretory activity has been observed after dietary supplementation with cellulose or wheat bran.[1,3,4] Quantitatively, enhancement of the relative number of goblet cells due to dietary fiber has been reported[3,4] but also lack of alteration or even a decrease.[3,4] In the colon, scanning electron microscopical observations[1,4] have indicated that the mucosal surface of rats fed wheat bran exhibits a larger amount of mucous producing cells compared to rats fed a fibre free diet.

In the present study, the increase in volume density of goblet cell mucin (Vg/Vmc), seen both in proximal and distal colon after both rye bran and soybean fibre supplementation, indicate that dietary fibre also induces a relative increase in the goblet cell mucin volume of the large intestine. A significant enhancement of total goblet cell mucin volume (Vg) was noted in the distal colon when the diet was supplemented with rye bran. This could be the result of an increase in either goblet cell size (hypertrophy) and/or number (hyperplasia). The mechanism seems complicated and can not be evaluated from the present data. It should be noted, however, that a close correlation between goblet cell mucin volume and number of goblet cells was observed in the small intestine from the same animals (unpublished data).

Water insoluble or poorly fermentable components of dietary fibre (e.g., cellulose and lignin) are related to the enhancement of fecal volume with dilution of toxic and carcinogenic substances.[1-8] Increased amounts of secreted mucins might also contribute to fecal bulk enhancement.[2] The increased goblet cell volume found in this study after rye bran supplementation may be related to such a possible mucous influence on fecal bulk. In this respect, the effect of both water insoluble and water soluble fractions of dietary fibre could be considered.

In conclusion, dietary supplementation with rye bran or soybean fiber did not result in qualitative alterations of goblet cell mucin in the hamster colon. Quantitatively, however, there was a general relative increase and in part a total enhancement of goblet cell mucin volume. Further studies are needed to demonstrate the influence of dietary fibre on mucin secretion, its significance and the possible mechanisms involved.

5. REFERENCES

1. M.M. Cassidy, F.G. Lightfoot, L.E. Grau, J.A. Story, D. Kritchevsky and G.V. Vahouny, Am.J.Clin.Nutr., 1981, 34, 218
2. L. Bustos-Fernandez, I. Ledsma-Paolo, J. Kofoed, E. Gonzalez, S. Hamamura, K. Ogawa-Furuya & C.Bernard, 'Colon, Structure and Function', Plenum Press, New York, 1983, p 65
3. G.V. Vahouny, T. Le, I. Ifrim, S. Satchithanandam, M.M. Cassidy, Am.J.Clin.Nutr., 1985, 4, 895
4. G.V. Vahouny, 'Physiology of the Gastrointestinal Tract', Raven Press, New York, 1987, p 1623
5. D. Kritchevsky, Cancer, 1986, 58, 1830
6. D.J.A. Jenkins, A.L. Jenkins, A.V. Rao and L.U. Thompson, Am.J.Gastroenterol., 1986, 81, 931
7. M.Estwood and W.G. Brydon, 'Physiological effects of dietary fibre on the alimentary tract', Academic Press, London, 1985, p105
8. L.R. Jacobs, Am.J.Clin.Nutr., 1988, 48, 775
9. D.G Sheahand and H.R. Jervis, Am.J.Anat., 1976, 146, 103
10. M.R. Neutra and J.F. Forstner, 'Physiology of the Gastrointestinal Tract', Raven Press, New York, 1987, p 975
11. M.I. Filip, Cell Pathol., 1979, 2, 195
12. D.C. Allen, N.S. Connolly and J.D. Biggart, Histopathology, 1988, 13, 399
13. C.B. Huang, J.Xu, J.F. Huang and X.R. Meng, Cancer 1986, 57, 1370
14. N.G. Asp, Molec.Aspects Med., 1987, 9, 17
15. R. Stenling and H.F. Helander, Cell Tissue Res.,1981, 217, 11
16. T.M. Mayhew and C. Middleton, J.Anat., 1985, 1, 141

THE DIETARY FIBRE MATRIX DURING GUT TRANSIT - MATRIX SOLUBILITY, PARTICLE SIZE AND FERMENTABILITY.

[1]J A Robertson [2]Sandra D Murison [2]A Chesson

[1]Institute of Food Research, Norwich Laboratory, Colney Lane, Norwich, NR4 7UA, United Kingdom; [2]Rowett Research Institute, Bucksburn, Aberdeen, AB2 9SB, United Kingdom.

1 INTRODUCTION

The physiological effects of dietary fibre (DF) are related to the behaviour of the DF matrix under conditions prevailing during gastrointestinal transit. Whilst methods have been developed for the analysis of DF in foods [1,2] less attention has been focused on its analysis from digesta during gut transit, although measurements of physicochemical properties and their relationship to physiological effect have been attempted [1,3,4]. To determine how fibre is affected by the environmental conditions in the gut lumen, samples of feed and its digesta recovered from specific points along the gastrointestinal tract of the pig, were analysed. A vegetable-based and a cereal-based diet were analysed for DF content and composition, particle size distribution and fermentability.

Materials and Methods

Pigs were individually housed and fed a weight maintenance diet based on either swede (Brassica napus) or cereal wheat bran to provide 240g/day vegetable fibre or 440g/day cereal fibre. Feed was presented twice daily and intakes recorded[5]. Digesta for analysis was recovered from animals sacrificed 3 weeks after acclimatization to diet. Potential fermentability and rate of fermentation were measured using pigs cannulated at the mid point in the caecum, and involved the suspension of a known weight of water-insoluble fibre in a nylon bag (5μm mesh) in the caecum for set time periods. Digesta samples recovered and residues from caecal fermentation were freeze dried prior to analysis.

<u>Table 1</u> Potential Fermentability of Water-Insoluble Swede Fibre and Wheat Bran in Feed and Ileal Digesta Samples Measured in the Pig Caecum and in Relation to Apparent Digestibility.

	SWEDE			BRAN		
	*Potential		#Apparent	*Potential		#Apparent
	Feed	Ileal	Faeces	Feed	Ileal	Faeces
'Fibre'[1]	93.1	90.5	87.8	45.2	45.3	41.5
Cellulose	93.5	94.0	91.7	23.1	29.6	24.1
Non-cellulosic neutral sugars	87.5	80.1	78.7	58.5	61.5	46.5
Uronic acid	96.5	81.3	88.6	45.0	27.9	39.2

[1]Non-starch polysaccharides and lignin. *Potential fermentability (% loss of component within 48hr) of feed and ileal digesta in the caecum. #Apparent digestibility measured in faeces (% loss of component during transit).

<u>Table 2</u> Fractional Rate of Fermentation of Swede Water-Insoluble Fibre and its Ileal Digesta and Bran Water-Insoluble Fibre and its Ileal Digesta.

	SWEDE				BRAN			
	Feed		Ileal		Feed		Ileal	
	k	$t_{1/2}$	k	$t_{1/2}$	k	$t_{1/2}$	k	$t_{1/2}$
Fibre'[1]	14.4 ±0.3	4.8	19.1 1.8	3.6	9.5 1.4	7.3	6.7 1.7	10.3
Cellulose	13.0 ±0.8	5.3	19.4 1.8	3.6	*8.8 4.8	7.9	*4.4 1.6	15.8
Non-cell. n. sugars	14.8 ±0.7	4.7	19.6 4.9	3.5	8.6 1.2	8.1	9.9 1.7	7.0
Uronic acid	18.4 ±1.6	3.8	25.8 2.7	2.7	*5.9 1.2	11.7	*2.9 1.2	23.9

[1]Non-starch polysaccharides and lignin. *No significant linear relationship. k = - Rate of fermentation (%h) (mean +/-standard error). $t_{1/2}$ = half life (h), i.e. time taken to ferment to 50% original (LN_e 2/K).

Chemical analysis was by standard methods and particle size analysed by wet sieving (stacked sieves, size range 4000 - 5μm; 10g sample dry weight; water flow rate 600 ml/min applied via spray nozzle; amplitude 1.5mm; time 30 min). Material retained on each sieve was recovered and represents water insoluble material of specific particle size.

Results

Estimation of potential fermentability (proportion of material lost during prolonged caecal incubation) showed swede fibre and its ileal digesta were rapidly fermented, with a potential fermentability over 90% achieved within 24h[5]. Similarly, bran samples were rapidly fermented but each had a potential fermentability of fibre of only around 45%[5]. Fermentability of fibre components varied within each fibre source and between swede and bran (Table 1). For both swede and bran samples apparent digestibility measured during gut transit agreed with component potential fermentability measured in the caecum. Rate of fermentation (Table 2) showed swede samples to be fermented faster than bran samples and swede ileal digesta to be fermented consistently faster than the feed. In bran, for only the non-cellulosic neutral sugars, was a linear and significant rate of fermentation found. Half life calculation from fermentation rates confirmd potential fermentability would be achieved within 24 hours. From chemical analysis of each sample (Figure 1) a proportionate decrease in water insoluble uronic acid was observed in swede during ileal transit. However, in faecal material uronic acid was still detectable despite the preferential fermentation of uronic acid in the caecum. In cereal bran no corresponding change in chemical composition was found during ileal transit and in faecal material each fibre component persisted, although apparent digestibility was 41.5%. Particle size analysis of feed and digesta (Figure 2) showed that little change in particle size distribution occurred in bran but in swede feed, particle size was reduced in the proximal gut. Stomach particle size and distribution was similar to that in feed and hence particle size reduction and change in distribution were not the result of mastication. A further reduction in swede particle size occurred in the large intestine and was accompanied by a loss of particulate material (87.8% apparent digestibility). Apparent digestibility of bran (41.5%) was not accompanied by a significant particle size reduction.

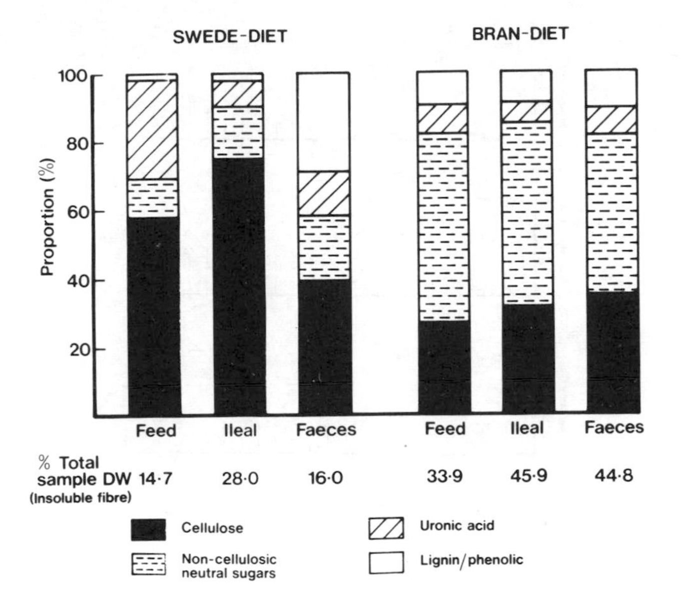

Figure 1. Proportionate composition of water-insoluble dietary fibre components recovered during gut transit in the pig.

Discussion

Swede and wheat bran comprise distinct tissue types and constituent cell wall biopolymers, which are cross linked to form the fibre matrix [1,6,7]. Persistence of cross links affect the solubility and structural integrity of the fibre matrix. Acid in the stomach may influence solublisation of pectic polysaccharides through displacement of calcium from constituent uronic acid residues involved in ionic cross link formation. Similarly, the mild alkaline conditions in the small intestine may encourage B-eliminative degradation of methyl esterified pectic polysaccharides[8]. Both processes will result in cell loosening/separation and hence particle size reduction, with a corresponding decrease in associated water insoluble pectic polysaccharides (uronic acid). The absence of pectic polysaccharides in bran

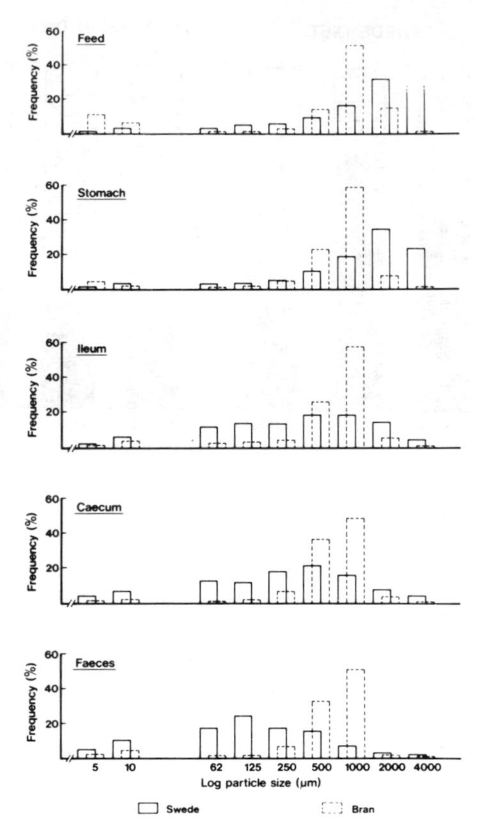

Figure 2. Particle size distribution (% insoluble organic material) of a cereal (Bran) - and a vegetable (Swede) - based diet measured during gut transit in the pig.

MEDIAN PARTICLE SIZE (μm)

	Feed	Stomach	Terminal Ileum	Caecum	Faeces
SWEDE	2398	2222	560	459	228
BRAN	1197	1110	1041	970	907

and the existence of a highly cross linked polysaccharide network involving phenolic esters and lignin will preclude a similar particle size reduction. However, selective fermentation of arabinoxylans, which can be attributed to be from aleurone layer cells[8] may result in a mechanical 'weakening' of the bran particle and a change in shape[9]. In swede, mainly parenchymatous tissue is found, also with selective fermentation of component polysaccharides,[10]. This tissue is readily fermentable and the result is the rapid and effective disappearance of swede cell wall material, ie particle destruction, during transit. In bran, fermentation is limited mainly to non-lignified tissues (endosperm and aleurone layer[8]) and hence rate and extent of fermentation is also limited. The resistance to fermentation of cross links in the lignified tissues of bran, maintains particle structure in bran during transit.

References

1 D A T Southgate and H Englyst, 'Dietary Fibre, Fibre Depleted Foods And Disease': [Ed H Trowell, D Burkitt and K Heaton], Academic Press Inc,(London) Ltd, 1985, Chapter 3, pp 31 - 55.
2 R R Selvendran, A V F V Verne and R M Faulks, 'Modern Methods of Plant Analysis New Series': [Ed H F Linskens and J F Jackson], Springer-Verlage, Berlin Heidelberg, 1989, 10, pp 234 - 259.
3 M A Eastwood and W D Mitchell, 'Fibre In Human Nutrition' [Ed G A Spiller and R J Amen], Plenum Press, New York, 1976, pp 109 - 130.
4. J A Robertson, Proc. Nutrition Society, 1988, 47, 143 - 152.
5. J A Robertson, S D Murison and A Chesson, J Nutrition, 1987, 117, 1402 - 1409.
6. R R Selvendran, B H J Stevens and M S DuPont, Advances In Food Research: 1987, 31, 117 - 209.
7. R R Selvendran and J A Robertson 1990 (These Proceedings).
8. B H J Stevens and R R Selvendran, Carbohydrate Research, 1988, 183, 311 - 319.
9. R Moss and D C Mugford, J Cereal Sci., 1986, 4, 171 - 178.
10. J A Robertson, S D Murison and A Chesson, J Sci Food Agric., 1986, 44, 151 - 166.

ESTIMATION OF THE DIGESTIBILITIES OF NSP FOR WHOLEMEAL BREAD AND HARICOT BEANS FED IN MIXED DIETS

F. B. Key* and J. C. Mathers

Department of Agricultural Biochemistry and Nutrition
University of Newcastle upon Tyne
Newcastle upon Tyne NE1 7RU

1 INTRODUCTION

It is now accepted that dietary non-starch polysaccharides (NSP) which escape small intestinal digestion are fermented within the large bowel of man,[1] that the extent of this fermentation varies considerably between foods and that the rat appears to be a reliable model for man in this respect.[2]

Most studies of NSP digestibility have investigated individual foods or food components and where mixed diets were eaten it has seldom been possible to obtain separate estimates of digestibility for each food. This study used a multiple linear regression (MLR) approach to test the hypothesis that the digestibility of NSP in foods is influenced by the dietary presence of other NSP sources using mixed diets containing wholemeal bread and cooked haricot beans which are commonly-eaten, rich sources of NSP in the UK diet. Separate estimates of digestibility for bread and beans were obtained and we found little evidence that beans affected the digestibility of bread NSP.

*Present address: MRC Dunn Nutritional Laboratory,
Milton Road,
Cambridge.
CB4 1XJ

2 METHODS

Four test diets were prepared each containing 500 g freeze-dried, ground wholemeal bread and 0, 150, 300 or 450 g cooked, freeze-dried and ground haricot beans (*Phaseolus vulgaris*)/kg diet with alterations in sucrose and casein to maintain similar energy and protein contents. Each diet was fed to groups of 6 male Wistar rats for 21 d with measurement of food intake and complete faecal collection over the final 7 d. NSP in diets and faeces were measured by the Englyst method.[3]

The digestibilities of NSP and its components in the bread and beans were calculated using a MLR model in which NSP output in faeces (Y) was regressed against intake of bread NSP (X_1) and beans NSP (X_2) for each individual animal as follows:

$$\text{Model 1} \quad Y = \alpha_1 X_1 + \alpha_2 X_2 + \alpha_3 X_1 X_3 \qquad 1$$

where X_3 has the value 0 or 1 when beans are absent or present in the diet respectively. α_1 and α_2 are the indigestibilities of the component in bread and beans respectively and α_3 the additional effect of presence of beans on bread indigestibility.

Where α_3 is non-significant, a simpler model is appropriate:

$$\text{Model 2} \quad Y = \beta_1 X_1 + \beta_2 X_2 \qquad 2$$

where β_1 and β_2 are the coefficients of indigestibility for bread and beans respectively.

Data were analysed using the STATGRAPHICS package (STSC Inc., Rockville, Maryland, USA).

3 RESULTS AND DISCUSSION

Whilst the digestibility of whole diet dry matter (DM) decreased linearly that of the NSP increased curvilinearly with increasing inclusion of beans in the diet (Figure 1).

MLR analysis (Model 1) showed that the presence of beans had little effect on the digestibility of NSP or any of its components in bread (α_3, not significant; $p>0.05$) so the simpler Model 2 was used to estimate digestibilities for bread and beans separately (Table 1).

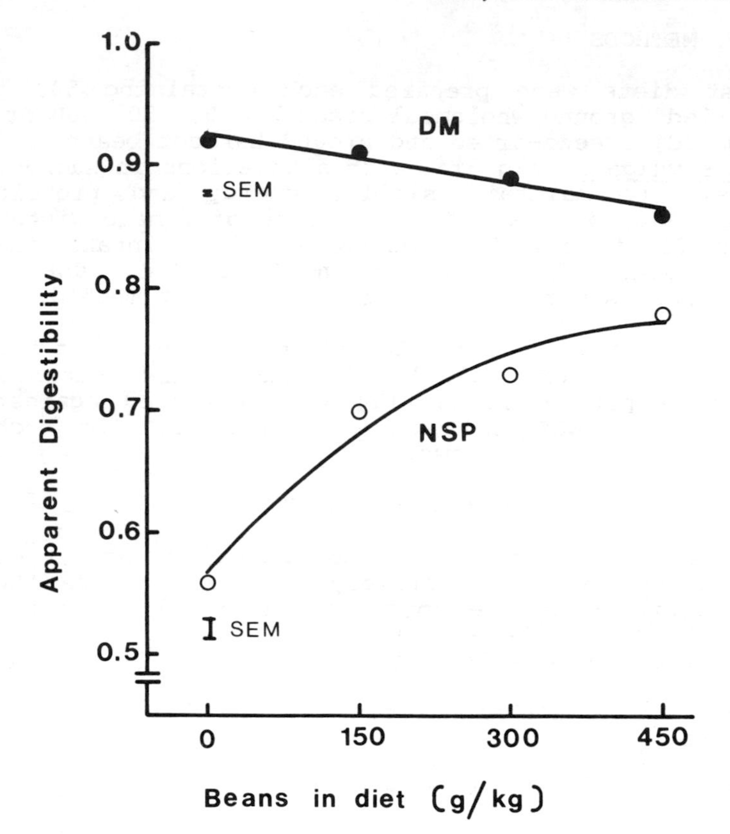

Figure 1 Digestibilities of whole diet dry matter (DM) and non-starch polysaccharides (NSP). Each point is the mean for 6 rats.

Table 1 Estimates by MLR using Model 2 of the digestibilities of NSP and its components for wholemeal bread and haricot beans fed in mixed diets

Component	Bread	SE	Beans	SE	R^2
NSP	0.56	0.023	0.86	0.017	0.99
NCP	0.64	0.020	0.92	0.013	0.99
Cellulose	0.22	0.038	0.23	0.045	0.98
Arabinose	0.56	0.022	0.98	0.015	0.98
Xylose	0.67	0.022	0.88	0.046	0.97
Mannose	0.53	0.088	0.85	0.039	0.86
Galactose	-0.04	0.185	0.78	0.050	0.91
Glucose	0.40	0.040	0.74	0.023	0.99
Uronic acids	0.99	0.008	0.99	0.001	0.89

NCP = non-cellulosic polysaccharides

Model 2 provided a very satisfactory description of the relationship between intake and ouput of NSP and its constituents (R^2 0.85 - 0.99) in these mixed diets. It is clear that the digestibility of beans NSP is much greater than that of bread and similar to values for peas.[4] Non-cellulosic polysaccharides (NCP) were considerably more digestible than was cellulose for both bread and beans.

Further studies should be carried out to determine to what extend food NSP digestibility is influence by other dietary components and we suggest that the MLR approach may have broad application in such studies.

We thank the AFRC for a Food Research Studentship (FBK) and Dr R. Henderson for advice on MLR analysis.

REFERENCES

1. J.H. Cummings and H.N. Englyst, Amer. J. Clin. Nutr., 1987, 45, 1243.

2. M. Nyman, N.-G. Asp, J.H. Cummings and H. Wiggins, Brit. J. Nutr., 1986, 55, 487.

3. H.N. Englyst and J.H. Cummings, Analyst, 1984, 109 937.

4. J.S. Goodlad and J.C. Mathers, Brit. J. Nutr., 1990 (submitted).

PHYSIOLOGICAL EFFECTS OF PEA DIETARY FIBRE IN THE RAT

M. FOCANT, A. VAN HOECKE, M. MEURENS and M. VANBELLE

Université Catholique de Louvain (AGRO-BNUT)
3, Place Croix du Sud, B-1348 Louvain-la-Neuve (Belgium)

1. INTRODUCTION

Most of the Western diets are dietary fibre deficient. One way of curing this is the consumption of foods including added dietary fibre. Tegument of pea (pisum sativum) is a usable by-product. Nevertheless, the fibre must be reduced in small particles in order to make it untasted.

It is generally assumed that the particle size *per se* does affect the susceptibility to bacterial degradation of the fibre but earlier studies are not consistent. A decrease in the susceptibility to bacterial fermentation has been found in man given coarse bran compared with fine bran[1], whereas a slight increase in the fermentation of cellulose has been reported in pigs given coarse bran[2]. In the rat, the particle size *per se* of wheat bran did not affect the faecal dry weight or the fermentability of the fibre[3]. No significant differences in mean retention time were observed among coarse and fine bran in the pig[2].

The present study was undertaken to provide information on whether particle size could change the fermentability and the mean retention time of the pea fibre. A rat experimental model was used.

2. MATERIAL AND METHODS

Two forms of tegument of pisum sativum (Fibragel, CASI Co, Schoten, Belgium) were used: a coarse form (Fibragel as it is) and a milled one. The geometric mean diameter was respectively 0.411 and 0.187 mm for the coarse and the milled fibre.

Three lots of 6 male Wistar rats (250 g mean live weight) were housed individually in metabolism cages. A control diet was composed of 4 % cellulose, 12.5 % casein, 40 % corn starch, 31 % sucrose, 7.5 % soja oil and 5 % of a mineral-vitamin mix. Two experimental diets were prepared by substituting 5 % sucrose by 5 % either coarse or milled pea fibre. The animals were fed 15 g of their respective diet once daily.

The pea fibre, the diets and the faeces were analysed for ash, Kjeldahl N, total dietary fibre (TDF)[4], neutral detergent fibre (NDF), acid detergent fibre (ADF) and lignin[5]. In addition diaminopimelic acid (DAPA) was analysed in the faeces[6]. The apparent digestive utilization of dry matter (DM), organic matter (OM), crude protein (CP, N x 6.25), NDF and ADF and the total mean retention time (MRT) of the indigestible dietary particles were determined[7] using chromic oxide as marker (Cr_2O_3)[8]. The results were submitted to analysis of variance.

3. RESULTS AND DISCUSSION

Chemical Composition of the Pea Fibre and of the Diets (Table 1).

On a DM basis, the tegument of pea was composed of 2.7 % ash, 8 % CP and 77.7 % TDF including 12.7 % hemicellulose (NDF-ADF), 54.8 % cellulose (ADF-lignin), 0.3 % lignin and 9.9 % soluble fibre (TDF-NDF). The main constituent of the TDF was water-insoluble fibre. As a reference, ADF content of wheat bran average 14 %.

The chemical composition of the 3 diets was similar for OM and CP. The TDF content was higher in the pea diets (9.3 % and 9.4 % vs. 5.3 %)

Table 1 : Chemical composition of the pea fibre and of the diets fed to rats (% of DM).

	PEA FIBRE	DIETS		
		CONTROL	COARSE PEA	MILLED PEA
ash	2.7	5.2	5.5	5.2
OM	97.3	94.8	94.5	94.8
CP	8.0	11.9	12.2	12.6
TDF	77.7	5.1	9.3	9.4
NDF	67.8	5.3	9.2	9.2
ADF	55.1	4.2	7.3	7.4
Lignin	0.3	0.4	0.7	0.9

Table 2: Chemical composition of the faeces

	CONTROL	COARSE PEA	MILLED PEA
Water (%)	47.7	43.4	46.4
Ash (%)	18.2	14.9	14.2
CP (%)	7.1	6.9	6.7
NDF (%)	20.4	28.6	27.5
ADF (%)	19.0	25.4	24.8
DAPA (mg/kg)	110.4	126.6	111.4

Chemical Composition of the Faeces (Table 2)

The addition of pea fibre to the diet was accompanied by a lower ash content and a higher fibre content. DAPA was used as bacteria marker. The lack of a significant increase of DAPA in stools demonstrated the resistance to fermentation of the pea fibre used.

The particle size had little effect on the faecal chemical composition. Only water content tended to be higher (46.4 vs. 43.4 %) with milled fibre when compared with coarse fibre.

Apparent Digestibility of the Diets (Table 3)

The apparent digestibility of both DM and OM in pea diets was 4 points lower than in control diet ($P<0.001$). The digestibility of CP in pea diets was 3 points lower ($P<0.001$). On the other hand, digestibility of NDF and ADF was not affected by the pea fibre and was low (respectively 16 % and 5 %). In rats, digestibility coefficients for DM and gross energy declined significantly with each increase in the ADF content from 0 % to 15 %[9]. On the other hand, fibre consumption causes a faecal N increase. Substitution of 10 % wheat bran in rats decreased apparent N digestibility from 93 % to 87 %[10]. Suggestions include a decrease in the true digestibility of protein, increased losses of protein from the tissues of the intestinal tract and increased microbial protein synthesis[11]. This latest suggestion can not be retained for our trial since DAPA was not increased (table 2). Two factors of considerable importance affecting the digestibility of dietary fibre are the chemical composition of the fibre and the transit time for passage through the alimentary tract. In our study, the main component of pea fibre was cellulose which is known to

Table 3: Apparent digestibility of the diets (%)

	CONTROL	COARSE PEA	MILLED PEA
DM	88.8^{a}	84.8^{b}	84.5^{b}
OM	92.3^{a}	88.2^{b}	87.8^{b}
CP	88.3^{a}	85.5^{b}	85.4^{b}
NDF	17.5	17.2	13.5
ADF	4.2	6.2	3.9

^{ab} : means with different superscripts differ significantly ($P<0.001$)

have a slow fermentative breakdown, and MRT was very rapid (table 4).

The particle size did not change significantly the fermentability of the fibre. This is in agreement with observations concerning wheat bran[3].

Faecal Bulk and Mean Retention Time of Chromium (Table 4)

Faecal bulk was increased in the rats fed the pea diets (3.7 and 4.0 vs. 3.0 g/d; $P<0.001$) and MRT was decreased from 18.4 h to respectively 14.1 h and 17.7 h for the coarse and the milled pea diets. When compared to coarse fibre, fine fibre leads to a higher faecal bulk (4.0 vs. 3.7 g/d) and to a decreased MRT (17.7 h vs. 14.1 h). It is generally assumed that higher faecal bulk shortens the retention time. This conviction is supported by the difference in MRT observed among control and pea diets but not when comparing both pea diets. In pigs, no significant differences in MRT were observed among coarse and fine bran[2].

Table 4: Faecal Bulk and Mean Retention Time of Chromium

	CONTROL	COARSE PEA	MILLED PEA
Faecal bulk (g/d)	3.0^{a}	3.7^{b}	4.0^{c}
MRT (h)	18.4^{a}	14.1^{b}	17.7^{ab}

^{ab}: values with different superscripts differ significantly ($P<0.05$).

4. CONCLUSION

Composed of more than 50 % cellulose on a DM basis, tegument of pea, as the other water insoluble dietary fibres decreased the apparent digestibility of OM and CP, increased the faecal bulk and shortened MRT when 5 % were substituted in the diet of rats. When compared to coarse pea fibre, fine pea fibre tended to increase the faecal water content, the faecal bulk and the MRT. As a decreased MRT is the wanted effect of insoluble dietary fibre in order to prevent constipation and to shorten the contact of carcinogens with the bowel inner side, more information is needed about the physiological properties of dietary fibre at reduced particle sizes.

5. REFERENCES

1. S.N. Heller, L.R. Hackler, J.M. Rivers, P.J. Van Soest, D.A. Roe, B.A. Lewis and J. Robertson, Am. J. Clin. Nutr., 1980, 33, 1734.
2. F.R. Ehle, J.L. Jeraci, J.B. Robertson and P.J. Van Soest, J. Anim. Sci., 1982, 55, 1071.
3. M. Nyman and N.G. Asp, Br. J. Nutr., 1985, 54, 635.
4. L. Prosky, N.G. Asp, I. Furda, J.W. Devries, T.F. Schweizer and B.F. Harland, J. Assoc. Off. Anal. Chem., 1988, 71, 1044.
5. P.J. Van Soest and R.H. Wine, J. Assoc. Off. Anal. Chem., 1967, 50, 50.
6. K. El Shasly and R.E. Hungate, Appl. Microbiol., 1966, 14, 27.
7. E. Sakagushi, H. Itoh, S. Uchida and T. Horigome, Br. J. Nutr., 1987, 58, 149.
8. H. Petry and W. Rabbs, Z. Tierphysiol. Tiernarhrg. Futtermittelkde, 1971, 27, 181.
9. M.V. Garrison, R.L. Rein, P. Fawley and C.P. Breidenstein, J. Nutr., 1978, 108, 191.
10. M.Nyman and N.G. Asp, Br. J. Nutr., 1982, 47, 357.
11. I.R. Davies, I.T. Johnson and G. Livesey, Int. J. Obesity, 1987, 11, 101.

This research was supported by grants from "La Région Wallonne".

DIETARY FIBRE PRODUCTS, THEIR CHARACTERISTICS AND GASTROINTESTINAL IMPLICATIONS

Inge Hansen*

Department of Animal Physiology and Biochemistry
National Institute of Animal Science
P.O. Box 39, 8800 Tjele, Denmark

1 INTRODUCTION

Dietary Fibre (DF) comprises a very complex and heterogeneous group of substances whose functional properties and nutritional effects vary with composition, structure and treatment of the fibre source[1,2,3].

In the present study 3 commercially available fibre products were examined: wheat bran, a long known fibre source used in numerous studies; oat bran , a rediscovered cereal fibre which has been claimed to have cholesterol-lowering effects, and pea fibre , a "novel" food ingredient whose nutritional relevance has only been studied in relation to blood glucose response[4].

The fibre sources were analysed chemically, and their physiological effects along the gastrointestinal tract examined in rats.

2. MATERIALS AND METHODS

The content of dietary fibre (DF), defined as non-starch polysaccharides (NSP)[5], was measured in wheat bran, oat bran, pea fibre, and a white wheat flour (control product).

---

* Present address: Grindsted Products
Edwin Rahrs Vej 38
8220 Brabrand, Denmark

The DF sources were mixed into diets with a total of 13% NSP and along with the control diet (low fibre, wheat flour diet) fed ad. lib. to rats with an initial body weight of about 280 g. Each diet was fed to 12 rats during a 2 weeks adaptation period and a 16 days balance period.

The faecal bulking defined as gram faecal dry matter (DM) per day was measured during the balance period.

The effects of the diets on the development of the gastrointestinal (GI) tract were measured on 6 rats per diet. After a total of 30 days on the diet the rats were killed, the entire GI tract removed and the length measured. The length of the small intestine (SI) and the large intestine (LI) was measured separately.

The passage rate of digesta was measured in 6 rats per diet after the termination of the balance period. Glass beads (diameter of 50 μm) were used as markers and administered on a single day over a 6 hour period. Total faecal collection followed over the subsequent 96 hours.

3 RESULTS

The total content of DF and the monomeric constituents of the DF sources differed widely. Wheat bran characterized by its high content of cellulose (6.4% of DM) and insoluble arabino-xylans (22% of DM) while oat bran contained a high amount of soluble betaglucans (9% of DM). Pea fibres comprised soluble and insoluble arabinans are (20% of DM) and insoluble cellulose (6% of DM) (Fig. 1). The solubility properties of the experimental diets varied from 65% in the oat bran diet, to 43% in the pea fibre diet, and to 19% in the wheat bran diet. The bulking effect of the diets differed significantly. In decreasing order the faecal bulking (g DM/d) of the diets were: wheat bran diet > oat bran diet > pea fibre diet > control diet (Fig. 2).

The length of the GI tract of the rats which were fed the experimental diets differed significantly. The elongating effects of oat bran were more pronounced than those of the wheat bran. Both cereal brans elicited effects on the small intestine (SI) as well as the large intestine (LI). The pea fibre had little effect on the length of the GI tract (Table 1).

Table 1 Length of gut segments (mean values and SD)

Dietary Group	Small Intestine		Large Intestine		Total Gut	
Control	111^{b}	7	15.3^{c}	1.4	130^{c}	7
Wheatbran	114^{b}	9	17.9^{ab}	2.0	138^{b}	7
Oatbran	128^{a}	12	19.1^{a}	1.5	147^{a}	13
Peafibre	114^{b}	16	17.2^{b}	2,9	135^{ab}	15

Column mean values with different superscript letters were significantly different ($P<0.05$).

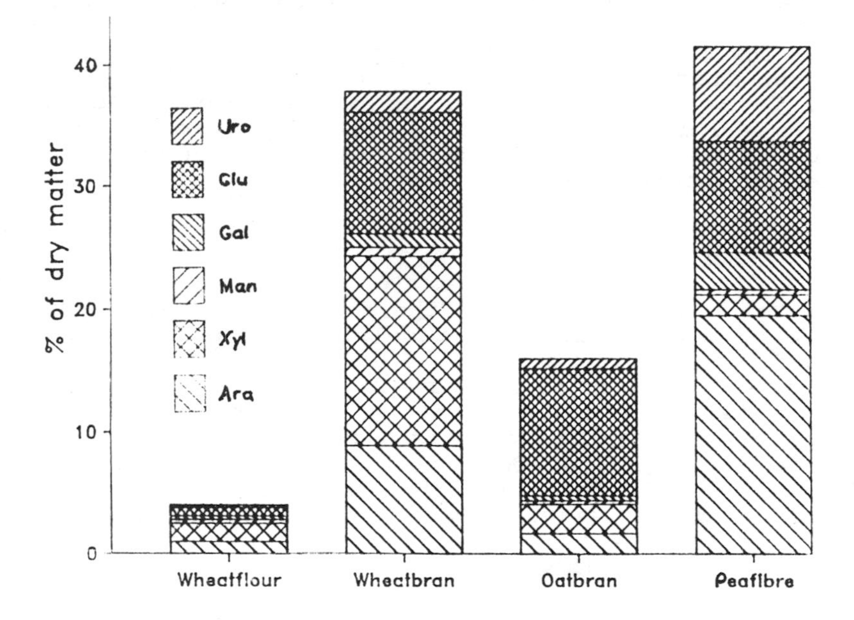

Figure 1 Content and compositions of non-starch polysaccharides (NSP).
Uro=Uronic acids, glu=glucose, gal=galactose, man=manmose, xyl=xylose, ara=arabinose

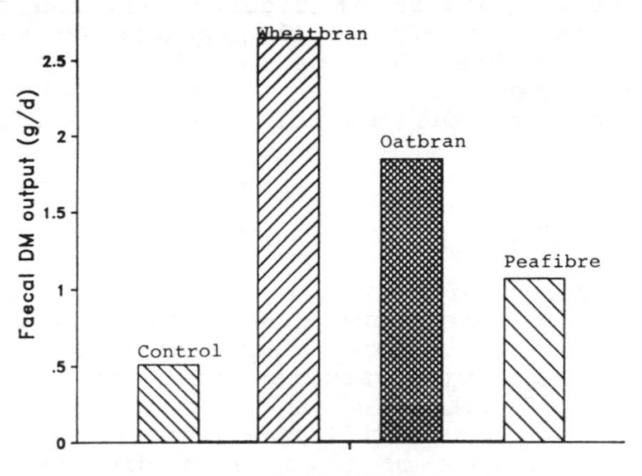

Figure 2 Faecal bulking (faecal DM/d) of diets

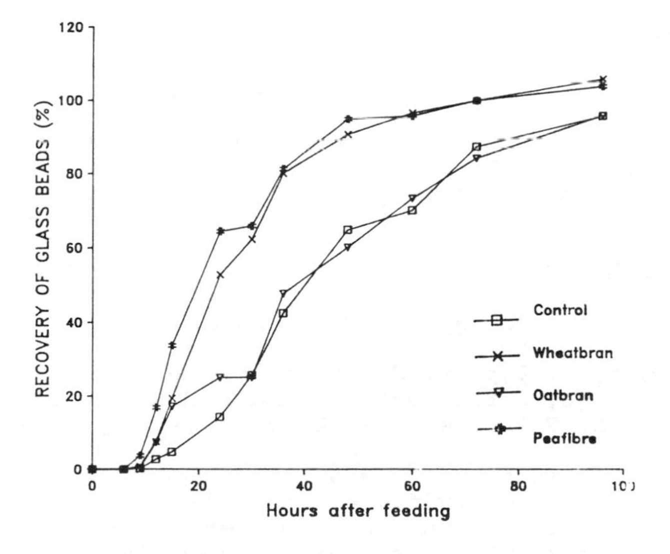

Figure 3 Rate of passage of digesta in rats fed different diets

The rate of passage of digesta differed significantly. The time for 50% of the markers to appear in the faeces was 22 hrs. for the pea fibre diet, 25 hrs. for the wheat bran diet, 37 hrs. for the oat bran diet, and 38 hrs. for the control (low fibre) diet (Fig. 3).

4. CONCLUSIONS

Analysis of three dietary fibre (DF) sources: wheat bran, oat bran and pea fibre showed considerable variation with regard to chemical composition and solubility properties. In vivo studies with rats confirmed that the mode of action of the DF sources along the gastro-intestinal tract differed greatly: bulking effects as well as rate of passage of digesta (transit time) were influenced to various degrees by the individual fibre sources. Also the direct effects of the DF on the development of the GI tract differed significantly which may be of importance in explaining long-term effects of DF.

In conclusion, the study provided evidence that dietary fibre sources of different origin elicit a range of effects along the GI tract. The mechanisms of action are complex, the knowledge of which is much needed to provide the consumer as well as the food producer with a soundly based guideline for the choice of acceptable and effective types of dietary fibre.

REFERENCES

1. R.R. Selvendran, Am. J. Clin. Nutr., 1984, 39, 320.
2. M. Nyman, I. Björck, H. Siljeström and N.-G.Asp, Proceedings, Cereal Science and Technology in Sweden (N.-G Asp, ed. **, Lund University, 1989, p. 40.
3. K.E. Bach Knudsen. I. Hansen, B. Borg Jensen, K. Østergård, Dietary fiber - new developments, physiological effects and physico-chemical properties, (J. Furda, C. Brine, eds.) Plenum Publ. Corp., 1990 (in press).
4. O. Hamberg, J.J. Rumessen, E. Gudmand-Høger, Am. J. Clin. Nutr., 1989, 50, 324.
5. H.N. Englyst, H.W. Trowell, D.A.T. Southgate, J.H. Cummings, Am. J. Clin. Nutr. 1987, 46, 873.

DIGESTIBILITY OF RAW LEGUME STARCH

J Taylor, K N Boorman, R J Neale and G Norton

Department of Applied Biochemistry and Food Science
School of Agriculture
Sutton Bonington
Leics LE12 5RD

1 INTRODUCTION

Legume seed starches have been found to exhibit considerable variation with respect to bioavailability (El-Harith *et al* 1976, Fleming and Vose 1979, Hoover and Sosulski 1985). No satisfactory explanation of this phenomenon has been forthcoming. A study has been initiated in this Department to assess the bioavailability of the starches of certain legume seeds and ultimately to define the physiochemical properties of the indigestible starch.

2 MATERIALS AND METHODS

The starch content of all food and faecal samples was estimated using a modification of the AOAC method 14.073/4 (1984).

Field bean (*Vicia faba*: varieties Maris Bead and Albinette), Lima bean (*Phaseolus lunatus*) and lentils (*Lens culinaris*) were obtained from the University of Nottingham Department of Agriculture or local health food stores. All were ground to pass through a 2mm sieve.

In all studies male Wistar rats were used (4 per group). Test animals were housed individually and allowed a 5 day adaptation period prior to the balance. Diets and water were supplied *ad libitum*. Feed intake, faecal output and weight gain were recorded in all cases.

Trial 1: Adult rats (250g) were offered diets containing 50% by weight of the above legumes. For comparative purposes one group of rats received a diet based on maize starch.

Trial 2: Weanling rats received diets containing field bean meal (variety, Maris Bead) at incorporation levels ranging from 340g/kg to 780g/kg. The diets were isonitrogenous (protein content = 195g/kg) and of comparable metabolisable energy (ME) values. The balance period lasted for 9 days.

Trial 3: Weanling rats received a series of diets containing 780g of Field bean meal (variety, Maris Bead) per kg of diet but with a range of ME values. At the expense of oil, silicic acid was included in the diets to give a range of ME values between 10.03 and 15.29 MJ/Kg.

3 RESULTS

All the legume seed starches tested were found to have lower apparent digestibilities ($p<0.001$) than the control diet (Trial 1 and Table 1). Although a considerable variation in the digestibility of Lima bean starch was obtained, all the faeces excreted were soft and pale. When viewed under the light microscope numerous intact starch granules could be seen.

The level of starch included in the diet did not significantly affect the proportion of starch digested overall or throughout the balance period(Trial 2 and Table 1). There was a significant difference ($p<0.05$) in food intake between those rats receiving low and high energy diets but this did not affect starch digestibility in any group (Trial 3 and Table 1).

4 DISCUSSION

It has been shown that the bioavailability of starch in raw legume seeds varies considerably. At this time the reason for the poor digestibility of these starches is obscure. Microscopic examination of the faeces from those rats which received a Lima bean based diet revealed large numbers of starch granules,some of which were intact while others appeared to have been partially digested. Thus it would appear that the somewhat lower digestibility of the Lima bean starch cannot be entirely ascribed to physical properties and chemical factors may be involved.

Studies on the bioavailability of starches of raw legume seeds may, in some cases, be complicated by the presence of anti-nutritional factors (ANF) (Gupta 1987). Such effects cannot, at this time, be completely overlooked in the Lima bean.

Although field bean starch was found to be more digestible than Lima bean starch, a constant fraction proved resistant to digestion irrespective of the amount consumed or the energy status of the diet.

The native starch escaping digestion or fermentation in the GI tract is termed refractive starch. Refractive starch can be distinguished from resistant starch which is formed during the processing of food by virtue of the fact that it can be assayed by conventional enzymatic procedures (Berry 1986). Refractive starch in raw or undercooked legume seeds can therefore be included in the fraction termed 'dietary fibre' since it should have similar physiological effects to other unavailable components of this fraction (Cummings and Englyst 1987). Refractive starch may, in the future, become of greater significance in this respect since legume consumption is increasing in both developed and less developed countries. This is particularly significant in many less developed countries where the beans are consumed uncooked.

TABLE 1

EFFECT OF DIETARY INCLUSION LEVEL AND ME CONTENT ON DIGESTIBILITY OF LEGUME STARCHES

LEGUME	% LEGUME STARCH IN DIET	ME VALUE OF DIET (MJ/kg)	% STARCH UNDIGESTED
LENTIL	24.6	15.42	1.34 ± 0.8
LIMA BEAN	21.5	15.22	9.38 ± 2.56
FIELD BEAN			
VARIETY ALBINETTE	17.5	15.19	2.05 ±0.42
VARIETY MARIS BEAD	14.22	15.26	2.05 ± 1.05
	23.43	15.18	2.21 ± 0.77
	32.63	15.29	2.77 ± 1.14
	32.63	12.15	3.75 ± 0.91
	32.63	10.03	4.17 ± 1.22

5 CONCLUSION

A variable proportion of the starch of different raw legume seeds has been found to be resistant to digestion in vivo. This fraction has been termed refractive starch and may be regarded as an additional component of the dietary fibre fraction of plant foods.

REFERENCES

1. El-Harith et al, 'On the Nutritive Value of Various Starches for the Albino Rat', J Sci Fd Agric 27, 1976
2. Fleming & Vose, 'Digestibility of Raw and Cooked Starches from Legume Seeds Using the Laboratory Rat', J Nutr 109, 1979
3. Hoover & Sosulski, 'Studies on the Functional Characteristics and Digestibility of Starches from Phaseolus vulgaris Biotypes' Starch 37, 1985
4. Gupta Y P, 'Anti-nutritional and Toxic Factors in Food Legumes: A Review', Plant Foods for Human Nutrition 37, 1987

5 Berry C S, 'Resistant Starch: Formation and Measurement of Starch that Survives Exhaustive Digestion with Amylolytic Enzymes During the Determination of Dietary Fibre', J Cer Sci 4, 1986
6 Cummings & Englyst, 'Fermentation in the Human Large Intestine and the Available Substrates', Am J Clin Nutr 45, 1987

ACKNOWLEDGEMENTS

J.T. is in receipt of a MAFF Studentship

The Effect of Isolated Complex Carbohydrates on Caecal and Faecal Short Chain Fatty Acids and Stool Output in the Rat.

Christine A. Edwards, Jacqueline Bowen and Martin A. Eastwood
Gastrointestinal Laboratory
Western General Hospital
Crewe Road
Edinburgh EH4 2XU

The mechanisms by which complex carbohydrates increase faecal output, and in particular faecal water, are not fully understood. Carbohydrates which are resistant to fermentation by bacteria in the colon are the most effective faecal bulkers probably because they retain their water holding capacity[1]. The extent to which a complex carbohydrate is fermented in the colon of an animal is dependent not only on the ease of fermentation but also the time it remains within the colon[2]. The factors which determine colonic transit time are not clear but propulsion may be increased by distension[3], bile acids[4] or by the stimulation of the mucosa with the edges of particulate matter[5]. The effect of fermentation products on motility has not been investigated thoroughly and short chain fatty acids (scfa) have been shown to inhibit motility in the caecum of the sheep[6] and the isolated rat colon[7], but to stimulate contractions in mid and distal rat colonic strips[8]. The major sites of fermentation are thought to be the caecum and proximal colon but it may be possible, if propulsion is stimulated before fermentation is complete, for fermentation of a complex carbohydrate to continue further along the colon. The absorption of scfa in the distal colon although very rapid may be less efficient than absorption in the proximal colon[9,10] and this could have important implications for colonic pH and the absorption of water. In this study, we have compared the caecal and faecal scfa of rats fed a variety of complex carbohydrates with a range of in vitro fermentabilities and have related this to the effects of the carbohydrates on faecal output and faecal water.

51 male Wistar rats (weight approx. 150g) were fed a fibre deficient diet (2.5% dietary fibre) for 28d. They were then divided into groups of 7 or 8 and their diet supplemented with 5% of guar, karaya, tragacanth, gellan, xanthan or ispaghula gum for a further 28d. At the end of this period faeces were collected over 2 days and the animals were killed and caecal contents obtained. Caecal and faecal wet and dry weight were measured and short chain fatty acid content determined by gas liquid chromatography[11].

Each carbohydrate had a different effect on caecal and faecal wet and dry weight and scfa content. However, they appeared to fall into three groups when compared with the fibre deficient group.

<u>Group 1</u>

Guar gum (G) was the only carbohydrate to increase the caecal scfa concentration (47% increase, $p<0.01$) but had no effect on faecal scfa or faecal output.

<u>Group 2</u>

Tragacanth (T) and karaya (K) significantly increased the total amount of caecal scfa (T 36% and K 89% $p<0.05$, $p<0.001$). Xanthan (X), however, decreased the total amount of caecal scfa (44%, $p<0.01$). All of these three gums significantly increased faecal scfa concentration (T 53%, K 55%, X 100%; $p<0.01$) and faecal wet weight (T 39%, K 47% and X 41% $p<0.05$) but had no effect on faecal dry weight.

<u>Group 3</u>

In contrast, ispaghula (I) and gellan (G) had very little effect on caecal scfa. Gellan increased scfa concentration per gramme caecal content dry weight but this was related to a decrease in dry weight and in fact resulted in a decrease in scfa concentration per gramme caecal content wet weight. Ispaghula and gellan also had no effect on faecal scfa concentration but did increase total daily scfa output (G 72%, I 114%). These carbohydrates significantly increased both faecal wet (G 88%, I 75% $p<0.001$) and dry weight (G 62%, I 33% $p<0.01$, $p<0.05$) and were the most effective faecal bulkers.

The caecum and proximal colon of the rat are thought to be the major sites of carbohydrate fermentation. If we can assume that comparison of caecal and faecal contents gives an indication of events occuring in the colon, these results indicate that the fermentation of some complex carbohydrates has a significant effect on luminal scfa content in more distal colon and this may be related to continued fermentation in more distal regions of the colon. In addition, an increase in the scfa concentration of faeces appeared to be related to an increased output of faecal water which may suggest that under some circumstances scfa absorption is less efficient and may play a role in determining faecal output.

Continued fermentation of complex carbohydrates in the distal colon of the rat has been indicated by pH measurements in other studies[12,13] and may explain the effects of some carbohydrates on the cellular proliferation of distal colonic epithelium[14].

In conclusion, although the knowledge that scfa are rapidly absorbed in the colon[9] has led us to believe that they play no role in determining faecal output, these results suggest that in some cases where carbohydrates are slowly fermented and increase faecal scfa the role of the scfa may need to be reassessed.

REFERENCES

1. M.I. McBurney, P.J.Horvath, J.L. Jeraci, and P.J.Van Soest. Brit.J.Nutr., 1985, 53, 17.
2. P.J.Van Soest, J.L.Jeraci, T. Foose, K.L.Wrick, and F.Ehle, 'Fibre in Human and Animal Nutrition', Eds. M.Wallace and L.Bell. Royal Society of New Zealand, New Zealand, 1983, p.75.
3. A.Chauve, G. Devroede, and E. Bastin, Gastroenterology, 1976, 70, 336.
4. C.A.Edwards, S.Brown, A.J. Baxter, J.J.Bannister, and N.W.Read. Gut, 1989, 30, 383.
5. J.Tomlin, and N.W. Read. Brit.Med.J., 1988, 297 1175.
6. P.Svendsen. Nord.Vet.Met., 1972, 24, 393.
7. C.A.Edwards. J.Physiol., 1989, 409, 59.
8. T.Yajima. J.Physiol., 1985, 368, 667.
9. N.I.McNeil, J.H.Cummings and W.P.T. James. Gut, 1978, 19, 819.
10. W.V.Englhardt and G. Rechkemmer. 'Intestinal Transport Fundamental and Comparative Aspects' Eds. M. Gilles-Ballier,and R. Gillers,

Springer Verlag, New York, 1985, p. 27
11. G.A.Spiller, M.C.Chernoff, R.A.Hill, J.E.Gates, J.J. Nassar, and E.A.Shipley. Am.J.Clin.Nutr., 1980, 33, 754.
12. D.F.Evans, J.Crompton, G.Pye, and J.D.Hardcastle. Gastroenterology, 1988, 94, 74.
13. J.R.Lupton, D.M.Coder, and L.R.Jacobs. J.Nutr., 1988, 118, 840
14. R.A.Goodlad, W.Lenton, M.A.Ghatei, T.E.Adrian, S.R.Bloom and N.A.Wright. Gut, 1987, 28, 171.

TIME COURSE OF THE EFFECTS OF DIETARY FIBRE ON ENERGY INTAKE AND SATIETY

V.J. Burley and J.E. Blundell

Department of Psychology
University of Leeds
Leeds LS2 9JT

INTRODUCTION

Satiety has previously been described as 'the state of inhibition over eating'[1]. The state of satiety may exist for a variable length of time which may be partitioned according to the mechanisms which are responsible for its maintenance. Figure 1 illustrates the proposed contribution of different processes to the intensity and time course of satiety. Thus, in the early phase of satiety, sensory and cognitive aspects of the food consumed may be responsible for activating and maintaining satiety. Post-ingestive properties such as the capacity of food to distend the stomach, its potential to evoke the release of gut hormones and the rate at which it is emptied from the stomach, are likely to be involved in prolonging the state of satiety in its mid and late stages. Late stages of satiety may be extended by the properties of food which decrease the rate of digestion and/or absorption. It follows therefore, that an assessment of the time course of satiety after a meal will reveal something of the underlying mechanisms responsible for its genesis and maintenance.

High fibre foods have been tested in relatively few short-term studies, and there is little information on how fibre influences the development and endurance of satiety. In this paper the results of two studies will be outlined which reveal something of the time course of action of high fibre meals on energy intake, appetite and satiety.

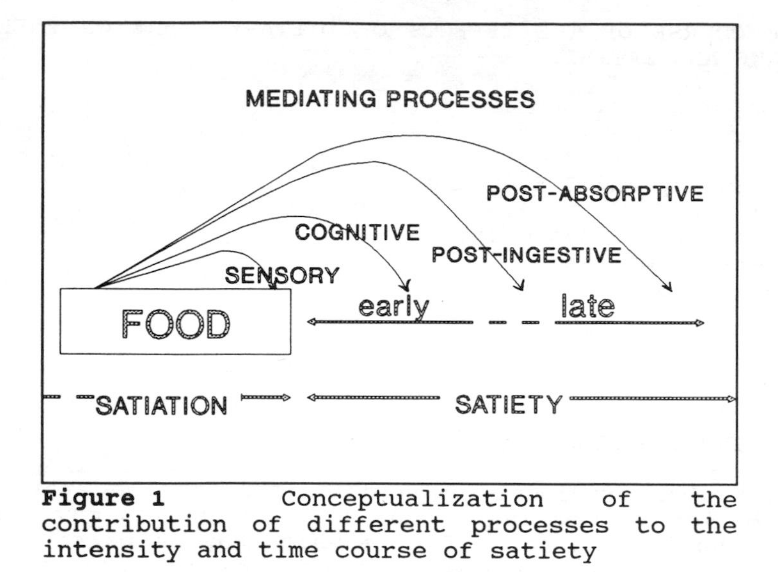

Figure 1 Conceptualization of the contribution of different processes to the intensity and time course of satiety

The Effects of a High and Low Fibre Lunches on Energy Intake and Satiety

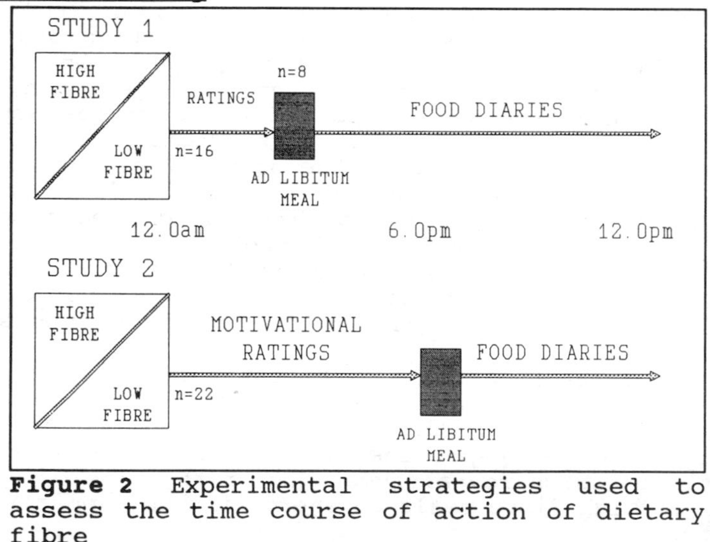

Figure 2 Experimental strategies used to assess the time course of action of dietary fibre

Using a within-subjects design, non-obese female volunteers consumed high and low fibre lunches after an over night fast. Lunches were prepared to be as similar as possible in terms of macronutrient content and palatability. Table 1 lists the components of each lunch, and Table 2 shows their calculated nutrient and fibre content.

Visual analogue ratings (100mm lines anchored with descriptors) of motivation to eat were completed by the subjects before and for 2½ hours after each meal in study 1, and for 6 hours after lunch in study 2. After 2½ hours, in study 1 half the subjects were offered an afternoon meal of sandwiches, biscuits and a beverage for the direct assessment of satiety. Following this, all subjects left the feeding area and maintained weighed diet records for the rest of each test day. Energy intakes were calculated from the afternoon meal intakes and at each meal recorded in the food diaries after each lunch by the use of a computerised version of British Food Tables. In study 2, after 6 hours a similar ad libitum meal was offered to all subjects, consisting of bread, spreads, ham, cheese, biscuits, salad and fruit. Energy intake was assessed with reference to food table values. Following this meal, subjects kept a food diary until 12.00 pm.

Table 1 Composition of the high and low fibre test lunches

HIGH FIBRE	LOW FIBRE
Lentil Soup	Chicken Soup
Wholemeal Bread	White Bread
Butter	Butter
Wholemeal Pasta	White Pasta
Spiced Red Kidney Beans	Bolognese Sauce
Cheddar Cheese	
Tinned Black currants	Fruit Yoghurt
Fruit Yoghurt	

Table 2 Energy (kcal) and Nutrient Content (g) of High and Low Fibre Lunches (* Southgate values)

	Study 1		Study 2	
	High	Low	High	Low
Energy kcal	795	790	762	764
Protein	33	33	34	34
Fat	28	29	25	26
Carbohydrate	106	106	106	106
Fibre*	30	3.3	30	4.1

<u>Study 1 - Results</u>. Analysis of the motivational ratings data showed that for 2½ hours following each lunch, there was no significant difference between the meals in terms of their effect on motivational ratings. Moreover, at the afternoon meal, consumed by half the subjects, energy intake did not differ significantly according to whether high or low fibre lunches had been consumed. However, energy intake calculated from food diaries, from 2½-3 hours after lunch until 12.0 pm was significantly lower after the high fibre lunch (967±532 vs 1559±930 kcal). Figure 3 illustrates cumulative energy intake following high and low fibre test lunches from food diaries (including energy intake of subjects who consumed the afternoon meal). This figure shows that until approximately 5-6 hours after lunch no effect of fibre on energy intake was apparent.

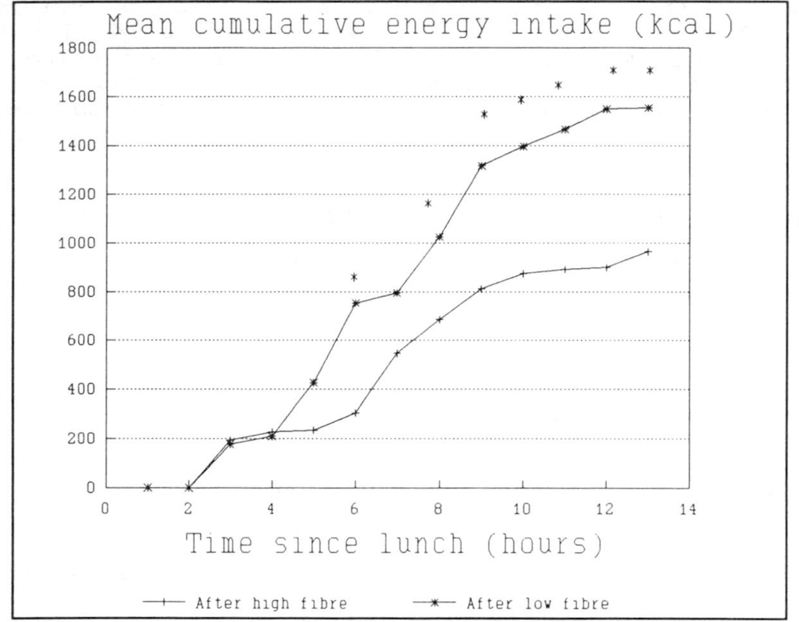

<u>Figure 2</u> Effect of High and Low Fibre Lunches on Cumulative Energy Intake (from food diaries) for 13 Hours After Consumption. *Means Significantly Different $p<0.03$

<u>Study 2 - results</u>. Analysis of visual analogue scale data revealed a greater reduction in motivation to eat following the high fibre lunch compared to the low fibre meal, a difference which persisted from 30 minutes

to 6 hours after eating. Figure 4 illustrates the mean prospective consumption rating (on a scale of 0-100mm asking the question, 'How much food do you think you could eat? Nothing at all - A large amount) following both meals.

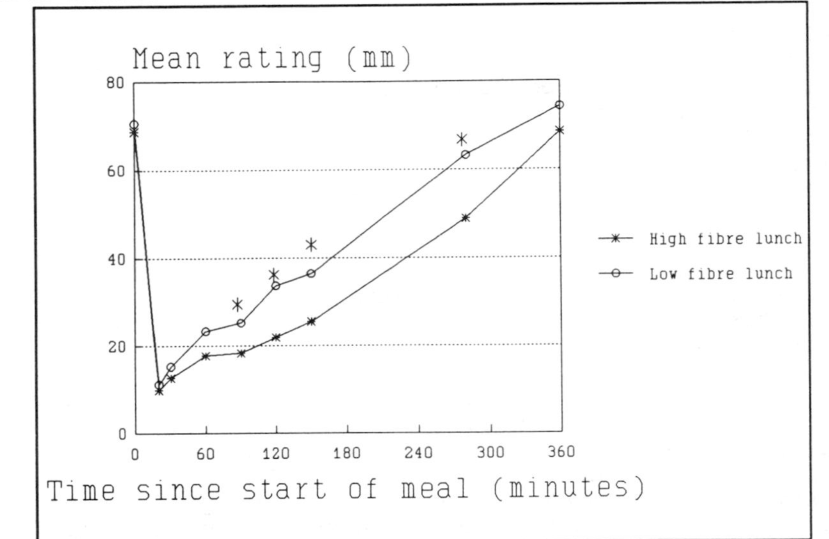

Figure 4 Effect of High and Low Fibre Lunches on Post-prandial Prospective Consumption Ratings. *Means Significantly Different $p<0.01$

Energy intake at the ad libitum meal presented 6 hours after lunch was lower by 86 ± 187 kcal after the high fibre meal ($p<0.04$), and intake after this meal, obtained from food diary analysis was also lower by a further 112 ± 528 kcal after fibre, though this difference failed to reach statistical significance.

Both these studies indicate that although there was some suggestion of increased satiety in the first 3 hours following the high fibre lunch in study 2, the primary impact of this meal was on later energy intake (5-6 hours after consumption). It is possible therefore that fibre consumed as high fibre foods reduces energy intake by an action on post-absorptive satiety.

1. J.E. Blundell and V.J. Burley, Int. J. Obesity., 1987, 11 (Suppl. 1), 9.

WORKSHOP REPORT: DIETARY FIBRE AND ENERGY BALANCE

G. Livesey

AFRC Institute of Food Research
Norwich Laboratory
Colney Lane
Norwich NR4 7UA

A wide range of topics was covered, aimed at understanding the extent of body weight regulation by 'fibre' and 'high-fibre' diets.

The question: ''Fibre', an adjunct to the treatment of obesity or not?' tended to be answered affirmatively but proof was lacking. Thirteen studies addressing this problem were reviewed by G. Livesey. These showed 3 with no effect and 10 each with an enhanced weight loss of about 2kg. Oddly the effect on absolute weight loss appeared independent of overall weight loss, duration of weight loss or the cumulative intake of 'fibre'. Methodological problems might explain the 2kg effect. No studies included 'high-fibre foods'. A 14th study, a 2yr coronary heart disease follow-up in men who had been advised to increase 'fibre' intake showed no effect on body weight, although fibre intake was assessed as increased. A potential problem was dietary advice from the media or from the researchers indirectly. One thought was that a study with 'high-fibre foods' is near impossible, on the other hand this undermined the original observations of Cleave, who must have considered subjects not eating fibre supplements. It was supposed fibre may behave like a drug, in which case there would be responders and non-responders. A study investing 'high-fibre foods' affecting weight loss was thought beyond the means of most (all) research groups. In the absence of such a study the discussion followed the approach that if all (or most) effects on the physiology were consistent with enhanced weight loss, the probability of an effect could be established on theoretical grounds.

'Caloric compensation with 'high- or higher-fibre diets' or satiety accompanied with lowered food intakes?' was the question addressed by V. Burley. Plenty of evidence was available for effects of fibre on satiety, but information linking 'fibre' with effects on calorie intake or intake of individual macro-nutrients was limited. High-fibre foods enhanced satiety at about the time of the next meal. This coincided with the onset of fermentation. It was suggested the effects of short-chain fatty acids on satiety be examined. Long-term studies of effects of fibre on food intake were absent and some thought these more important than studying the mechanisms in short-term effects. Methodology for assessing intake in the long-term was poor however, while in animals adaptation occurred so fibres eventually had no effect on intake. Short-term effects could be useful, e.g. when reintroducing subjects to conventional foods after procedures for severe body weight reduction. Furthermore there may be critical periods in life when a short-term effect would help, e.g. with dangerously obese subjects.

Are there significant energy losses from the terminal ileum other than 'dietary fibre'? was addressed by N. Read who gave evidence from ileostomists, intubated subjects and breath hydrogen studies. Ileostomists fed a control diet with and without 10g guar gum in an acute study showed terminal losses increasing from about 100 kcal to 180 kcal daily with only about a half of the increase that could be due to guar. The losses with the control diet and the increased losses due to guar were due to fat, protein and glucose (G_n). The elevated losses due to guar were thought by the workshop as an upper limit to what might be expected with other fibres which show less effects and adaptation might occur. It was considered valuable to obtain more information on the ileal losses of energy substrates.

John Mathers gave an account of an empirical model which can be used to calculate losses and gains of energy from fermentation of materials entering the colon from the small bowel. The major factor influencing energy salvage is the extent of fermentation of the organic materials supplied in ileal effluent. Assuming a value of 0.8 for the latter and making reasonable assumptions about i) the yield of bacterial biomass, ii) colonic SCFA pattern and iii) the proportion of SCFA absorbed he calculated a salvage of only 34-37% of energy from carbohydrate entering the colon of man. No single experiment emerged that could test the calculated result.

Information on the effects of protein and fat entering the colon was lacking and studies were needed to assess this.

Species differences in the extent of fermentation in man, pig and rat was addressed by I. Hansen. It was concluded that rat was similar to humans in the extent of fermentation of a whole range of materials from poorly fermented Solka floc up to easily fermented soluble fibres. Possibly greater variation occurred with poorly fermented fibres and between dosage with fibre within the same species. The similarity between man and rat in this respect made the rats a useful model for quantitative studies.

The energy values of 'dietary fibre' was a subject addressed by the British Nutrition Foundation Task Force and there being no time to continue with discussions individuals were suggested to consult the Task Force report when this becomes available. Briefly, here it can be stated that the Task Force concluded a value of 2 kcal/g was appropriate if a caloric conversion factor was needed for a mixed meal, with 'unavailable carbohydrate' having a fermentability of about 70%. The same could be true for supplements except it could be as high as 3 kcal/g and as low as 0 kcal/g when fermentability is about 100% and about 0% respectively. With diets high in cereal grain problems remain with quantification of carbohydrate that is unavailable _in vivo_.

Dietary Fibre and Lipid Metabolism

DIETARY FIBER IN LIPID METABOLISM

David Kritchevsky

The Wistar Institute of Anatomy and Biology
3601 Spruce Street
Philadelphia, Pennsylvania 19104
U.S.A.

There are many facets of research involving lipid metabolism and dietary fiber. Data have been collected concerning the influence of fibers on serum or plasma lipid levels in man or animals; usually the emphasis is on cholesterol and often on cholesterol and its principal metabolites, the bile acids. Since bile acids are related closely to cholesterol absorption, effects of fibers on bile acid binding or turnover are also of interest. As an extension of interest in cholesterol metabolism, there have been investigations into the effects of fibers on experimental atherosclerosis. Studies of fiber effects in animals almost always involve the use of purified or defined sources of fiber. In man, effects of both identifiable fibers and fiber-rich foods have been studied. In a space constrained exposition such as this, it is not feasible to offer in-depth discussion or exhaustive citations of the literature, but rather one is limited to high-lighting studies which have been of a pioneering nature or which have influenced research directions.

Lipid Metabolism in Animals

Portman observed that diet affected bile acid turnover and excretion in rats. Compared to a commercial ration, a semipurified diet increased cholic acid turnover by 110% and reduced the body pool by 38%.[1,2] These experiments represent one of the earliest studies of metabolic effects of fiber.

Ershoff and his colleagues carried out studies in which rats were fed fiber-free diets containing cholesterol and compared the effects with those observed in rats fed the cholesterol diet with added fiber. Pectin, guar gum, locust bean gum and carageenan all lowered serum and liver cholesterol, but agar and cellulose did not.[3,4] Pectin has also been shown to be hypocholesterolemic in rabbits[5,6], chickens[7], and swine[8]. Cellulose appears to be hypercholesterolemic in rats[3,9] and rabbits[6] and also increases total body cholesterol in rats.[10] Soluble fibers such as pectin or guar gum also exert beneficial effects on plasma lipoproteins and apolipoproteins in rats.[11,12] A comprehensive compilation of fiber effects on blood and liver cholesterol levels in a number of animal species has been published.[13]

Since bile acids are a metabolites of cholesterol and bile salts are required for the micelle formation which is necessary for lipid absorption, it is logical to study fiber effects on bile acid metabolism. Eastwood and Hamilton[14] studied the binding *in vitro* of cholic and taurocholic acid to a number of plant substances at pHs of 3.9 or 8.0. They concluded that the binding was a hydrophobic phenomenon most probably due to the lignin content of the plant materials used. Story and Kritchevsky[15] studied binding of cholic, chenodeoxycholic, deoxycholic acids and their taurine and glycine conjugates to alfalfa, bran, cellulose and lignin. Each substrate was bound to a different extent by each of the fibers used. Cellulose bound very little of any of the substrates. Studies with lipid micelles show that different fibers can also bind cholestesrol and lecithin.[16] Taurocholic acid is bound to a number of spices such as curry powder, cloves and oregano.[17]

A study in which a variety of fibers were fed to rats for four weeks showed that almost all of them increased fecal steroid excretion (mg/day) over that seen in rats fed a fiber-free diet but fecal steroid concentration (mg/gm dry weight) was usually lower.[18] The insoluble fibers caused a consistent increase in fecal neutral steroids and a variable increase in acidic steroid excretion. The percentage of fecal acidic steroids present as primary bile acid is increased in almost all cases suggesting reduced metabolic activity by colonic bacteria. Results of the fecal steroid analyses are summarized Table 1. Eastwood and Boyd[19] had observed earlier that addition of cellulose to a stock diet increased excretion of trihydroxy bile acids at the expense of dihydroxy

Table 1. Fecal Steroids in Rats Fed Fiber*

Fiber (%)	Fecal Dry Weight	Steroids		
		Total	Neutral (OL/ONE)[a]	Acidic (P/S)[b]
None	(100)	(100)	(100) (3.65)	(100) (0.26)
Cellulose (10)	349	161	151 (3.65)	186 (0.42)
Wheat Bran (10)	218	146	154 (3.10)	125 (0.38)
Alfalfa (10)	245	149	145 (4.03)	157 (0.53)
Pectin (7)	77	68	86 (7.33)	66 (0.31)
Guar Gum (5)	196	180	161 (6.14)	113 (0.32)
Psyllium (10)	220	188	182 (3.55)	204 (0.41)

* After ref. 18

a Cholesterol + cholestanol + coprostanol/coprostanone
b Primary/secondary bile acids

bile acids. The various fibers also affected the liver phospholipid spectrum; principal differences were in levels of lecithin (7-25% above control level) and sphingomyelin (20-53% below control level).[20]

Investigation of mechanisms of action of fiber in rats has shown that recovery of cholesterol in lymph four hours after feeding a test meal is significantly lower in rats fed either soluble or insoluble fiber, but at 24 hours lymphatic absorption in rats fed insoluble fibers begins to approach that of the fiber-free control but recovery in lymph of rats fed soluble fibers is still reduced significantly. Recovery of oleic acid is unaffected in 4 hour lymph of rats fed insoluble fiber and at 24 hours there are no significant differences in recovery except for rats fed psyllium.[21]

Recovery of cholesterol (endogenous or exogenous) from serum and tissues of rabbits fed a semipurified diet is significantly higher than in those fed a commercial preparation; recovery from the feces is lower.[22] The data are consistent with enhanced transit and excretion of cholesterol in rabbits fed a diet high in crude fiber.

Experimental Atherosclerosis

In the late 1950s, Lambert et al.[23] and Malmros and Wigand[24] reported that they had produced atherosclerosis in rabbits by feeding them a cholesterol-free semipurified diet containing high levels of saturated fat. These findings appeared paradoxical since other investigators[25-27] had found that addition of saturated fat to commercial ration did not render that diet atherogenic. A summary of existing literature[28] showed that addition of saturated fat to a semipurified diet (in which the fiber was usually cellulose) rendered that diet atherogenic but addition of the same fat to a stock diet was without effect. It was hypothesized[28] that a factor determining a diet's atherogenic potential was its fiber content. Another view was that the small amount of polyunsaturated fat present in commercial ration was sufficient to negate the effects of saturated fat.

To compare these hypotheses, an experiment was conducted in which rabbits were fed a semipurified diet containing coconut oil, the same diet plus 2% of the fat (iodine value 115) extracted from commercial ration, or the extracted residue plus 14% saturated fat. Addition of the unsaturated fat did not affect cholesterolemia or severity of atherosclerosis. The extracted residue plus 14% coconut oil or stock ration plus 14% coconut oil led to significantly lower serum cholesterol levels and atherosclerosis of 54-71% lower severity.[29,30] In a somewhat similar study, Moore[31] fed rabbits 20% butter fat in diets in which the fiber (19%) was wheat straw, cellulose, cellophane or cellophane-peat 14:5. Cholesterol levels were highest and atherosclerosis most severe in the rabbits fed cellophane. Rabbits fed wheat straw exhibited significantly lower serum cholesterol values (by 47%) and 66% less severe atherosclerosis. When the diet contained corn oil rather than butter fat, cellophane was still significantly more cholesterolemic and atherogenic than wheat straw. Dilution of an atherogenic regimen by addition of alfalfa (1:9)[32] or stock diet (1:1)[33] will also reduce cholesterolemia and severity of atherosclerosis in rabbits.

Fiber may also influence the effects of other components of an atherogenic diet. Thus, when effects of casein or soy protein were compared in diets containing cellulose, the former was more cholesterolemic and atherogenic; when the dietary fiber was wheat straw, casein was still more cholesterolemic than soy protein but the atherogenicity was about the same; alfalafa in the diet rendered effects of the two proteins virtually equal.[34]

Aortic sudanophilia in Vervet monekys fed semipurified diet containing wheat straw was significantly less severe than that seen in monkeys fed the same diet with cellulose.[35] When the diet contained 0.1% cholesterol, aortic sudanophoilia was similar in monkeys fed cellulose or pectin.[36] Chickens fed 0.6% cholesterol and 3% pectin exhibit lower cholesterol levels and less severe atherosclerosis than those fed cholesterol and cellulose.[37]

Human Studies

Walker and Arvidsson[38] suggested in 1954 that the low levels of cholesterol in black Africans and their low incidence of coronary disease could be due, in part, to the high levels of fiber in their diet.

Keys et al.[39] found that pectin lowered cholesterol levels in man, whereas cellulose has no effect. The effects of dietary fiber on serum or plasma cholesterol have been summarized by Kay and Truswell[40] and by Schneeman and Lefevre.[41] In general, insoluble fiber such as cellulose or wheat bran has no effect on either cholesterol or lipoprotein levels. Oat bran is hypolipidemic probably because it contains oat gum.[42] Soluble fibers have been shown to have hypolipidemic properties particularly guar gum[43], pectin[44], and locust bean gum.[45] Diets rich in fiber[46] or legumes[47,48] are hypocholesterolemic. A thorough review of dietary fiber effects on lipid metabolism has appeared recently.[49]

Effects of a vegetarian life style on lipidemia can be studied in Seventh Day Adventists, some of whom are vegans and most of whom are lacto-ovo vegetarians. Hardinge and his colleagues[50,51] showed that vegans whose fiber intake was judged to be significantly higher than that of the general population had significantly lower levels of plasma cholesterol. A later study[52] confirmed this observation and found that the major difference in fiber intake between Seventh Day Adventist vegans, other Seventh Day Adventist groups and the general population was in pectin consumption which was almost twice as high in vegans as in the other groups. Intake of other fibers were comparable. Vegetarians exhibit lower plasma levels of those apolipoproteins associated with increased risk of atherosclerosis.[53]

In summary, soluble (or gelling) fiber affect lipidemia and atherosclerosis in animals and man. The effect may be due, in part, to binding of bile acids but this property is not always reflected in bile acid excretion. Fiber can affect intestinal morphology[54] and activity of intestinal[54] and pancreatic[55] enzymes but the precise relation of these effects to lipid metabolism has not been established. Effects on lipid absorption and distribution into lipoproteins have been demonstrated.[21] The role of the products of colonic fermentation of fiber, namely, the short chain fatty acids, remains to be clarified.

References

1. O.W. Portman and P. Murphy, Arch. Biochem. Biophys., 1959, 76, 367.
2. O.W. Portman, Am. J. Clin. Nutr., 1960, 8, 462.
3. A.F. Wells and B.H. Ershoff, J. Nutr., 1961, 74, 87.
4. B.H. Ershoff and A.F. Wells, Proc. Soc. Exp. Biol. Med., 1962, 110, 580.
5. L.M. Berenson, R.R. Bhanadaru, B. Radhakrishnamurthy, S.R. Srinivasan and G.S. Berenson, Life Sci., 1975, 16, 1533.
6. R.M.G. Hamilton and K.K. Carroll, Atheroscler., 1976, 24, 47.
7. M.J. Fahrenbach, B.A. Riccardi and W.C. Grant, Proc. Soc. Exp. Biol. Med., 1966, 123, 321.
8. H.D. Fausch and T.A. Anderson, J. Nutr., 1965, 85, 145.
9. S. Kiriyama, Y. Okazaki and A. Yoshida, J. Nutr., 1969, 97, 382.
10. M.A. Mueller, M.P. Cleary and D. Kritchevsky, J. Nutr., 1983, 113, 2229.
11. W.J.L. Chen and J.W. Anderson, Proc. Soc. Exp. Biol. Med., 1979, 162, 310.
12. B.O. Schneeman, J. Cimmarusti, W. Cohen, L. Downes and M. Lefevre, J. Nutr., 1984, 114, 1320.
13. D. Kritchevsky and J.A. Story, "CRC Handbook of Dietary Fiber in Human Nutrition", ed. G.A. Spiller, CRC Press, Boca Raton, FL, 1986, Chapter 4.3, p. 129.
14. T.L. Cleave, J. Royal Naval Med. Service, 1956, 42, 55.
15. J.A. Story and D. Kritchevsky, J. Nutr. 1976, 106, 1292.
16. G.V. Vahouny, R. Tombes, M.M. Cassidy, D. Kritchevsky and L.L. Gallo, Lipids, 1980, 15, 1012.
17. J.A. Story and D. Kritchevsky, Nutr. Rep. Int., 1975, 11, 161.
18. G.V. Vahouny, R. Khalafi, S. Satchithanandam, D.W. Watkins, J.A. Story, M.M. Cassidy and D. Kritchevsky, J. Nutr., 1987, 117, 2009.
19. M.A. Eastwood and G.S. Boyd, Biochim. Biophys. Acta, 1967, 137, 393.
20. D. Kritchevsky, S.A. Tepper, S. Satchithanandam, M.M. Cassidy and G.V. Vahouny, Lipids, 1988, 23, 318.

21. G.V. Vahouny, S. Satchithanandam, I. Chen, S.A. Tepper, D. Kritchevsky, F.G. Lightfoot and M.M. Cassidy, Am. J. Clin. Nutr., 1988, 47, 10.
22. D. Kritchevsky, S.A. Tepper, H.K. Kim, D.E. Moses and J.A. Story, Exp. Mol. Pathol., 1975, 22, 11.
23. G.F. Lambert, J.P. Miller, R.T. Olsen and O.V. Frost, Proc. Soc. Exp. Biol. Med., 1958, 97, 544.
24. H. Malmros and G. Wigand, Lancet, 1959, 2, 749.
25. D. Kritchevsky, A.W. Moyer, W.C. Tesar, J.B. Logan, R.A. Brown, M.C. Davies, and H.R. Cox, Am. J. Physiol., 1954, 178, 30.
26. E.F. Hirsch and R. Nailor, Arch. Pathol., 1955, 59, 419.
27. A. Steiner, A. Varsos and P. Samuel, Circ. Res., 1959, 7, 448.
28. D. Kritchevsky, J. Atheroscler. Res., 1964, 4, 103.
29. D. Kritchevsky and S.A. Tepper, Life Sci., 1965, 4, 1468.
30. D. Kritchevsky and S.A. Tepper, Atheroscler. Res., 1968, 8, 357.
31. J.H. Moore, Br. J. Nutr., 1967, 21, 207.
32. F.B. Cookson, R. Altschul and S. Fedoroff, J. Atheroscler. Res., 1967, 7, 69.
33. A.N. Howard, G.A. Gresham, I.W. Jennings and D. Jones, Prog. Biochem. Pharmacol., 1967, 2, 117.
34. D. Kritchevsky, S.A. Tepper, D.E. Williams and J.A. Story, Atheroscler., 1977, 26, 397.
35. D. Kritchevsky, L.M. Davidson, D.A. Krendel, J.J. Van der Watt, D. Russell, S. Friedland and D. Mendelsohn, Ann. Nutr. Metab., 1981, 25, 125.
36. D. Kritchevsky, L.M. Davidson, G.T. Goodman, S.A. Tepper and D. Mendelsohn, Lipids, 1986, 21, 338.
37. H. Fisher, W.G. Soller and P. Griminger, J. Atheroscler. Res., 1966, 6, 292.
38. A.R.P. Walker and U.B. Arvidsson, J. Clin. Invest., 1954, 33, 1358.
39. A. Keys, F. Grande and J.T. Anderson, Exp. Biol. Med., 1961, 106, 555.
40. R.M. Kay and A.S. Truswell, "Medical Aspects of Dietary Fiber", Plenum Medical Book Co., New York, 1980, Chapter 9, p. 153.
41. B.O. Schneeman and M. Lefevre, "Dietary Fiber: Basic and Clinical Aspects", ed. G.V. Vahouny and D. Kritchevsky, Plenum Press, New York, 1986, Chapter 20, p. 309.
42. R.W. Kirby, J.W. Anderson, B. Sieling, E.D.

Rees, W.J.L. Chen, R.E. Miller and R.M. Kay, Am. J. Clin. Nutr., 1981, 34, 824.
43. S.J. Gatenby, "Dietary Fibre Perspectives", Vol. 2, ed. A.R. Leeds, John Libbey and Co., Ltd., London, 1990, Chapter 7, p. 100.
44. D.J.A. Jenkins, A.R. Leeds, C. Newton and J.H. Cummings, Lancet, 1975, 1, 1116.
45. K.M. Behall, K.H. Lee and P.B. Moser, Am. J. Clin. Nutr., 1984, 39, 209.
46. F. Grande, "Sugar in Nutrition", Academic Press, New York, 401.
47. K. Mathur, M.A. Khan and R.D. Sharma, Br. Med. J., 1968, 1, 30.
48. J.W. Anderson, L. Story, B. Sieling, W.L. Chen, M.S. Petro and J. Story, Am. J. Clin. Nutr., 1984, 40, 1146.
49. S. Pilch ed., "Physiological Effects and Health Consequences of Dietary Fiber", FASEB, Bethesda, MD, 1987.
50. M.G. Hardinge and F.J. Stare, Clin. Nutr., 1954, 2, 83.
51. M.G. Hardinge, A.C. Chambers, H. Crooks and F.J. Stare, Am. J. Clin. Nutr., 1958, 6, 523.
52. D. Kritchevsky, S.A. Tepper and G. Goodman, Am. J. Clin. Nutr., 1984, 40, 921.
53. J. Burselm, G. Schonfeld, M.A. Howard, S.W. Werdman and J.P. Miller, Metabolism, 1978, 27, 711.
54. G.V. Vahouny and M.M. Cassidy, "Dietay Fibre: Basic and Clinical Aspects", ed. G.V. Vahouny and D. Kritchevsky, Plenum Press, New York, 1986, Chap. 13, p. 181.
55. B.O. Schneeman, "Dietary Fiber in Health and Disease", ed. G.V. Vahouny and D. Kritchevsky, Plenum Press, New York, 1982, Chapter 7, p. 73.

DO OATS LOWER BLOOD CHOLESTEROL?

E. K. Lund, C. A. Farleigh and I. T. Johnson

Department of Nutrition, Diet and Health
AFRC Institute of Food Research
Norwich Laboratory
Colney Lane
Norwich NR4 7UA

1 INTRODUCTION

Oats and oat bran have recently received considerable attention due to their alleged cholesterol lowering properties. The active constituent is generally assumed to be the soluble fibre fraction or oat gum which is predominantly β-glucans. We have previously shown that rats fed *ad libitum* a diet based on 50% oats (ca. 2% gum) have significantly lower cholesterol levels compared to cellulose-fed controls[1]. Food intake remained the same in the two groups and both diets contained less than 350 mg of cholesterol per Kg of diet.

2 THE HYPOCHOLESTEROLAEMIC EFFECT OF OAT GUM IN RATS

Hypocholesterolaemic properties have been suggested not only for oat gum but also for the lipid and protein fractions, so we have now looked for a dose-response effect of isolated oat gum. Figure 1 shows that no significant effects on total, LDL or HDL cholesterol were found when the same control diet as that used above, but with ; 0.5, 1.0, 2.0 and 4.0 % gum substituted for cellulose, was fed for 4 weeks. However the control cholesterol level was already reduced by placing all rats on restricted food intake (18g/day) to obtain equal energy intake. A reduction in cholesterol in rats on restricted intakes has been reported previously[2] but has largely been ignored in studies on soluble fibre. Where rats are fed 10% guar gum or pectin, food consumption is reduced. This in itself will lower their blood cholesterol level, and may account for much of the hypocholesterolaemic affect of these diets.[3]

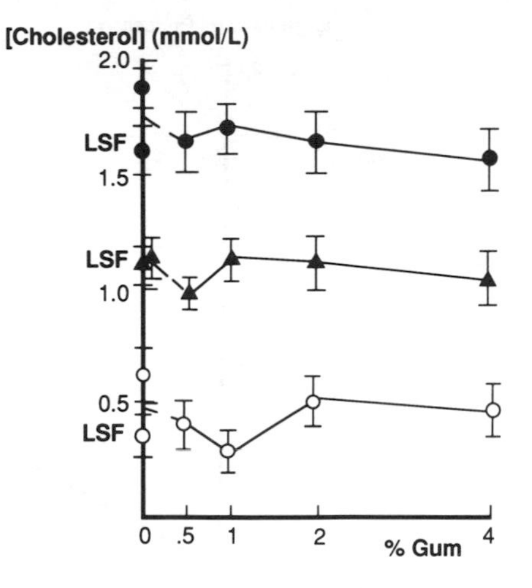

Figure 1 Plasma cholesterol levels vs. oat gum concentration. (LSF. 6% solkafloc control. o-Total Cholesterol Δ HDL-Cholesterol o-LDL-Cholesterol (mean ± SEM n=6)

3 CHOLESTEROL REDUCTION IN MAN

These problems, of separating the effects of reduced food intake from any true hypocholesterolaemic value of oats, are also relevant to human studies. We have recently conducted a study on 35 healthy subjects mean age 36 (22-54). They had a relatively low mean cholesterol level (4.9 ± 0.2 m.mol/L) at the beginning of the trial which, after eating 50g of oats/day for 42 days, dropped by 1.4% to 4.8 ± 0.2 m.mol/L ($p>0.05$). This statistically insignificant reduction is consistent with many other studies using volunteers with mean cholesterol levels close to the accepted maximum desirable level of 5.2 m.mol/L[4-7]. When hypercholesterolaemic subjects have been fed oats or oat bran, large statistically significant, reductions in cholesterol levels have been reported[6-9]. Figure 2 shows that a dose-dependent effect of oats is only apparent in hypercholesterolaemic groups of subjects. In many of these studies a small weight loss was observed and, therefore, as with rats, reduced food intake may have been a complicating factor. This however seems unlikely to account for all the hypocholesterolaemic effect.

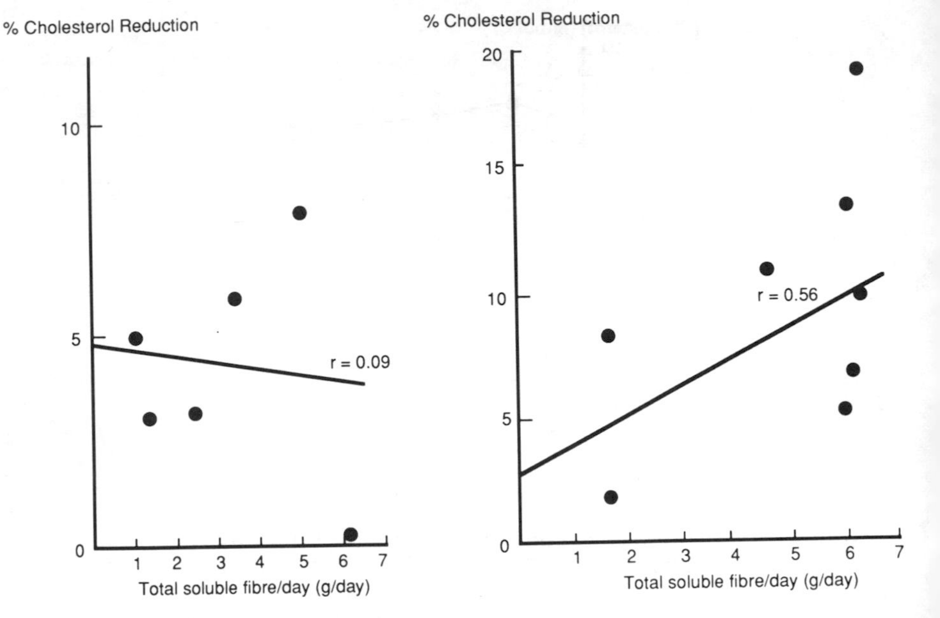

Figure 2 Percent reduction in mean plasma cholesterol in normocholesterolaemic (a) and hypercholesterolaemic (b) subjects from 14 studies.

3 DISCUSSION

Oats probably do reduce LDL-cholesterol when fed to hypercholesterolaemic rats or human subjects but this is hard to distinguish from effects due to reduced saturated fat and total food intake. A further complicating factor is that the percentage cholesterol reduction is dependent on the baseline level. The correlation between % reduction and starting cholesterol is stronger than that between % reduction and the amount of gum fed (r=0.74 and r=0.53 respectively, n=14). It is still not certain that soluble fibre is the key factor in lowering serum cholesterol but recent studies have shown that oat gum can reduce plasma levels in cholesterol fed rats. Present studies using hypercholesterolaemic rats but with no cholesterol in the diet should resolve some of these points.

REFERENCES

1. E.K. Lund and I.T. Johnson, Proc.Nutr.Soc., 1990, (In Press).
2. K.R. Feingold, G. Zsigmond, S.R. Lear and A.H. Moser. Am.J.Physiol., 1986, 251, G362-G369.
3. J.M. Gee and I.T. Johnson. Fibre 90.
4. T.R. Gormley, J. Kevany, B. O'Donnell and R. McFarlane. Irish J. Food Sci.Technol., 1978, 2, 85-91.
5. M.R. Gould, J.W. Anderson and S. O'Mahony. 'Cereals for Foods and Beverages', Academic Press, New York, 1980.
6. K.V. Gold and D.M. Davidson. West.J.Med., 1988, 148, 299-302.
7. J.F. Swain, I.L. Rouse, C.B. Curley and F.M. Sacks. New Eng.J.Med., 322, 147-152.
8. W.H. Turnbull and A.R. Leeds. J.Clin.Nutr. Gastroenterol., 1987, 2, 177-181.
9. R.W. Welsh. Proc.Nutr.Soc., 1990, In press.
10. C.D. Jennings, K. Boleyn, S.R. Bridges, P.J. Wood and J.W. Anderson. Proc.Soc.Exp.Biol.Med., 1988, 189, 13-20.

MECHANISMS WHEREBY FIBRE COULD LOWER PLASMA CHOLESTEROL

D. L. Topping, R. J. Illman, Kerin Dowling and R. P. Trimble

CSIRO Division of Human Nutrition, Glenthorne Laboratory, O'Halloran Hill, SA 5158, Australia.

1. INTRODUCTION

Fibre preparations such as oat bran, guar gum and pectin have gained attention through their potential for lipid lowering with reported reductions of 10-20% in plasma cholesterol in man.[1] Oat bran has been examined most thoroughly and recent reports generally have confirmed more modest (but useful) reductions than earlier studies.[2,3] In an investigation of three cereal brans in mildly hypercholesterolaemic men, oat bran reduced cholesterol by 6% relative to wheat bran.[3] We have screened eight commercially available Australian oat brans in rats and found that they all lower cholesterol. These rat data have been confirmed in man[4] for one product (The Uncle Toby Company) with similar reductions to the earlier study of Kestin *et al.*[3] Foods like oat bran and barley flour are believed to lower cholesterol through their water-soluble non-starch polysaccharide (NSP) content.[5] This is because very similar compounds (such as guar gum) in isolated form also lower plasma cholesterol.[6] Three mechanisms of action have been proposed: alteration of digestion through viscosity in aqueous solution; inhibition of hepatic cholesterogenesis by the propionate produced by large bowel bacterial NSP fermentation; and, enhanced sterol excretion through a specific binding by NSP. However, none has been established beyond doubt and one (inhibition of cholesterol synthesis) seems incorrect.

2. EFFECTS OF NSP VISCOSITY ON PLASMA CHOLESTEROL

Oat and barley β-glucans and gums such as guar are highly viscous in aqueous solution. This viscosity can effect cholesterol reduction by slowing gut transit and reducing absorption of fat and cholesterol.[5] Delayed transit from the stomach to the small intestine is a well-documented effect of NSP.[7] High viscosity

Table 1. Plasma cholesterol concentrations of rats fed a low-cholesterol diet containing different cereal preparations.

Diet	Plasma Cholesterol* (mM)
Wheat bran	3.22 ± 0.11^{a}
Oat bran	2.59 ± 0.10^{b}
Barley flour (unmalted)**	2.75 ± 0.13^{b}
Barley flour (malted)**	2.55 ± 0.10^{b}

*Mean±SEM of 6 observations per group. Values with the same superscript are not significantly different.
**From George Weston Foods.

guar gum lowers plasma cholesterol more than low viscosity guar gum in hypercholesterolaemic men.[8] This is consistent with an effect of viscous drag on transit and absorption. The viscosities of cereal β-glucans and "high viscosity" guar gum are similar and both lower cholesterol in rats fed hypercholesterolaemic diets.[5] However, the cholesterol loads in such diets are excessive compared to most human diets and in rats fed a low-cholesterol diet, plasma cholesterol is unaffected by NSP viscosity in the range up to and including that of guar gum.[9] Moreover, in rats fed malted barley (in which β-glucans are greatly modified) plasma cholesterol is the same as with the unmalted product. (Table 1).[10]

Malting has similar effects to treatment of cereals with β-glucanase which in chickens abolishes the hypocholesterolaemic effects of barley.[5] Thus either the same mechanism does not operate in different species or effects of NSP may be influenced by other factors such as diet. The latter seems to be a real possibility. In the rat, NSP modify lymphatic fat and/or cholesterol transport, depending both on the type of NSP and fat.[11] We have found that in rats fed fish oils with rice bran, plasma lipids were lower than with wheat bran and the down-regulation of the liver low density lipoprotein receptor seen with the latter bran was prevented.[12] Finally, a recent study has shown that oat bran does not enhance the cholesterol reduction of a "prudent" low-fat diet in normocholesterolaemic men.[13] Thus, the degree of cholesterol lowering could depend on the type of fat as well as the source of NSP and the cholesterol status of the subject.

3. VFA AND HEPATIC CHOLESTEROL SYNTHESIS

The fermentation of NSP by the hindgut microflora of omnivores yields VFA in significant quantities and it has been proposed that one acid (propionate) mediates the cholesterol reduction by such NSP.[7] Propionate at 1-5% of the diet lowers plasma cholesterol in animals. Propionate also inhibits hepatic cholesterogenesis in vitro but at concentrations 10-15 times higher than those found in the portal vein in vivo during NSP fermentation.[14] Moreover, propionate arising from NSP fermentation in the hindgut is absorbed with a totally different time course to that consumed in the diet. In the pig, portal venous concentrations rise and fall rapidly after propionate feeding with a peak concentration far below that needed to inhibit hepatic cholesterol synthesis. The rise in propionate formed by fermentation is delayed for several hours until food reaches the hindgut. In rats the effects of NSP on plasma cholesterol appear unrelated to propionate production with some preparations raising both caecal propionate levels and plasma cholesterol.[15] On balance, propionate does not seem to be the agency for cholesterol lowering by NSP.

4. NSP AND BILE ACID AND NEUTRAL STEROL EXCRETION

The third alternative mechanism for cholesterol reduction is binding of bile acids by NSP and is supported by observations in rats that faecal bile acid excretion is greater with oat bran than with cellulose.[16] Neutral sterol excretion is also increased. This could simply be another aspect of effects of NSP on small intestinal digestion but the changed molar ratios of the bile acids suggests more specificity. It is an attractive mechanism, resembling that of therapeutic bile acid sequestrants such as cholestyramine. NSP preparations that lower cholesterol generally do increase steroid excretion.[17] Such binding also could account for the effects of NSP on fat and cholesterol absorption and transport described by Ikeda et al, especially as chitosan (an anion-exchange resin) was most effective in reducing absorption.[9]

However, it is hard to discern how neutral NSP could effect such specific binding. Possibly, the NSP have domains where bile acid micelles might become sequestered but this mechanism cannot apply to acidic NSP (such as pectin) as both they and bile acids are ionized at the mildly alkaline pH values of the small intestine and therefore would be mutually repulsive. Additionally, we have found no relationship between steroid excretion and plasma cholesterol in rats fed different oat brans.

5. NON-FIBRE COMPONENTS AND PLASMA CHOLESTEROL: POSSIBLE INTERACTIONS BETWEEN NSP AND LIPIDS

Investigations into the hypolipidaemic effect of plant foods precedes the relatively recent interest in NSP. For example, Judd and Truswell concluded from their own work and earlier animal studies that cholesterol reduction by oats in man was due to both the lipids and NSP.[18] It is generally assumed that the important lipids are the unsaturated triacylglycerols present in oats at 8% (or more). More recently Qureshi et al have reported that minor components of barley, the tocotrienols, directly inhibit hepatic cholesterol synthesis in chickens.[19] We have found that delipidation of oat bran and relipidation with polyunsaturated oils abolishes the cholesterol lowering in rats fed a low cholesterol diet. These effects were unrelated to the tocotrienol content of the diet.

6. CONCLUSIONS

NSP isolates may lower cholesterol by their physicochemical properties such as viscosity in solution. For neutral NSP (e.g. cereal glucans) this could involve interaction in the gut either with dietary or biliary lipids. Interaction with bile acids seems unlikely for acidic NSP under physiological conditions. Cereals such as oats and barley, may lower plasma cholesterol through both their NSP and lipid components. VFA production by NSP fermentation in the large bowel does not seem to relate to cholesterol reduction in animals.

ACKNOWLEDGMENTS

The financial support of Tip Top Bakeries (a Division of George Weston Foods) and The Uncle Toby Company is gratefully acknowledged.

REFERENCES

1. B.O. Schneeman, Food Technol., 1986 40. 104.
2. L.V. van Horn, K. Liu, D. Parker, L. Emidy, L. Liao, W. H. Pan, D. Giumetti, J. Hewitt and J. Stamler, J. Am. Diet. Ass., 1986, 86, 759.
3. M. Kestin, R. Moss, P. M. Clifton and P. J. Nestel, Am. J. Clin Nutr., 1990, In the Press.
4. J. Whyte, P. M. Clifton and P. J. Nestel & D. L. Topping, Unpublished observations.
5. K. Newman, C. W. Newman & H. Graham, Cereal Foods World, 1989, 34, 883
6. L. A. Simons, S. Gayst, S. Balasubramaniam & J. Ruys, Atherosclerosis, 1982, 45, 101.
7 J. W. Anderson & W.-J.L. Chen in 'Oats Chemistry and Technology', F. H. Webster, editor, American Association of Cereal Chemists, St. Paul, 1986, Chapter 11, p.309.

8. H. R. Superko, W. L. Haskell, L. Sawrey-Kubicek and J. W. Farquhar, Am. J. Cardiol., 1988, 62, 51.
9. D. L. Topping, D. Oakenfull, R.P. Trimble & R. J. Illman, Br. J. Nutr., 1988, 59, 21.
10. K. Jackson, R. J. Illman, R.P. Trimble & D. L. Topping, Unpublished observations.
11. I. Ikeda, Y. Tomari and M. Sugano, J. Nutr., 1989, 119, 1383.
12. D. L. Topping, R. J. Illman, P. D. Roach, R. P. Trimble, A. Kambouris and P. J. Nestel, J. Nutr., 1990, Accepted for Publication.
13. J. F. Swain, I. L. Rouse, C.B. Curley and F. M. Sacks, New Engl. J. Med., 1990, 322, 147.
14. R. J. Illman, D. L. Topping, G. H. McIntosh, R. P. Trimble, G. B. Storer, M.N. Taylor and B.-Q. Cheng, Ann. Nutr. Metab., 1988, 32, 97.
15. B.-Q. Cheng, R.P. Trimble, R. J. Illman, B.A. Stone and D. L. Topping, Br. J. Nutr., 1987, 57, 69.
16. R. J. Illman and D. L. Topping, Nutr. Res., 1985, 5, 839.
17. M. Stasse-Wolthuis, World Rev. Nutr. Diet., 1981, 36, 100.
18. P. A. Judd and A. S. Truswell, Am. J. Clin. Nutr., 1981, 34, 2061.
19. A. A. Qureshi, W. C. Burger, D. M. Peterson and C. E. Elson, J. Biol. Chem., 1986, 261, 10544.

BENEFICIAL EFFECT OF WHEAT GERM ON PLASMA LIPIDS AND LIPOPROTEINS IN HYPERCHOLESTEROLEMIC SUBJECTS.

L. CARA, P. BOREL, M. ARMAND, H. PORTUGAL[1], C. LACOMBE[2], H. LAFONT and D. LAIRON.

Unité 130-INSERM (Institut National de la Santé et de la Recherche Médicale), 18 AV. Mozart, 13009 Marseille, France, and [1]Laboratoire Central, Hopital Ste-Marguerite, 13009 Marseille, and [2]Unité 317-INSERM, Université Paul Sabatier, 31400 TOULOUSE, France.

1 INTRODUCTION

According to several epidemiological studies, the concept that a high dietary fibre intake may protect against coronary heart disease becomes more and more plausible (1,2,3). It is generally assumed that hypercholesterolemia and especially elevated LDL and VLDL-cholesterol , hypertriglyceridemia and hypertension are major risk factors for cardiovascular diseases.

In addition to wheat bran, another by-product of wheat milling is wheat germ, which contains about 8-12 % total dietary fibres, most of them being hemi-celluloses and cellulose as in wheat bran. We have already shown in the adult rat (4) that the addition of 7 % wheat germ to a high fat-cholesterol diet significantly decreased the VLDL-cholesterol and the VLDL-triglycerides and increased the HDL-cholesterol after 7 weeks feeding.

Given the beneficial effects previously observed in the rat on the liver and blood parameters, the present study was performed in hypercholesterolemic subjects in view to experiment the effects of a 30 g daily intake of wheat germ on various parameters of lipid metabolism.

2 SUBJECTS AND METHODS

Subjects

Ten adult volunteers participated in the study. Free-living subjects (8 females and 2 males) were from 35 to 68 years old. Based on body mass index values, none was obese. None was a cigarette smoker. None had received hypolipidemic agents in the 3 months before the study.

All volunters had serum cholesterol concentrations exceeding normal value, from 6.5 to 9.46 mM cholesterol. Fasting serum triglycerides were variable, with normal values in 4 subjects and above normal values in 6 other ones (1.7 to 5.0 mM). Glycemia were in the normal range.

Experimental design and diets

The protocol consisted of 1 week basal diet, 4 weeks experimental diet (basal diet plus 30 g/d wheat germ) and a 4 weeks follow-up coming back to basal diet. The wheat germ used was Supergermes coming from Diepal, France and contained 27.5 % protein, 27.2 % carbohydrate, 11.0 % fat and 9.7 % total dietary fibre . Its pancreatic lipase inhibitory capacity was 497 inhibitory units per g, as measured in vitro (5).

The usual diet (basal diet) of each subject was monitored during the first week and for 3 days during the last week of the experimental period (basal diet plus wheat germ). During the control week, complex and refined carbohydrates, fat and proteins represented 41.0 %, 8.7 %, 35.9 % and 14.4 % of the total energy intake, respectively. The mean daily intake of dietary fibres was 13.6 g.

When comparing the basal diet plus wheat germ with the basal diet consumed during the week before supplementation, only the protein daily intake was significantly higher ($p \leq 0.05$). The mean daily intake of dietary fibres increased from 13.6 to 15.1 g.

Plasma Analysis

Blood samples were taken after an overnight fast at baseline (first week of basal diet), after the 4 experimental weeks and after the 4 week follow-up period. The lipoprotein classes (VLDL: $d \leq 1.006$; LDL: $1.006 \leq d \leq 1.060$; HDL: $1.060 \leq d \leq 1.21$) were separated from 1.5 ml plasma by ultracentrifugation (38,000 rpm at 15 ° for 24 h in a Beckman SW41 rotor) on a KBr discontinuous density gradient. Total and free cholesterol, triglycerides, phospholipids and glucose were measured by enzymatic procedures.

Student's t test for paired values was used to assess the statistical significance of the differences observed between the experimental periods at the probability level of 95 %.

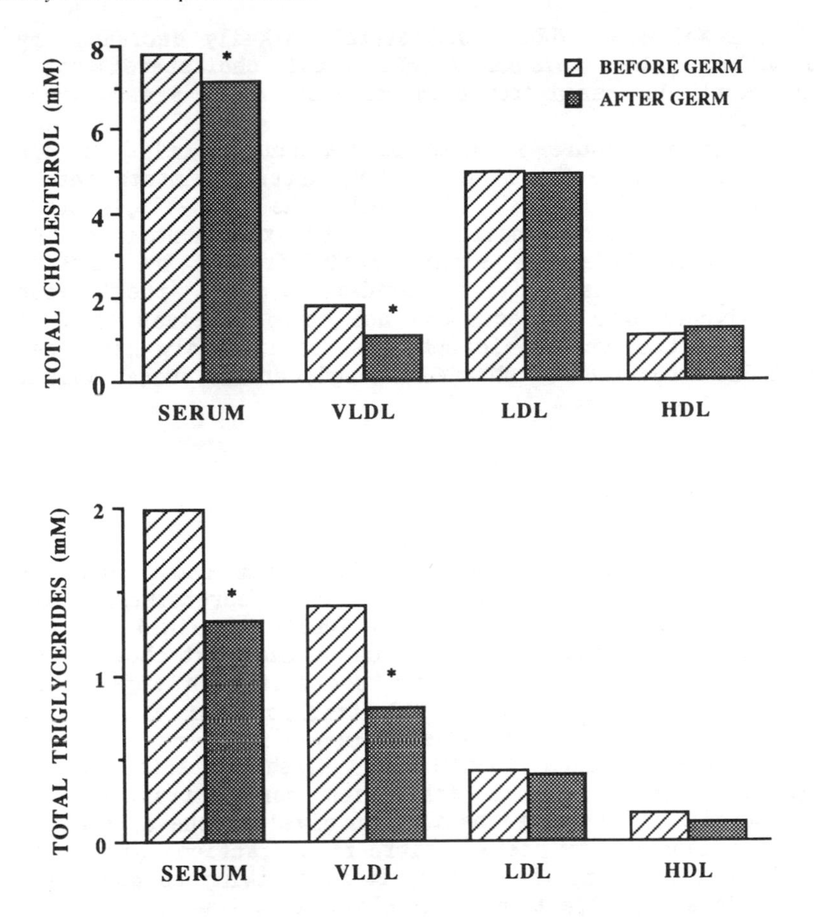

<u>Figure 1</u> Changes in plasma or lipoprotein cholesterol and triglycerides after supplementing diet with 30 g/d wheat germ for 4 weeks. Paired columns (mean values of 10 subjects) bearing an asterisk are significantly different ($p \leq 0.05$).

3 RESULTS

Normal glycemia did not change after wheat germ supplementation. The changes observed in plasma lipids and lipoproteins are shown in Figure 1. The intake of 30 g per day wheat germ significantly decreased plasma total cholesterol by 8.3 %. HDL cholesterol tended to increase moderately (+ 15.2 %) while LDL cholesterol did not noticeabily change

(- 1.2 %). The VLDL cholesterol markedly decreased by 40.6 %.Thus, the plasma / HDL total cholesterol ratio significantly decreased from 8.74 to 7.24 after 4 week wheat germ intake.

As shown in Figure 1, mean plasma triglycerides of ten subjects markedly decreased (- 33.7 %) after one month intake of wheat germ. The LDL and the HDL-triglyceride concentrations were not modified over the experimental period, while VLDL-triglycerides significantly dropped by 43.3 %. Indeed, the mean triglyceridemia of the four normotriglyceridemic subjects did not change whereas that of the six hypertriglyceridemic subjects significantly decreased from 2.68 to 1.64 mM. VLDL-triglycerides were mainly affected.

4 DISCUSSION

The data obtained during this short-term study done on subjects with type IIa and IIb hyperlipoproteinemia are strikingly comparable to those previously obtained in the adult rat (4). Indeed, supplementing a high fat-cholesterol diet with 7 % wheat germ for 7 weeks resulted in significant decreases in VLDL-cholesterol, VLDL-triglycerides and in the plasma / HDL total cholesterol ratio in the rat.

Numerous studies have been already performed to investigate the effects of fibre-rich cereal fractions on lipid metabolism but the mechanisms involved in the alteration of lipid metabolism by dietary fibre-rich fractions are still debated(6). Concerning wheat germ, it is striking to observe a parallel lowering effect on both elevated plasma cholesterol and triglycerides in the rat and in humans; this can account for the drop in the Very Low Density Lipoproteins. In fact, we have already demonstrated in the rat (5,7) that addition of 10 % wheat germ into fatty test-meals lowers fat lipolysis and thus the intestinal uptake and the output in the blood stream of dietary lipids and cholesterol.

The respective role of different wheat germ components may be briefly discussed. The addition of 2.9 g dietary fibres provided daily by wheat germ did not significantly increase the fibre intake and it seems unlikely that this low amount of wheat germ fibres, mainly composed of cellulose and hemi-celluloses, plays a key role in lipid metabolism. (8,9).

The protein intake significantly increased because the high protein content of wheat germ. Various vegetable proteins have plasma cholesterol lowering properties and proteins isolated from wheat germ have the ability to inhibit pancreatic lipase activity in vitro (8,10) and in vivo in the rat gut (5).

The total amount of phytosterols (98 mg/30 g) provided by wheat germ could also contribute to the lowering of plasma cholesterol. Phytosterols are well known to inhibit the intestinal absorption of cholesterol and indeed wheat germ was shown to decrease the intestinal uptake dietary cholesterol in the rat (7).

Since the accumulation of apo-B containing lipoproteins is well known to increase the risk for atherosclerothic lesions, wheat germ supplementation could represent a preventing factor.

Acknowledgements. We are grateful to Drs. Vigneron, Ode, Chanal, Bergier and Chardon for their medical help. This work was supported by grants from M.R.T. (N° 88-G-0894) and from I.N.R.A.

REFERENCES

1. D. Kromhout, E.B. Bosschieter and C. de Lezenne Coulander, Lancet, 1982, 2, 518.
2. M.J. Lichtenstein, M.L. Burr, A.M. Fehily, J.W. Yarnell, J. Epidemiol. Comm. Health., 1986, 40, 330.
3. K.T. Khaw and E. Barrett-Connor, Amer. J. Epidemiol. 1987, 126, 1093.
4. D. Lairon, C Lacombe, P. Borel, G. Corraze, M. Nibbelink, M. Chautan, F. Chanussot and H. Lafont, J. Nutr., 1987, 117, 838.
5. P. Borel, D. Lairon, M. Senft, M. Chautan and H. Lafont, Am. J. Clin. Nutr., 1989, 49, 1192.
6. J.W. Anderson and J. Tietyen-Clark, Am. J. Gastroenterol. 1986, 81, 907.
7. P. Borel, M. Armand, L. Cara, H. Lafont and D. Lairon. J. Nutr. Biochem., 1990, in press.
8. D. Lairon, P. Borel, E. Termine, R. Grataroli, C. Chabert and J.C. Hauton, Nutr. Rep. Int., 1985, 32, 1107.
9. P. Borel, D. Lairon, M. Senft and H. Lafont, Ann. Nutr. Metab., 1989, 33, 237.
10. P. Borel, D. Lairon, E. Termine, M. Martigne, H. Lafont, Plant Foods Hum. Nutr., 1989, 39, 339.

PRUNES AS A SOURCE OF DIETARY FIBER

Barbara O Schneeman*, Lesley Tinker*, Paul Davis* and Dan Gallaher+

*University of California
Davis, CA 95616

+University of Minnesota
St. Paul, MN 55108

Prunes are a partially dried fruit derived from a variety of plums. We were interested in studying prunes as a source of dietary fiber for several reasons. Analysis of the fiber content of prunes (32% moisture) indicated that they provide 6-7 g of total dietary fiber per 100 g, an amount that is typically higher than most fresh fruits. An analysis of the carbohydrate components of prunes conducted by John Labovitch, Ph.D. (Department of Pomology University of California, Davis) indicated that approximately 60% of the total dietary fiber in prunes is pectin. Consumption of pectin has been associated with a decrease in plasma cholesterol in human clinical trials and in experimental animal studies (1-7). The dose of pectin fed in the human studies ranges from 9-50 g per day with 15 g as a typical daily dose. The hypocholesterolemic effect of pectin has been associated with an increase in fecal bile acid excretion, although this response has not been observed in all studies. Stasse-Wolthuis et al. (7) have reported that increasing the amount of fruits and vegetables in experimental diets in order to provide an additional 8 g of pectin daily lowers plasma cholesterol in human volunteers. For modification of diets consumed by the population as a whole it is clearly important to test the efficacy of pectin-containing foods in lowering plasma cholesterol.

The objective of the study was to determine if adding 100 g of prunes to the diet daily would be associated with some of the potential benefits of dietary fiber, including a lowering of plasma cholesterol and an increase in fecal bulk.

Male subjects with plasma cholesterol levels of 5.7-6.0 mmol/L were recruited as volunteers. Each subject participated in the study for 10 weeks which included a 2-week training period, 4 weeks of consuming 100 g of prunes and 4 weeks of consuming 240 mL of grape juice. Grape juice (GJ) served as the control to provide a similar intake of energy from carbohydrate. A plasma sample and 3-day fecal collection were taken at the end of each block. Food records were kept by each subject throughout the experimental period.

The consumption of either prunes or grape juice, which supplied 240 kcal primarily from carbohydrate, was associated with a slight reduction in the percentage energy from fat and a slight increase in the percentage of energy from carbohydrate. Substitution of prunes for GJ lead to an average increase in fiber intake of about 6 g per day. Both plasma cholesterol and LDL-C tended to be higher in the GJ period than the baseline. However, consumption of prunes resulted in levels lower than the GJ control.

There was no difference in fecal wet or dry weight between baseline values and the GJ control. However, both wet and dry weights were significantly higher after the prune period than the GJ control. The increased fecal weight due to prunes indicated that subjects complied with the consumption of the dietary supplements.

In the feces total bile acid excretion did not differ among the experimental periods. However, the fecal concentration of bile acids was altered by supplementation with prunes. After the prune period the fecal concentration of secondary bile acids, and specifically the concentration of lithocholic and deoxycholic acid, was lower than after the GJ period. There were no differences in fecal bile acid concentrations between the GJ and baseline periods.

These results indicate that increasing fiber intake by about 6 g from a dried fruit was adequate to increase fecal weight and lower fecal bile acid concentration. Although the changes in plasma and LDL-cholesterol levels were not large, analysis of our results indicates that the change in fiber intake due to the prune supplement was associated with a change in plasma cholesterol and suggest that a larger increase in fiber intake may have a greater effect on plasma lipids. Given the overall design of our study, our results indicate that the addition of sources of fiber to the diet has benefits beyond that of simply lowering total fat intake and that prunes can contribute effectively to total dietary fiber intake.

Supported by the California Prune Board, NIH grant DK 20446 and the Clinical Nutrition Research Unit (DK 35747).

REFERENCES

1. P.N. Durrington, C.H. Bolton, A.P. Manning and M. Hartog, Lancet, 1976, Aug 21, 394.
2. D.J.A. Jenkins, A.R Leeds, C. Newton and J.H. Cummings, Lancet, 1975, 1116.
3. P.A. Judd and S. Truswell, Br. J. Nutr. 1982, 48, 451.
4. R.M. Kay and A.S. Truswell, Am. J. Clin. Nutr., 1977, 30, 171.
5. A. Keys, F. Grande and J.T. Anderson, Proc. Exp. Biol. Med., 1961, 106, 555.
6. T.A. Miettenen and S. Tarpila, Clin. Chim. Acta, 1977, 79, 471.
7. M. Stasse-Wolthuis, H.F.F. Albers, J.G.C. van Jeveren, J.W. de Jong, J.G.A.J. Hautvast, R.J.J. Hermus, M.B. Katan, W.G. Brydon and M.A. Eastwood, Am. J. Clin. Nutr., 1980, 33, 1745.
8. S. Pilch (ed.), Physiological Effects and Health Consequences of Dietary Fiber. Life Sciences Research Office, Fed. Am. Soc. Expt. Biol., Bethesda, MD, 1987.

EFFECTS OF RYE BRAN, OAT BRAN AND SOY BEAN FIBRE ON LIPID AND BILE METABOLISMS, AND GALLBLADDER MORPHOLOGY IN MALE SYRIAN HAMSTERS

J. X. Zhang, E. Lundin, C. O. Reuterving, G. Hallmans, R. Stenling, E. Westerlund* and P. Åman*

Departments of Pathology and Nutritional Research
University of Umeå
S-901 87 Umeå, Sweden
*Department of Chemistry
Swedish University of Agricultural Sciences
S-750 07 Uppsala, Sweden

1 INTRODUCTION

Food sources of fibre are complex, and their physiological responses are different. Oat bran which is rich in water soluble, viscous fibre has been shown to have hypocholesterolemic effects.[1,2] Soy bean fibre, alhough containing less water soluble fibre, has also cholesterol lowering effects.[3] Rye bran contains considerable amounts of water soluble, highly viscous fibre[4] but little is known about its physiological effects. Some isolated fibre components prevent cholelithiasis in animals by influence on the bile metabolism[5] but little information is available on the effects of natural fibre sources. A lithogenic diet causes changes in morphology of the gallbladder in mice,[6] but whether dietary fibre has an effect on the morphology of gallbladder is still obscure.

The purpose of this study was to investigate the effects of rye bran, oat bran and soy bean fibre on serum cholesterol level, bile composition, gallstone formation, and the morphology of the gallbladder in hamsters.

2 MATERIALS AND METHODS

Male Syrian hamsters (age 4-6 wk) were randomized into 7 dietary groups, 8-14 animals in each group. Two basal fibre free diets were used in this experiment: The O1 diet is a stone provoking diet based on glucose and casein,[7] and the non-stone provoking O2 diet is based on

Table 1 Diet composition and nutrient content (%,w/w)[a]

	O1	O-O1	S-O1	R-O1	O2	S-O2	R-O2
Glucose	72.7	31.5	61.9	54.4	0	0	0
Casein	20.2	8.6	17.1	15.1	0	0	0
Wheat starch	0	0	0	0	38.2	32.6	26.2
Milk protein	0	0	0	0	37.6	29.0	23.9
Corn oil	0	0	0	0	15.9	13.1	11.9
Vitamins[b]	1.0	0.4	0.9	0.7	1.2	1.0	1.0
Minerals[b]	5.1	2.2	4.3	3.7	6.1	5.1	4.9
Gelatine	1.0	1.0	1.0	1.0	1.0	1.0	1.0
Oat bran	0	56.3	0	0	0	0	0
Soy fibre	0	0	14.8	0	0	18.2	0
Rye bran	0	0	0	25.1	0	0	31.1
Protein	18.0	17.9	16.9	17.5	26.1	22.1	21.6
Fat	0.2	4.6	0.2	1.0	16.4	13.6	13.3
Carbohydrates	73.3	56.9	63.8	62.0	40.4	34.1	33.1
Dietary fibre[c]							
soluble	0	3.4	1.3	2.5	0	1.6	3.0
insoluble	0	8.6	10.7	9.5	0	13.2	11.9

[a]fresh wt; [b]reference (7); [c]excluded from carbohydrates.

starch, milk protein and corn oil. The diets of O-O1, S-O1 and R-O1 were prepared by supplementation of the O1 diet with oat bran, soy bean fibre and rye bran. The S-O2 and R-O2 diets were the O2 diet supplemented with soy bean fibre and rye bran. Diet composition and nutrient content are listed in Table 1. The animals were fed respective diets for 6 weeks before sacrifice. The methods for sampling and analysis of bile and serum, and the evaluation of the gallstone formation were essentially performed as previously described.[8] The bile phospholipid was determined according to the method of Qureshi et al.[9] The lithogenic index was calculated according to Thomas et al.[10] The morphological analyses were carried out on 5 μm paraffin sections of the gallbladder wall stained with H & E or Van Gieson's method. The point counting method was used for the stereological volumetric analysis.[11]

For statistical evaluation the X^2-test was used for the frequency of gallstones and Student's t-test for determining the significance of difference of mean value. The results in tables 2-5 are reported as the mean ± S.E. The symbols *, ** and *** denote the following significance levels as compared with the fibre free control diets: $P<0.05$, $P<0.01$ and $P<0.001$.

3 RESULTS AND DISCUSSION

Oat bran and soy bean fibre have repeatedly been shown to exert hypocholesterolemic effects.[1,2,12,13] Results in this experiment are also consistent with these former findings (Table 2). The cholesterol lowering effect was also observed in the animals fed O1 and O2 diets supplemented with rye bran. The water soluble fraction of fibre in the oat is supposed to be responsible for the hypocholesterolemic effect. Although diets supplemented with rye bran and soy bean fibre contain less water soluble fibre than the oat diet it may still play a role in lowering serum cholesterol. In rye, the dominant part of dietary fibre is pentosans. The water soluble part of pentosans is known to have high viscosity.[4] Whether this plays a role in lowering serum cholesterol should be considered.

Some isolated dietary fibre components prevent cholesterol gallstone formation in animal experiments.[5,14,15] Although the mechanism is still obscure changes in bile metabolism and a lowered lithogenic index are supposed to be the main reasons. In this experiment, rye bran supplementation reduced the incidence of gallstone formation significantly (Table 3) as compared to the control, O1 diet group. The concentration of bile acids was significantly higher in groups fed the rye bran diets. (Table 4), but the

Table 2 Effect of dietary fibre supplementation on serum cholesterol level (mmol/L)

Dietary group	Serum cholesterol	Dietary group	Serum cholesterol
O1(14)	4.19±0.24	O2(9)	4.03±0.15
O-O1(10)	2.91±0.16***	S-O2(8)	3.10±0.29***
S-O1(10)	3.28±0.20*	R-O2(8)	2.90±0.15***
R-O1(10)	2.73±0.16***		

Table 3 Effect of dietary fibre supplementation on gallstone formation

Dietary group	Number of animals with gallstones	Dietary group	Numberof animals with gallstones
O1(14)	8	O2(10)	1
O-O1(10)	3	S-O2(8)	1
S-O2(10)	4	R-O2(8)	0
R-O2(10)	1*		

Table 4 Effects of dietary fibre supplementation on bile composition (mmol/L) and lithogenic index

Dietary group	Bile acids	Phospho-lipids	Chole-sterol	Lithogenic index
O1(14)	10.6±0.4	5.6±0.4	1.5±0.2	1.1±0.1
O-O1(10)	11.6±0.4	5.5±0.4	1.0±0.2*	0.5±0.1*
S-O1(10)	9.8±0.7	5.3±0.4	1.3±0.1	0.9±0.1
R-O1(10)	14.5±1.1*	6.0±0.5	1.7±0.1	0.8±0.1
O2(9)	11.8±1.2	9.6±1.0	1.2±0.1	0.7±0.1
S-O2(8)	15.8±2.1	8.3±2.4	1.1±0.1	0.5±0.1
R-O2(8)	17.0±1.1**	9.3±2.1	0.8±0.1	0.3±0.0***

lithogenic index was lower only for the hamsters fed the O2 diet supplemented with rye bran as well as in those fed the O1 diet supplemented with oat bran. Most gallstones found in this experiment were of the cholesterol and pigment mixed stones. Their frequency did not correlate well to the lithogenic index. This is inconsistent with previous findings where the stone provoking diet produced only cholesterol gallstones. The reason for this may be the varying response in different animal strains or some other unknown factors.

When mice were fed a lithogenic diet, an increase in volume density of gallbladder epithelium was observed (6). In this experiment the O1 diet seemed to have a similar effect on hamsters. Whether this change is associated with the formation of gallstones is not clearly known. Rye and oat bran supplementation to the lithogenic, O1 diet decreased the volume density of the epithelium (Table 5). In addition, supplementation with rye bran, oat bran and soy bean fibre to the O1 diet

Table 5 Effects of dietary fibre supplementation on volume densities of epithelium and smooth muscle of the gallbladder (volume percentage to gallbladder wall)

Dietary group	Epithelium	Smooth muscle
O1(13)	36.0±1.1	33.4±1.2
O-O1(10)	32.4±1.0*	40.8±2.4**
S-O1(9)	34.0±1.1	41.9±2.3**
R-O1(10)	32.8±0.8*	38.0±1.2*
O2(8)	33.5±1.7	32.7±1.8
S-O2(8)	33.8±1.0	38.7±2.3
R-O2(8)	29.8±1.2	34.2±1.9

increased the volume density of the smooth muscle layer. The mechanism for this is unclear and should be studied further.

In conclusion, rye bran as well as oat bran and soy bean fibre were found to have serum cholesterol lowering effects in hamsters. Rye bran increased the concentration of bile acids in the bile and prevented gallstone formation in the hamsters. These fibre sources also exert effects on the morphology of the gallbladder wall.

REFERENCES

1. W.J.L. Chen, J.W. Anderson and M.R. Gould, Nutr.Rep.Int., 1981, 24, 1093.
2. L. Van Horn, K. Liu, D. Parker, L. Emidy, Y. Liao, W.H. Pan, D. Giumetti, J. Hewitt and J. Stamler, J.Am.Diet.Assoc., 1986, 86, 759.
3. G.S. Lo, A.P. Goldberg, A. Lim, J.J. Grundhauser, C. Anderson and G. Schonfeld, Atherosclerosis, 1986, 62, 239.
4. D. Pettersson and P. Åman, Acta Agric.Scand., 1987, 37, 20.
5. D. Kritchevsky, S.A. Tepper and D.M. Klurfeld, Experientia, 1984, 40, 350.
6. T. Wahlin, Virchows Arch.[Cell Pathol.], 1976, 22, 273.
7. H. Dam and F. Christensen, Acta Pathol.Microbiol. Scand., 1952, 30, 236.
8. J.X. Zhang, F. Bergman, G. Hallmans, G. Johansson, E. Lundin, R. Stenling, O. Theander and E. Westerlund, Acta.Pathol.Microbiol.Immunol.Scand., (in press)
9. M.Y. Qureshi, G.M. Murphy and R.H. Dowling, Clin. Chim.Acta, 1980, 105, 407.
10. P.J. Thomas and A.F. Hofmann, Gastroenterol., 1973, 65, 698.
11. E.R. Weibel, W. Stäubli, H.R. Gnägi, F.A. Hess, J.Cell Biol., 1969, 42, 68.
12. J.W. Anderson, L. Story, B. Sieling, W.J.L. Chen, M.S. Petro and J. Story, Am.J.Clin.Nutr., 1984, 40, 1146.
13. J. Sasaki, M. Funakoshi, K. Arakawa, Ann.Nutr. Metab., 1985, 29, 274.
14. F. Bergman and W. Van der Linden, Z.Ernährungswiss, 1975, 14, 218.
15. O.D. Rotstein, R.M. Kay, M. Wayman and S.M. Strasberg, Gasteroenterol., 1981, 81, 1098.

METABOLIC EFFECTS OF RAW AND PROCESSED CARROTS IN HUMANS

E. Wisker[1], W. Feldheim[1] and T.F. Schweizer[2]

[1] Institute of Human Nutrition, University of Kiel, D-2300 Kiel (Federal Republic of Germany)

[2] Nestlé Research Centre, Nestec Ltd., Vers-chez-les-Blanc, CH-1000 Lausanne 26 (Switzerland)

INTRODUCTION

It is often stated that the processing of dietary fibre sources could induce physical and chemical changes in the dietary fibres and alter their physiological effects. However, very few human studies have actually addressed processing effects. With their comparatively thin and little lignified cell walls vegetables appear to be attractive study materials for this purpose. In addition, many vegetables can be easily consumed both as raw or processed foods. Yet, vegetables received little attention in comparison with cereal fibres or purified fibre sources.

Carrots were chosen for this study which followed a similar protocol as previously applied to assess the faecal bulking capacity and other properties of a number of fibre sources[1-4]. Two hundred grams of raw carrots per day have been reported to lower serum cholesterol by 11% after three weeks and to increase faecal weight, faecal bile acids and faecal fat[5], but fiber-mediated effects of processed carrots have not been investigated.

EXPERIMENTAL

Twelve healthy females (age 21-27) volunteered for this study which comprised four experimental periods each of three weeks in a latin square design. These periods were separated from each other by at least three weeks in order to avoid carry-over effects. During all periods the subjects consumed the same, strictly controlled, basic diet which provided about 15 g dietary fibre per day. During three periods this diet was supplemented with either raw, blanched or canned carrot cubes of equal size, in amounts providing additional 15 g dietary fibre. Blood samples were drawn at the beginning and end of each period, all stools and urines were collected during the third week of the periods.

RESULTS AND DISCUSSION

Raw, blanched and canned carrots increased mean stool weights similarly, from 93 g/day during the basal periods to 129-151 g/day during carrot periods. Dry weights increased by about 30%, faecal water by 43-72%. The faecal bulking of carrots was comparable to fine wheat bran[2].

Faecal fibre increased from an average 6 g/day (basal diet) to 8 g (carrot periods). Thus apparent carrot fibre digestibilities were around 85%, with no difference between the different carrots. Such high carrot fibre fermentabilities have recently also been found in rats[6,7].

Serum cholesterol decreased during all carrot periods, by 14, 19 and 11% for raw, blanched and canned carrots, respectively. There were no significant differences between the different carrots, but a 12% decrease was also observed after the basal diet alone, making a mainly carrot-mediated cholesterol-lowering effect unlikely.

The mean daily faecal bile acid excretion and the extent of conversion of primary to secondary bile acids were similar in all periods. Faecal bile acids were more diluted after carrots than after the basal diet, but the interindividual variation was considerable.

CONCLUSIONS

a) Dietary fibres in carrots are highly fermentable and yet have good stool bulking ability in man.
b) In spite of appreciable effects of processing (blanching and canning) on carrot texture and microscopic structure (C. Schlienger & C. Probst, personal communication), the physiological effects of raw and processed carrots were very similar.
c) There appears to be good agreement between the faecal parameters measured in this human study and the results obtained with similarly processed carrots in rats[6,7].

This finding further strengthens the rat model for the predictive screening of dietary fibre sources.

REFERENCES

1. E. Wisker, A. Maltz and W. Feldheim, J.Nutr., 1988, 118, 945.
2. E. Wisker, U. Krumm and W. Feldheim, Akt.Ernähr., 1986, 11, 208.
3. E. Wisker and W. Feldheim, Akt.Ernähr., 1983, 8, 200.
4. T.F. Schweizer, A. Bekhechi, B. Koellreutter, S. Reimann, D. Pometta and B.A. Bron, Am.J.Clin.Nutr., 1983, 38, 1.
5. J. Robertson, W.G. Brydon, K. Tadesse, P. Wenham, A. Walls and M.A. Eastwood, Am.J.Clin.Nutr., 1979, 32, 1889.
6. M. Nyman, T.F. Schweizer, S. Tyrèn, S. Reimann and N.-G. Asp, J.Nutr., 1990, in press.
7. M. Nyman, T.F. Schweizer, K.-E. Pålsson and N.-G. Asp (submitted).

WORKSHOP REPORT: DIETARY FIBRE AND STEROL METABOLISM - WHAT SHOULD WE MEASURE APART FROM PLASMA TOTAL CHOLESTEROL?

C.S. Berry

Flour Milling and Baking Research Association
Chorleywood
Rickmansworth
Herts WD3 5SM

Suppose that a particular fibre has been shown to lower plasma total cholesterol (TC). What additional data should be gathered that is of potential benefit to the consumer at large, the physiologist or even the food manufacturer?

In his opening remarks the Chairman (Dr. C. Berry, Chorleywood) noted that lowering of TC per se cannot be assumed to confer reduced risk of CHD if we are ignorant of the fate of that cholesterol. Dr. David Kritchevsky supported this point by citing instances of 2 cholesterol-lowering strategies - colestipol/nicotinic acid and partial ileal bypass, which increased rather than reversed atherosclerosis in a minor but nevertheless substantial proportion of patients (e.g. ~14%). The Chairman suggested that monitoring of faecal total sterols (acidic and neutral) remained an important non-invasive method for confirming that surplus plasma cholesterol was indeed lost from the body, but that it should be done in a way that recognised the existence of steady-state constraints. Thus the decrement in plasma cholesterol may appear in stool as a short-lived pulse in faecal sterols; the latter may then settle back to their original value even though the diet remained supplemented with fibre. Failure to detect increased faecal sterol output at a single fixed arbitrary sampling time was not inconsistent with any of the suggested mechanisms of cholesterol-lowering, including sequestration by fibre of sterols in the gut. Additionally, careful measurement of the final steady state value of faecal sterol excretion, relative to its initial value before feeding fibre, may give important clues to effects of the fibre and associated substances on rates of de novo cholesterol synthesis. The latter may be increased, for example, by homeostatic mechanisms (relief of feedback inhibition etc.).

If this is shown to occur, then a search might usefully be made for synergistic agents in diet that, when co-ingested with fibre, might prevent or attenuate the rebound effect.

Elizabeth Lund presented a survey of literature data showing that the hypocholesterolaemic effect of oat bran diminished with initial plasma TC concentration. Dr Kritchevsky and others considered that more weight should be given to fibres that lowered TC that was not initially greatly elevated.

Dr Gur Ranhotra (American Institute of Baking) reported that he had lowered plasma TC and apolipoprotein B in human subjects by increasing soluble dietary fibre supplied primarily in the form of refined cereal products. The precise mechanism of the effect is unknown.

Dr John Mathers (Newcastle) reported that including peas (a rich source of fermentable fibre and slowly digestible starch) in the diet of rats increased significantly the activity of hepatic HMG Co A reductase. Similarly, feeding guar gum increased *in vivo* hepatic cholesterol synthesis measured by the tritiated water incorporation method. However orally administered SCFA given in the amounts expected to be produced from the guar gum fermentation had no effect on liver cholesterol synthesis. Thus the data suggest that the rat liver readily compensates with increased synthesis when sterol excretion is raised and that propionate is unlikely to be responsible for cholesterol lowering.

Dr Linda Morgan (Surrey) reported studies in normal and diabetic subjects in which effects of guar on apolipoproteins were monitored. She recommends measurement of the ratio of apolipoproteins A_1/B as being of superior predictive value for CHD than total cholesterol, or ratios of its complexes with lipoproteins.

Dr David Topping (CSIRO Australia) presented data showing synergism between fish oils and dietary fibres (wheat and oat brans) on plasma TC. Whilst omega-3 fatty acids on their own are not hypocholesterolaemic in man, they are potentially valuable as synergists by their ability to prevent the down-regulation of the LDL-receptor. Effects of this nature will no doubt be of great potential for both food manufacturers and the general public.

Dr Zhang (Umea, Sweden) had been observing hypocholesterolaemic effects of oat bran and also, interestingly, a source of insoluble fibre (spent brewers grains) in ileostomy subjects, which reinforces data from Dr Mather and others that SCFAs are not obligatory for cholesterol-lowering.

In the short time available at the end for discussion, it was agreed that fibre effects on sterol metabolism were so diverse and complex that it was not sensible at the present time to recommend rigid test protocols, even for routine documentation. This area must still be approached with an open mind.

Dietary Fibre and the Food and Pharmaceutical Industries

RAW MATERIALS NATURALLY RICH IN FIBRE VERSUS FIBRE CONCENTRATES/ISOLATES IN FOOD PRODUCTION

N.-G. Asp

Department of Applied Nutrition and Food Chemistry
University of Lund, Chemical Centre P.O. Box 124
S-221 00 Lund, Sweden

1 INTRODUCTION

The tremendous interest in dietary fibre, both among nutritionists and in the general public, has stimulated food producers to upgrade fibre rich waste products - hitherto used mainly as cattle feed - to food ingredients. The recommendations on dietary fibre intake issued recently in several countries - 3 g/MJ in the Scandinavian countries (1) - has reinforced the possibility to enrich foods with high fibre ingredients, in order to help in reaching these recommendations.

The use of gums and other soluble, gel-forming polysaccharides as consistency aids in foods preceeds the dietary fibre era, but it is now realized that their use may contribute in providing nutritionally beneficial properties attributed to this type of dietary fibre.

This paper addresses the question to what an extent physiological properties of foods naturally rich in fibre can be restored by addition of fibre concentrates or isolates.

2 CLASSIFICATION AND NUTRITIONAL PROPERTIES OF FIBRE PREPARATIONS

A large number of fibre concentrates or isolates are now available. A recent overview (2) listed about 20 different sources of fibre, many of them available from several producers. Another special issue (3) listed 128 different food fibre preparations available in the United States in 1987.

Table 1 attempts to classify various fibre sources.

As for protein preparations, it might be useful to distinguish between concentrates that contain appreciable amounts of non-fibre material, and isolates of more or less pure fibre polysaccharides.

A number of raw materials with very high fibre content are available, that have not earlier been used in foods. These include oat hulls, straw, and cellulose. The dietary fibre in these materials is almost completely insoluble and difficult to ferment in the intestine, resulting in a good fecal bulking capacity (4). Their content of available nutrients is negligible.

Cereal bran from wheat, rye and oats can be regarded nutritionally as flours, that are enriched in dietary fibre and nutrients, roughly parallelly. However, the phytic acid, located mainly in the aleuron cells that contain high concentrations of nutrients as well, is also enriched.

Table 1 Classification of fibre concentrates and isolates

Concentrates

1. Husk, straw etc
2. Bran - wheat, oats, rye, barley, maize, rice, soy, pea
3. Cotyledon - soy, pea
4. Tuber pulp - beet, carrot, potato
5. Fruit pulp - orange, apple etc.

Isolates

1. Purified, natural polysaccharides
   - pectins, gums, alginates, carragheenan etc

2. Processed, natural polysaccharides - microcrystalline cellulose
   - carboxymethyl cellulose etc
   - modified food starches
   - resistant starch (retrograded amylose)

3. Related undigestible poly- or oligosaccharides
   - polydextrose
   - inulin, fructo-oligosaccharides
   - soy α-galactosides

Brans are used either raw or processed. In the case of oat bran, heat processing to inactivate lipases is necessary for an acceptable shelf life. Other enzymes, including the phytase naturally occurring in oats as well as in other cereals (5), are also inactivated. Therefore, enzymic degradation of phytate is less likely during processing of oat bran containing foods. This problem needs further consideration in view of the present emphasis on the consumption of large amounts of oat bran. The bioavailability of zink in oat porridge has been demonstrated to be very low (6).

As illustrated in Fig. 1, rye and barley contain as much soluble fibre as oats. Rye bread gave lower post-prandial glycemic response than the corresponding wheat bread (7), an effect most probably due to the soluble fibre content of rye. A further glucose lowering effect is obtained by baking with whole grains of rye (8).

Fibre preparations from leguminous seeds are available both from the outer layers (bran) and from the cotyledon, with distinctly different composition and properties. The bran preparations contain mainly insoluble fibre, whereas the cotyledon fibre preparations are rich in pectins and hemicelluloses, partly soluble. The considerable amounts of protein in these preparations must be considered when evaluating their physiological effects, for instance on serum cholesterol levels.

Fibre preparations derived from the pulp of sugar beets, carrots or potatoes have a relatively high soluble fibre content. Therefore, they have been regarded as promising concerning effects on serum cholesterol levels and postprandial glucose response.

The fibre isolates can be divided into 1) Purified, natural polysaccharides traditionally used as food additives - pectins, gums, alginates, carragheenan etc and 2) Modified materials, e.g. microcrystalline cellulose and carboxymethyl cellulose. Chemically modified food starches and resistant starch (retrograded amylose) should also be regarded in this context. The physiological properties vary widely due to the different physico-chemical properties. A third group, not analysed as dietary fibre with any current method, include polydextrose, inulin and fructo-oligosaccharides, and soy a-galactosides. These are not absorbed in the small intestine and can therefore be expected to have physiological properties generally attributed to dietary fibre. The reason why they are not analysed as dietary fibre is their solubility in 78 - 80 %

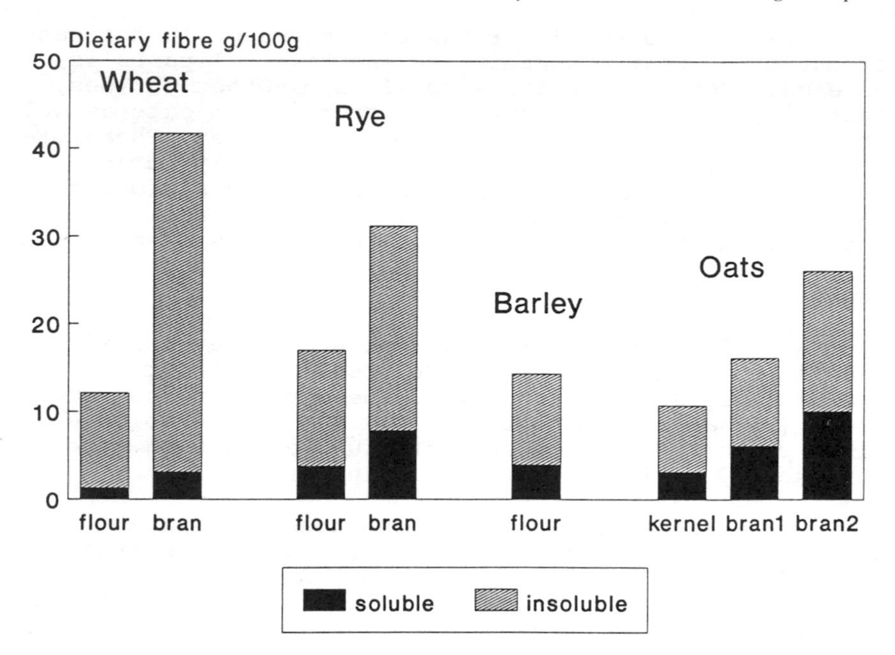

Figure 1 Soluble and insoluble dietary fibre in cereals analysed according to Asp et al. (9). Data from 4,10-12. The extraction rate of the barley flour corresponds to about 90 %.

ethanol. Polydextrose, and also chemically modified food starches (13) are alcohol soluble after degradation to low molecular weight fragments by amylase used in dietary fibre analysis.

3 USE OF FIBRE CONCENTRATES AND ISOLATES

Dietary fibre concentrates or isolates can be used in different ways (Table 2): 1) As functional ingredients at a low concentration (usually 0.5 - 2 %) or 2) To increase the dietary fibre content. In the first instance, physiological effects of the fibre per se can hardly be expected. However, the fibre addition can have important indirect effects, enabling the formulation of foods with reduced fat and energy content.

Only a few types of foods are suitable for incorporation of large amounts of fibre: Bread, breakfast

Table 2 Fibre concentrates/isolates can be used in different ways

	Functional ingredient	Dietary fibre enrichment
Typical food	Meat prod Jam Ketchup Bread	Bread Breakfast cereals Bars
Typical amount	0.5 - 2 %	5 - 20 %
Dietary fibre/ serving	< 1 - 2 g	2 - 10 g

cereals, other cereal products, bars etc. Only foods containing appreciable amounts of fibre should be marketed with claims on fibre content. The Association of Swedish Food Industries recommends (14) that only foods containing at least 10 g dietary fibre/1000 kcal or 2 g/serving should be claimed as "containing fibre". The corresponding limits for claiming "rich in fibre" are 15 g/1000 kcal or 3 g/serving.

4 PHYSIOLOGICAL EFFECTS OF ADDED FIBRE

The physiological effects of dietary fibre in foods are related partly to properties of the fibre polysaccharides themselves, and partly to structural features of plant cell walls, attained by the fibre. Addition of fibre preparations can restore some, but not all of these effects. Recent studies with fibre preparations, especially sugar beet fibre (Fibrex[R]), will be reviewed to illustrate this point.

Glycemic response

A number of food properties - in addition to the content of soluble, viscous types of dietary fibre - influence the glycemic reponse (15,16), as shown in Table 3. Gross and cellular structure, as well as degree of starch gelatinisation (17) seem especially important.

Incorporation of 12 g dietary fibre (about 3 g soluble fibre) from Fibrex into a breakfast meal decreased postprandial glucose response in type 2 diabetics significantly (18). In healthy controls, however, only the insulin response was decreased (19). In spite of a sizable increase in soluble and total fibre, these effects were

Table 3 Food properties influencing the rate and extent of carbohydrate digestion and absorption

- Gross and cellular structure
- Dietary fibre
- Starch gelatinisation
- Amylose/amylopectin ratio
- Starch/lipid complexes
- Starch/protein interactions
- Starch retrogradation
- Amylase inhibitors

rather small compared to the large variation in glycemic indices reported for different foods (16). A similar experiment with a potato fibre preparation containing comparable amounts of total and soluble fibre gave no significant effect on the glycemic response in type 2 diabetics (Hagander et al., to be published).

An 8+8 weeks cross-over study in well controlled type 2 diabetics with 30 g sugar beet fibre incorporated into various foods in a conventional diabetic diet (percent of energy from protein 20, fat 35, and carbohydrates 45; dietary fibre in basic diet g/day) failed to show any significant improvement of the glycemic control. The blood pressure, however, was lowered significantly (19).

Two studies have been reported in which guar bread was given to type 2 diabetics providing 5 or 6 g guar gum/day with improvements of serum cholesterol and glycosylated hemoglobin values (20,21). The palatability of guar bread can be questioned, however.

In summary, incorporation of suitable dietary fibre preparations into bread and other cereals may help to reduce the glycemic response with possible favorable long term effects. However, other factors such as the form of the food, its cellular structure and the degree of starch gelatinisation seem to be more important in this respect.

Serum cholesterol level

It is well documented that soluble, gel-forming types of dietary fibre lowers the serum cholestrol, and that the decrease is due to a fall in LDL-cholesterol. Pectin and guar gum have been most extensively studied in doses generally around 10-15 g/day, i.e. considerably more than present in a normal diet. In studies with guar gum, 5 g/day is the lowest dose reported to lower the serum cholesterol significantly (for review see 22).

Sugar beet fibre (22-26 g/day) has shown significant lowering effect on total and LDL cholesterol in two different studies (23,24), whereas the 8+8 weeks study on diabetics (19) did not show any significant influence on plasma lipids.

The cholesterol lowering effect of oat bran has been much emphasized recently and attributed to its ß-glucan content. Several studies have demonstrated a 3 - 5 % cholesterol lowering effect of oat bran in addition to that obtained by a fat modified diet acording to current recommendations. The recent study by Swain et al (25) is compatible with ealier studies in that it does not exclude a 4 % cholesterol lowering effect of the oat bran fibre, although the fat displacement effect of adding a large amount of any cereal is emphasized.

The contributions to the serum cholesterol lowering effect of various dietary changes in a typical Western diet according to current dietary guidelines suggested by Kay and Trushwell (26) still seem valid. These authors expected about half of the effect to be due to fat modifications, whereas decreased dietary cholesterol, increased dietary fibre, and increased ratio vegetable/ animal protein was judged to be of roughly equal importance in explaining the rest of the change. Thus, a lowering of the fat content by incorporation of dietary fibre may be more important in influencing plasma lipids than specific effects of the fibre in itself.

Fecal bulking

The fecal bulking capacity of dietary fibre is inversely related to the fermentation. For example, oat hulls are very difficult to ferment and have a high bulking capacity, whereas oat bran is more extensively fermented with a lower bulking capacity (5). There is no evidence that added fibre preparations have different bulking effect than corresponding naturally occurring fibre, although particle size and heat treatment have been reported to influence the bulking effect of wheat bran.

Sugar beet fibre increased fecal dry matter in rats by 0.6 g/g fibre (27) - a figure similar to that obtained for both raw and variously processed vegetables (28). Human studies also show similar fecal bulking capacity of fibre from sugar beet and carrots (W. Feldheim, personal communication; E. Wisker, W. Feldheim, T.F. Schweizer: Metabolic effects of raw and processed carrots in humans, this symposium).

Table 4 Relationship between properties of dietary fibre and physiological effects

Main effects	Solubility	Viscosity	Monomeric comp.	Lignification	Functional groups/associated compounds
Fecal bulking	- - -			+ + +	
Fermentation	+ + +			- - -	
Glycemic response	+	+ + +	+		+
Lipoprotein levels	+ +	+			+
Mineral binding					+ + +

5 PREDICTION OF PHYSIOLOGICAL EFFECTS BY IN VITRO ASSAY

In view of the highly variable physiological effects of different dietary fibre preparations, methods capable of predicting such effects are desirable. Table 4 summarizes properties of dietary fibre related to physiological effects.

Fecal bulking and fermentation are properties highly related to solubility and lignification, and glycemic response is influenced by viscous fibre. Solubility is also related to effects on plasma lipids. Monomeric composition has some relationship to effects on glycemic response and plasma lipids, in that pectins with galacturonic acid as main monomeric constituent is one class of dietary fibre polysaccharides with such effects. In general, however, the monomeric composition is a poor predictor of physiological effects. There is a great need for further method development for prediction of physiological response.

6 CONCLUSIONS

Dietary fibre concentrates or isolates can be used as functional components in small amounts, or as dietary fibre enrichment/fortification. Structural effects of dietary fibre may be nutritionally important and cannot be restored by addition of fibre concentrates or isolates. Indirect effects due to lowered fat and energy content may be more important nutritionally than physiological effects of the fibre per se. Cereal brans can be regarded nutritionally as enriched flours in that nutrients are concentrated roughly proportionally to the fibre. Even detailed chemical analysis of fibre polysaccharides does not predict physiological function. So far, physiological effects have to be documented in vivo.

REFERENCES

1. Anon, Nordiska Näringsrekommendationer (Nordic Nutrition Recommendations) PNUN-rapport 1989, 2.
2. C. Andres, Food Processing, August 1989.
3. Anon, Dietary Fiber Guide. Cereal Foods World, August 1987.
4. M. Nyman and N-G. Asp, Am.J.Clin.Nutr., 1988, 48, 274.
5. W. Frölich, M. Wahlgren and T. Drakenberg, J. Cereal Sci., 1988, 8, 47.
6. B. Sandström, et al., J. Nutr., 1987, 117, 1898.
7. B. Hagander, I. Björck, N-G. Asp, S. Efendic, J. Holm, P. Nilsson-Ehle, I. Lundquist and B. Schersten, Diabetes Research & Clinical Practice, 1987, 3, 85.
8. D. Jenkins, T. Wolever, A. Jenkins, C. Giordano, S. Giudici, L. Thompson, J. Kalmusky, R. Josse and G. Wong, Am.J.Clin.Nutr., 1986, 43, 516.
9. N-G. Asp, C-G. Johansson, H. Hallmer and M. Siljeström, J.Agric.Food Chem., 1983, 31, 476.
10. C-G. Johansson, P. Åman, N-G. Asp and O. Theander, Swedish J.Agric.Res., 1982, 12, 157.
11. M. Nyman, M. Siljeström, B. Pedersen, K.E. Bach-Knudsen, N-G. Asp, C-G. Johansson and B.O. Eggum, Cereal Chem., 1984, 61, 14.
12. N-G. Asp and C-G. Johansson, In: W.P.T. James, O. Theander (eds.): The analysis of dietary fiber in food. Marcel Dekker Inc., New York, 1981, 173.
13. K. Östergård, I. Björck and A. Gunnarsson, Stärke, 1988, 40, 58.
14. Anon, Märkning av färdigförpackade livsmedel. SLIM, Box 5501, Stockholm, Sweden.

15. J. Holm, N-G. Asp and I. Björck, In: I.D. Morton (ed.): Cereals in European Context, Ellis Horwood, Chichester, G.B., 1987, 169.
16. P. Würsch, World Review of Nutrition and Dietetics, 1989, 60, 199.
17. J. Holm, I. Lundquist, I. Björck, A-C. Eliasson and N-G. Asp, Am.J.Clin.Nutr., 1988, 47, 1010.
18. B. Hagander, N-G. Asp, S. Efedic, P. Nilsson-Ehle, I. Lundquist and B. Scherstén, Diabetes Research, 1986, 3, 91.
19. B. Hagander, N-G. Asp, R. Ekman, P. Nilsson-Ehle and B. Scherstén, European J.Clin.Nutr., 1989, 43, 35.
20. J.P. McNaughton, D.D. Morrison, L.J. Huhner, M.M. Earnest, M.A. Ellis and G.L. Howell, Nutr.Rept.Int., 1985, 31, 505.
21. D.B. Peterson, P.R. Ellis, M.M. Baylis, P. Fielden, J. Ajodhia, A.R. Leeds and E.M. Jepson, Diabetic Medicine, 1981, 4, 111.
22. S.J. Gatenby, In: A.R. Leeds (ed.): Dietary Fibre Perspectives, Reviews & Bibliography, John Libbey, London, 1990, 7.
23. B. Israelsson, G. Järnblad and K. Persson. Serum cholesterol reduced with Fibres a sugar-beet fibre preparation, 1990 (in publication).
24. Z.T. Cossack, M.O. Musaiger, Fiber supplementation lowers LDL-cholesterol and triglycerides in hypercholesterolemic desert nomads, 1990 (in publication).
25. J.F. Swain, I.L. Rouse, C.B. Curley and F.M. Sacks, New Engl.J.Med., 1990, 322, 127.
26. R. McPherson Kay, A.S. Truswell. In; G.A. Spiller, R. McPherson Kay (ed.), Medical Aspects of Dietary Fiber, Plenum Medical Book Company, New York, 1980, p. 153.
27. M. Nyman and N-G. Asp, Br.J.Nutr., 1982, 47, 357.
28. M. Nyman, T.F. Schweizer, S. Tyren, S. Reimann and N-G. Asp, J.Nutr., 1990, 120, in press.

DIETARY FIBRE ENRICHMENT, A SIGN OF QUALITY: POSSIBILITIES IN BREAD.

U. Pechanek*, H. Wutzel**, W. Pfannhauser*

Research Institute of the Food Industry,*
A-1190 Vienna, Blaasstrasse 29.
Association for Advancement of Baker's Business,**
A-1080 Wien, Florianigasse 13.

1. INTRODUCTION

The aim of this research project was to find out the limitations of the enrichment of bread with dietary fibres, maintaining still a good sensorial quality of the products. To meet this goal we tried to find out new sources of dietary fibre which give a better acceptance and possibly higher content of dietary fibre than wheat bran.

Table 1: Contents of dietary fibre of some materials.

	dietary fibre (% dry matter)		
	soluble	insoluble	total
Soya bran	9.6	59.6	62.2
Pea bran	9.3	74.1	83.4
Cacao hulls	22.6	47.2	69.8
Coffee hulls	9.8	54.2	64.0
Sugarbeet fibre	26.1	26.0	52.1

We analyzed a great amount of different materials (some are given in the table) for their content of dietary fibre and their rheological properties with respect to bread making.

It turned out that soja bran and especially pea bran can be employed for this purpose. Their acceptance was tested also regarding their rheological properties in dough as well as their baking properties.

The breads were analyzed for their sensorial quality and their volume. The results pointed out that an enrichment up to more than 20% dietary fibre (on dry matter basis) does not result in breads with acceptable sensorial quality.

WORKSHOP REPORT: SOURCES AND PROCESSING OF DIETARY FIBRE

T. Galliard

RHM Research Ltd
The Lord Rank Research Centre
Lincoln Road
High Wycombe HA2 3QR

1 INTRODUCTION

The main theme of the workshop was the practical, rather than nutritional, aspects of high-fibre foods; i.e. what could be produced commercially and accepted by the consumer at large, taking account of legislative requirements around the world.

Fibre concentrates or isolates are generally by-products of the food industry. A wide range of sources mentioned included cereal and legume brans (husks), sugar beet fibre and by-products of fruit processing, e.g. from citrus fruits, apples, grapes and olives. Other potential sources from waste products of vegetable preparation were identified; from a compositional point of view, these could be good sources of dietary fibre, but need to be studied further with respect to commercial applications. The nature of the non-starch polysaccharides in the different sources of fibre varies widely, e.g. in solubility, degree of substitution, cross-linking etc. Even in a given class, (e.g. pectins) chemical and physical properties can be very different. Thus, apart from nutritional considerations, the resultant textural and organoleptic characteristics of a food product will determine the type of fibre incorporated.

Some constraints on the food manufacturer producing fibre-enriched products were discussed; these included:-

- the cost, availability and potential variability in raw materials.
- functional effects on product quality; e.g. sugar beet

fibre can be used in pasta and some types of bread, but not in others.
- palatability; since the population at large regard eating as a pleasurable and social activity. Fibre-enriched foods must be enjoyable (not medicine!). It was suggested that palatability and nutritional studies on new types of food should be run in parallel.

The processes used to produce beet fibre and fruit pectin were described. For many types of fibre, processing is necessary for food safety, product shelf-life or organoleptic reasons. However, the type of processing can have important effects on the fibre quality. Examples quoted were the particle size of cereal brans and the type of hydrothermal treatment used; for instance, conventionally cooked cornflakes have much higher levels of resistant starch than do extruded cornflakes. Soluble fibre imparts high viscosity and, whilst this may be beneficial nutritionally, it limits applications in, e.g. drinks; partial hydrolysis reduces viscosity but raises questions about efficacy and definition of fibre.

Finally, legislative and labelling constraints in different countries were discussed. There are wide discrepancies, even within Europe. In the U.K. and Sweden, "wholemeal" means 100% and 70% whole grain, respectively; in Norway, food containing fibre isolates cannot be labelled as 'fibre-enriched'. Some rationalisation is needed, but this was outside the remit of this workshop!

PHARMACEUTICAL ASPECTS OF DIETARY FIBRE

Professor N.W. Read

Sub-Department of Gastrointestinal
Physiology and Nutrition,
K Floor,
Royal Hallamshire Hospital
Sheffield S10 2JF

Purified soluble complex polysaccharides have been advocated as medicines to treat diabetes mellitus, hypercholesterolaemia and dumping syndrome. Some have been used as slimming agents, others, particularly isphagula and methyl cellulose, are used as bulk laxatives in patients with constipation and Irritable Bowel Syndrome.

1 DIABETES MELLITUS

The administration of viscous polysaccharides with a meal or with a drink of glucose reduces postprandial hyperglycaemia [1-6] and reduces the plasma insulin response. Does this matter? It is generally believed that the microvascular and macrovascular complications of diabetes are directly related to hyperglycaemia. Stabilising the blood glucose should help to delay the onset of these complications.

Viscous polysaccharides can also reduce reactive hypoglycaemia that is thought to occur when the 'too rapid' absorption of glucose causes and excessive secretion of insulin pushes the plasma glucose concentration down below the fasting level [7]. Thus, viscous polysaccharides may help to prevent hypoglycaemic episodes in patients taking a high carbohydrate, low fat diet especially with brittle 'type 1' diabetes. It has even been suggested that the increased plasma levels of acetate following colonic fermentation of polysaccharides may reduce the effects of hypoglycaemia by providing an alternative energy source for the brain [8].

Mode of action

Viscous polysaccharides are thought to reduce plasma glucose and insulin by delaying absorption of glucose in the small intestine. This action may have three components; 1: delay of the delivery of carbohydrate from the stomach to the small intestine, 2: delay in the ingestion of starch and 3: delay in the access of the products of digestion to the epithelial surface [9]. Viscous polysaccharides appear to work better when they are mixed more intimately with the food material than when they are given in a capsule form or even as a drink before the meal [3]. It is not always possible to predict the effect of these agents on blood glucose levels in vivo from measurements of viscosity in vitro, since dilution with acid and alkaline digestive juices can radically alter the viscous properties of some polysaccharide solutions [10].

Ingestion of viscous polysaccharides do not just reduce plasma glucose and insulin levels after the meal in which they are incorporated. They also influence postprandial glycaemic responses after the following meal [4]. There are at least two possible mechanisms; the first is that the delay in absorption of nutrients, particularly fat, causes an increased load to the ileum which then slows the emptying of the subsequent meal from the stomach and reduces the glycaemic response [11]. The second is that the short chain fatty acids released into the peripheral blood as a result of fermentation of the polysaccharide, may reduce hepatic glycogenolysis and enhance insulin sensitivity [12].

Is the prolonged use of viscous polysaccharides useful in treatment of type 2 diabetes?

Although the acute administration of viscous polysaccharides with a meal may reduce postprandial glycaemia, this does not mean that chronic administration will necessarily influence the disease process. Aro and his colleagues [13] have shown a reduction in fasting plasma glucose on long term guar with little effect on postprandial measurements. The success indices in long term management of type 2 diabetes are fasting plasma glucose, glycosylated haemoglobin levels, plasma lipids and weight. Studies by Simpson and his colleagues showed that a diet high in carbohydrate caused lower plasma glucose levels even when fibre intake remained the same [14]. But when leguminous fibre was added, there were improvements in both fasting and postprandial blood glucose levels [15-17]. How does fibre work long term? Is the effect due to a direct pharmacological effect in delaying absorption or is it caused by a compensatory reduction in other components of the diet? We attempted to answer this question by treating a group of 24 obese 'type 2' diabetics with a sensible high carbohydrate, low fat reducing diet, either by itself or together with supplements of guar gum or bran [18]. Our results showed that all three groups of subjects lost weight and showed an improvement in glycoslated haemoglobin levels and fasting plasma glucose and there was no significant difference between each group of subjects. We concluded that if the energy content of the diet was carefully controlled, the addition of viscous polysaccharide did not contribute significantly to diabetic control. This did not necessarily mean that the viscous polysaccharide is not effective in the overall management of diabetes. On the contrary, it may be effective but when taken ad libitum its mode of action may be to assist the patient in reducing energy and losing weight.

Type 2 diabetes is very much related to the obese state. Anything that causes weight loss in these patients is likely to result in improved diabetic control. Several studies have demonstrated that the administration of viscous polysaccharides does cause a modest weight loss in the diabetic patient and this can be associated with an improvement in diabetic control.

2 HYPERCHOLESTEROLAEMIA

Despite the doubts that blood cholesterol levels may not be the best discriminators of risk of coronary artery disease, there is nevertheless a very marked association between mortality from coronary artery disease and cholesterol levels [19]. A large number of studies have demonstrated that viscous polysaccharides, in particular guar gum [20], cause a reduction in total cholesterol and LDL-cholesterol, most of the studies have been conducted over a short term, but two recent studies have shown that the effects of guar gum on reducing cholesterol could be sustained for over 12 months [21, 22].

Mode of action

It is unlikely that the hypocholesterolaemic

action of viscous polysaccharides occurs by direct reduction in cholesterol absorption. Some complex polysaccharides bind bile acids, impairing their absorption. Under these circumstances, cholesterol could be used to synthesise the bile acid lost in the faeces or the bile acid pool could be reduced to such an extent that insufficient bile acid could be available for lipid solubilisation and absorption. This hypothesis was not supported by Gallagher and Schneeman [23], who recently showed that although guar gum bound significant amounts of bile acid in the small intestine it did not reduce the amount of solubilised lipid. Other studies have suggested that viscous polysaccharides could trap fat in the intestinal lumen or delay the diffusion of micelles to the cell surface [24]. Another theory is that hepatic cholesterol synthesis is reduced by reduced postprandial insulin and glucose levels; hyperinsulinaemia may be a primary risk factor for the development of hypotrigliceridaemia and coronary artery disease [25]. Plasma cholesterol may also be reduced by increased portal blood levels of propionic acid produced by bacterial fermentation of polysaccharide in the colon. Finally, there are significant correlations between obesity and hypercholesterolaemia and between obesity and arteriosclerosis. Several studies have shown that reduction in plasma cholesterol induced by viscous polysaccharides is related to a loss in weight [26].

3 OBESITY

When concentrates of viscous polysaccharides like guar gum and pectin are added to test meals, they slow down gastric emptying and increase feelings of satiety [27]. When taken before a meal, however, the reduction in hunger levels by guar gum is very brief [personal observations]. Ingestion of crackers rich in isphaghula immediately before a meal caused a small but significant reduction in total energy intake over 2 week period [28]. Can this effect extend over a much longer time period? Most controlled studies have shown an advantage of supplements of dietary fibre over placebo if a sufficiently high dose is taken [29-34].

4 DUMPING SYNDROME

Some patients experience severe postprandial symptoms of nausea and faintness after gastric surgery. These symptoms are thought to be explained by two mechanisms. The early symptoms are thought to be related to the rapid entry of food material into the small intestine. This can distend the small intestine and also cause profound osmotic shift of fluid from the extracellular fluid space into the gut and consequent hemoconcentration. The more delayed symptoms are thought to be related to the rapid absorption of glucose which causes an excess insulin secretion and a reactive hypo-glycaemia. Supplements of viscous polysaccharides taken with meals have been shown to reduce the reactive postprandial hypoglycaemia and to depress dumping symptoms [35].

5 IRRITABLE BOWEL SYNDROME

The popularity of wheat bran as a natural way to manage IBS has made it the first line treatment for this condition for over 10 years [36]. But is it an effective treatment for IBS? Manning and Heaton showed that the addition of 7 grams of fibre in the form of wheat bran for 6 weeks to the diet of patients with IBS resulted in significant improvement in symptoms [37]. The possibility that the control diet in Manning and Heaton's study had little placebo effect [38] make the results of the bran diet difficult to interpret in a condition that shows such a marked response to placebos [39]. Other placebo controlled trials of bran in IBS [39, 40, 41, 42]. have failed to show any convincing effect of the fibre on overall symptom patterns. The most recent study [42] compared the effect of supplementing the daily diet of 44 IBS patients with either 12 bran biscuits (12.8 grams fibre) or 12 placebo biscuits (2.5 grams fibre). Patients were randomly allocated to receive either high fibre or placebo biscuits for 3 months and then to cross over to the alternative biscuit for another three months. The patients were not told which type of biscuits they had been given, though some may have guessed! The results showed that both placebo and bran groups experienced similar improvements in overall symptom scores. Moreover, the beneficial effects of bran were independent of any change in stool weight. The paucity of constipated patients may have influenced the conclusions from this study.

Cann and his colleagues [39] compared the independent responses of a number of typical IBS symptoms to bran with the responses to placebo tablets in 38 patients. Eighteen patients (47%) said they had improved on bran treatment, but only five of these said that they were entirely satisfied and did not require further treatment. Eleven patients (29%) experienced no change in their symptoms, and nine patients (24%) found that their symptoms were exacerbated by a high fibre diet. When the characteristics of the patients who said they had improved on bran were compared with those who said that their symptoms had not changed or become worse, the bran responders had smaller, harder, less frequent stools and longer colonic transit times upon entering the trial than the non-responders. In other words, it was the more constipated patients that responded to bran. This impression was confirmed by analysis of the responses to individual symptoms; constipation was the only symptom that showed a significantly greater response to bran than placebo. This doesn't mean that none of the other symptoms responded to bran. They did, but the responses were no greater than the responses to placebo. This study failed to confirm the suggestion that ingestion of wheat bran in patients with diarrhoea delays rapid intestinal transit [43, 44]. Instead, wheat bran tended to accelerate transit in everybody and exacerbate symptoms of diarrhoea.

So is there any point in advising patients to take bran for their irritable guts? It would, in my opinion, be a mistake to reject the idea completely. What is needed is a sense of realism. Some IBS patients, particularly those who are constipated, undoubtedly obtain symptomatic relief from taking bran in their diet and should continue to do so. Others find that bran induces or exacerbates symptoms of distension, flatulence, diarrhoea and abdominal pain. Their hypersensitive and irritable guts are probably irritated by the particulate nature of the bran, its bulk and the products of its fermentation by colonic bacteria. In the current climate of bran for everything, judicious reduction in fibre intake in these patients can prove very useful. For the remainder, then bran is a less toxic 'placebo' than many drugs and is better than placebo for constipation associated with IBS.

Viscous polysaccharides

Are other forms of dietary fibre more effective in IBS? Ritchie and Truelove [45] showed that ispaghula was better than wheat bran when both were given in combination with a psychotropic agent and an antispasmodic. A large controlled trial, published last year [46], Prior and Whorwell reported overall symptomatic improvement in 82% patients, who had been taking ispaghula compared with 53% patients who had been taking placebo. Many of their patients had constipation and in these ispaghula accelerated transit time and increased stool weight. There was no evidence that ispaghula improved patients with diarrhoea.

Kumar and his colleagues [47] carried out a dose ranging study in a group of male Asian patients to determine the optimum dose of ispaghula husk in IBS and to assess the correlation between the relief in patients' symptoms and colonic function. Thirty grams of ispaghula caused a significant improvement in patients symptoms, accompanied by an increase in stool weight but no significant change in transit time. Surprisingly, patients reported improvement in diarrhoea, even though stool weight increased.

6 CONSTIPATION

Meta-analysis of the data from 20 studies [48] showed that (i) stool output and transit times responded less to fibre supplementation in constipated patients than in normal subjects, and (ii) that treatment with fibre of constipated patients often failed to return transit time and stool output to normal. There is therefore little evidence to support the contention that constipation in all patients is wholly caused by fibre deficiency and constipated patients should not be blamed for non-compliance if dietary advice fails. Constipation should probably be regarded as a disorder of colonic or anorectal motility that may respond to the mild laxative action of complex polysaccharides rather than simply the result of a fibre deficient diet [49].

How do complex polysaccharides work on the colon

Wheat bran and other bulk laxatives containing complex polysaccharides are not digested in the small intestine, but act on the colon to increase stool weight and frequency to make the stool softer and bulkier and to reduce whole gut transit time [36, 50]. The mechanism of the laxative action of bran and other types of fibre on the colon is not established. The most popular theory relates to their bulk. Fibre contains plant cell walls that may resist breakdown by bacteria; and the associated complex polysaccharides adsorb and retain water. Complex polysaccharides also stimulate microbial cell growth, resulting in a greater faecal bacterial cell mass. Bannister and his colleagues [51] recently showed that small hard spheres mimicking faecal pellets are much more difficult to expel from the rectum than large spheres. Fermentation of fibre release gases which may be trapped in colonic contents, contributing to their bulk and plasticity. The increased colonic bulk promote colonic propulsion, which leads to reduced water absorption by the colon and the easier passage of bulkier and softer stools. In a recent study comparing the action of a number of complex polysaccharides with their fermentation characteristics, Tomlin [52] suggested that polysaccharides that resisted breakdown increased stool weight, while those that were fermented, accelerated colonic transit. The best bulk laxatives appeared to be those, like bran and ispaghula, that retained their structure but were also fermented. An increase in their volume of colonic contents will distend the colon, stimulating colonic propulsion and secretion [53].

There may, however, be other mechanisms of action. It is possible, for example, that the lignified particles of bran irritate the colonic epithelium, activating nervous reflexes that cause colonic secretion and propulsion. This may explain why coarse bran is a more potent laxative than the same amount of finely milled bran [54]. In support of this idea, Tomlin has recently observed [55] that the addition of 15g/day of small segments of polyvinyl tubing to the diet increases stool mass and frequency, improves stool consistency and accelerates whole gut transit to the same extent as the addition of an equivalent weight of coarse wheat bran.

REFERENCES

1. Jenkins, D.J.A. et al. (1978) Br. Med. J. 1: 1392-1394.
2. Jarjis, N.A. et al. (1984) Br. J. Nutr. 51: 371-378.
3. Fuessl, H.S. et al. (1987) Diabetic Med. 4: 463-468.
4. Jenkins, D.J.A. et al. (1980) Diabetologia 19: 21-24.
5. Karlstrom, B. (1988) Acta Univ. Upsaliensis 153: 1-50.
6. Vinnik, A.I. and Jenkins, D.J.A. (1988) Diabetes Care 11: 160-173.
7. Hoffman, W.P.M. et al. (1988) Gut 29: 930-934.

8. Akanji, A.O., Bruce, M.A. and Frayn, K.N. (1989) Eur. J. Clin. Nutr. 43: 107-115.
9. Edwards, C.A. and Read, N.W. (1989) Fibre and small intestinal function. In: Dietary Fibre Perspectives 2 - Reviews and Bibliography. Ed. A. Leeds, John Libbey, London, Paris.
10. Edwards, C.A. et al. (1987) Am. J. Clin. Nutr. 46: 72-77.
11. Welch, I.Mc., Cunningham, K.M., Read, N.W. (1988) Gastroenterology 94: 401-404.
12. Orskov, H. et al. (1982) Acta. Endocronol. 99: 551-558.
13. Aro, A. et al. (1981) Diabetologia 21: 29-33.
14. Simpson, H.C.R. et al. (1982) Diabetologia 23: 235-239.
15. Simpson, H.C.R. et al. (1981) Lancet 1: 1-5.
16. Kinmouth, A.L. et al. (1982) Arch. Dis. Childhood 57: 187-194.
17. Monnier, L.H. et al. (1981) Diabetologia 20: 12-17.
18. Beattie, V.A. et al. (1988) Br. Med. J. 296: 147-149.
19. Shaper, A.G. (1988) Coronary heart disease: risk and reasons, Current Medical Literature, in association with Duncan, Flockhart and & Co. Limited.
20. Gatenby, S.J. (1989) Guar gum and hyperlipidaemia - a review of the literature. In: Dietary Fibre perspectives 2, Ed. A. Leeds, J. Libbey.
21. Uusitupa, M. et al. (1989) Am. J. Clin. Nutr. 49: 345-351.
22. Simons, L.A. et al. (1982) Atherosclerosis 45: 101-108.
23. Gallagher, M.D., Schneeman, B.O. (1986) Am. J. Physiol. 250: G420-G426.
24. Phillips, D.R. (1986) J. Sci. Food Agricul. 37: 548-552.
25. Coulston et al. (1984) Am. J. Clin. Nutr. 59: 163-165.
26. Tuomilehto, J. et al. (1980) Act. Med. Scand. 208: 45-48.
27. DeLorenzo, C. et al. (1988) Gastroenterology 95: 1211-1215.
28. Stevens, J. et al. Am. J. Clin. Nutr. 46: 812-817.
29. Krotkievsky, M. (1984) Br. J. Nutr. 52: 97-105.
30. Rossner, S. et al. (1988) Acta Med. Scand. 223: 353-357.
31. Walsh, D.F. et al. (1984) Int. J. Obesity 8: 289-93.

32. Ryttig, K.R. et al. (1984) Tidssk Nor Laegeforen 104: 898-91.
33. Solum, T.T. et al. (1987) Int. J. Obesity 11: 7167-71.
34. Blundell, J.E. and Burleigh, V.J. (1987) Int. J. Obesity 11 (1): 9-25.
35. Jennings, D.J.A. et al. (1977) Gastroenterology 72: 215-217.
36. Manning, A.P. and Heaton, K.W. (1976) Lancet i: 588.
37. Manning, A.P. et al. (1977) Lancet i: 417-418.
38. Heaton, K.W. (1985) Role of dietary fibre in the treatment of irritable bowel syndrome. In: Irritable Bowel Syndrome Ed. N.W. Read, Grune and Stratton, Orlando, FL. pp. 203-222.
39. Cann, P.A., Read, N.W. and Holdsworth, C.D. (1984) Gut 24: 168-173.
40. Soltoft, J. et al. (1976) Lancet ii: 270-272.
41. Arffman, S. et al. (1985) Scand. J. Gastroenterol. 20: 295-298.
42. Lucey, M.R. et al. (1987) Gut 28: 221-225.
43. Harvey, R.F., Pomare, E.W. and Heaton, K.W. (1973) Lancet i: 1278-1280.
44. Payler, D.K. et al. (1975) Gut 16: 209-213.
45. Ritchie, J.A. and Truelove, S.C. (1979) Br. Med. J. 1: 376-378.
46. Prior, A. and Whorwell, P.J. (1987) Gut 28: 1510-3.
47. Kumar, A. et al. (1987) Gut 28: 150-155.
48. Muller-Lissner, S.A. (1988) Br. Med. J. 296: 615-617.
49. Cummings, J.H. (1984) Postgrad Med. J. 60: 811-819.
50. Burkitt, D.P., Walker, A.R.P. and Painter, N.S. (1972) Lancet ii: 1408-1412.
51. Bannister, J.J. et al. (1987) Gut 28: 1246-1250.
52. Tomlin, J. and Read, N.W. (1988) Br. J. Nutr. 60: 467-475.
53. Chauve, A., Devroede, G., Bastin, E. (1976) Gastroenterology 70: 336-344.
54. Brodribb, A.J.M., Groves, C. (1978) Gut 19: 60-63.
55. Tomlin, J., Read, N.W. (1988) Brit. Med. J. 297; 1175-1176.

BEET FIBRE PRODUCTS: NEW TECHNOLOGIES AND PRACTICAL APPLICATIONS TO DIET THERAPY

E. Del Toma, A. Clementi, C. Lintas* and G. Quaglia*

Nutrition and Diabetes Management Unit, Ospedale Forlanini, 00149 Roma
* National Institute of Nutrition, 00178 Roma

1 INTRODUCTION

As an alternative to guar and its products, already successfully experimented in some diabetic foods [1,2] but of limited palatability and acceptability, sugar beet pulp appears an attractive source of dietary fibre (DF).[3,4] It is in fact available in large amounts and contains high levels of fibre. Unlike guar, beet fibre is obtained from a local crop, of common use in Italy, more economical and also well accepted. This aspect was found of particular interest. Only with a good compliance it can be hoped, in fact, that diabetics will respect dietetic recommendations not merely for a few days but rather for the rest of their life, as needed for these patients.

The Aim of the present study was to investigate the metabolic impact of pasta enriched with beet fibre, natural or enzymatically modified, in comparison with commercial pasta or rice, traditional Italian dishes. Furthermore, the opportunity of integrating the DF intake with other beet fibre-enriched products (extruded and baked products), and its limitations, was evaluated.

2 MATERIALS AND METHODS

Ten non insulin-dependent diabetic (NIDDM) patients, mean age 59.8±2.2 years, were recruited from the metabolic day care unit at the Ospedale Forlanini. The patients, all treated by diet alone, were in good metabolic control and homogeneous for weight (67.0 ± 3.5 kg), BMI (24.0±1.7 kg/m^2), basal blood glucose (117.3±3.8 mg/dL) and HbA_{1c} (5.9±0.1 % Hb). All the patients were free from other major illness and not treated with any drug for the whole duration of the experiment.

After giving their informed consent to the study, the patients were summoned to the metabolic day care unit for a basal control and kept resting until lunch time. At that time, over a 30-min period each patient ate one of the three test meals, received in random order during three test periods at least 2-wk apart, under the supervision of dietitians.

The test meals, calculated to be isoglucidic and isocaloric, consisted of commercial pasta (100 g), beet fibre enriched pasta* (15% fibrex in the dough, 130 g) and rice (regular long grain white rice, 95 g), all served with the same tomato sauce. The ingredients of the sauce were the following: fresh ripe tomatoes, 250 g; celery, carrot and olive oil, 20 g each; onion and basil leaves, 10 g; salt. Duplicate samples of each meal were sent to the Food Chemistry Unit of the I.N.N. for composition analysis. Official AOAC methods were utilized for the major nutrients. DF was determined according to the method of Prosky *et al.* [5] As shown in Table 1, the actual meals were not isoglucidic, as calculated from tabular data.[6]

In vitro digestibility of starch was evaluated on the test meals, with and without sauce, and also on some other fibre-enriched products according to the method of Wong *et al.* [7] Sugars in the enzymic digesta were quantified by ionic chromatography.

Table 1 Composition of test meals (g/ration)

Nutrient	Pasta + F	Pasta	Rice
Protein (Nx5.7)	16.9	16.1	9.8
Lipid*	18.5	22.5	23.9
Soluble carbohydrates	9.9	11.8	11.8
Starch	76.5	65.4	51.6
Ash	6.0	5.1	5.9
Dietary fibre	14.3	7.5	4.8
soluble	3.8	2.8	1.5
insoluble	10.5	4.7	3.3
Energy Kcal	587	581	511
Kj	2456	2430	2139

*acid hydrolysis

* The sugar beet fibre enriched pasta was kindly provided by SIRC S.p.A., Milano

Venous blood was collected *via* an indwelling needle before eating and 15, 30, 45, 60, 90, and 120 min after the beginning of the meal. Blood glucose was determined by a glucose oxidase method and serum insulin by RIA. Glycosylated Hb was tested by ion-exchange chromatography. Results are expressed as mean±SEM. Significance of differences between means was determined by the paired *t*-test (two-tailed).

3 RESULTS

The changes in plasma glucose concentration after the oral ingestion of the three meals are presented in Figure 1. As it can be observed, the post-prandial glucose response to the pasta + fibrex meal was lower than those of the other two meals. However, a statistically significant difference ($p<0.03$) could be observed only when comparing fibre-enriched pasta with rice.

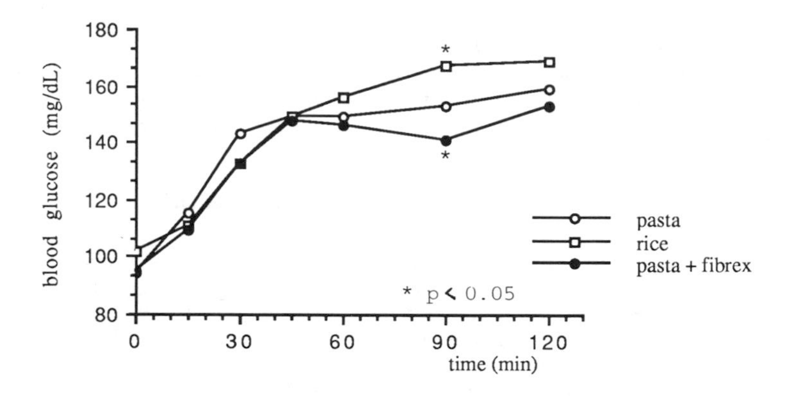

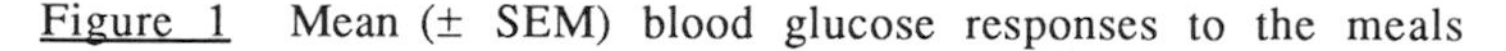
Figure 1 Mean (± SEM) blood glucose responses to the meals

Regarding hormonal changes the fibre-enriched pasta meal tended to produce a mean insulin rise intermediate between those of rice and pasta, the difference being statistically not significant.

The *in vitro* starch digestibility confirmed the results of *in vivo* absorption: both pasta meals showed lower digestibility (and consequently a more modulated absorption) than rice (Figure 2). The difference in starch digestibility was significant when comparing rice vs pasta ($p<0.05$), while in the case of rice vs fibre-enriched pasta more significant differences were observed in absence than in presence of the sauce ($p<0.01$ vs $p<0.05$, respectively).

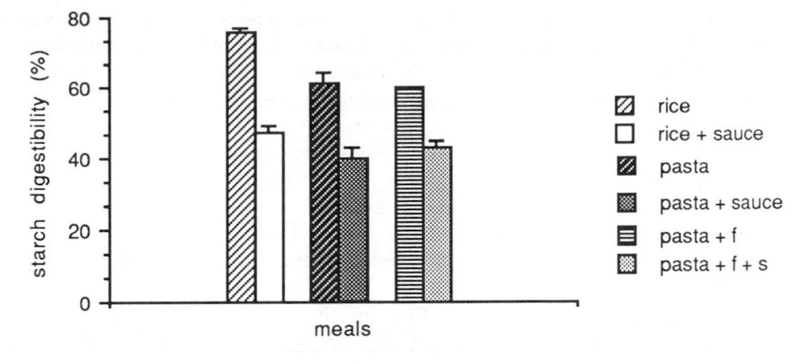

Figure 2 In vitro starch digestibility of the meals

4 DISCUSSION

Our results confirm that the rate of starch hydrolysis *in vitro,* as already reported,[8] is a useful mean to assess the degree of gelatinization of the starch in a product and to predict bioavailability of starch *in vivo.*

In vivo, Although the fibre-enriched pasta had a total DF content twice that of pasta (14.3 vs 7.5 g, respectively), their glycemic responses were not significantly different. It is well known however that only, or mainly, soluble fibre is metabolically active. Thus, probably the increment in soluble fibre of the fibre-enriched pasta (1.O g) was not sufficient to induce a postprandial glycemic response significantly different from that of pasta. On the other hand, its increment vs rice (2.3 g) was sufficient to induce a significant difference.

Consequently, in view of the need of increasing the soluble DF content of fibre sources, a study has been undertaken to modify the soluble/insoluble DF ratio of beet pulp fibre by enzymatic treatment. The modified beet fibre has then been utilized in the formulation of extruded products (pasta and cracotte). Such products appear very promising, in terms of processing, DF content and *in vitro* starch digestibility, but their palatability still needs improvement.

To maximize the effect of soluble fibre on metabolic parameters without affecting patients' acceptability, it would in fact be helpful to have other fibre-enriched products. In the present study the fibre-enriched pasta and consequently there was very much liked by the patients and consequently there was no problem at all with compliance. For daily use, traditional products such as for example pasta in Italy, are preferable to pharmacological fibre supplements (guar, xantham gum), particularly since the fibre incorporated in a food is more effective than fibre administered separately.[10]

REFERENCES

1. D.B. Peterson, J.I. Mann, Diabetic Med., 1985, 2, 345.
2. E. Gatti, G. Catenazzo, E. Camisasca, et al., Ann. Nutr. Metab., 1984, 28, 1.
3. B. Hagander, N.-G. Asp, S. Efendic, P. Nilsson-Ehle, I. Lundquist and B. Scherstén, Scand. J. Gastroenterol., 1987, 22 (suppl. 129), 284
4. L.M. Morgan, Y.A. Tredger, C.A. Williams, W. Marks, Proc. Nutr. Soc., 1988, 47, 185A.
5. E. Carnovale and C.F. Miuccio, "Tabelle di composizione degli alimenti", I.N.N. and M.A.F., Italy, 1989.
6. L. Prosky, N.-G. Asp, T. Schweizer, J. De Vries and I. Furda, J. AOAC, 1988, 71, 1017.
7. S. Wong, S. Traianedes and K. O'Dea, Am. J. Clin. Nutr., 1985, 42, 38.
8. F.R.J. Bornet, A.-M. Fontvieille, S. Rizkalla, P. Colonna, A. Blayo, C. Mercier and G. Slama, J. Am. Clin. Nutr., 1989, 50, 315.
9. E. Del Toma, C. Lintas, A. Clementi and M. Marcelli, Eur. J. Clin. Nutr. 1988, 42, 313.
10. G.V. Vahouny, D. Kritchevsky, "Dietary fiber: basic and clinical aspects", Plenum Press, New York, 1986, p. 343.

SUGAR BEET FIBRE: A CLINICAL STUDY IN CONSTIPATED PATIENTS

A. Giacosa, S. G. Sukkar, F. Frascio and M. Ferro

Nutritional Unit
National Institute for Cancer Research
Genova
Italy

1 INTRODUCTION

During the last decades various epidemiological studies suggested a correlation between decreased ingestion of dietary fibres and some diseases mostly affecting Western countries (1,2).

These clinical conditions are represented by gastrointestinal diseases (constipation, diverticulosis, irritable bowel syndrome, cholelithiasis) and dismetabolisms (dislipidemia, diabetes, overweight and obesity) and cardiovascular pathologies (1,3).

The clinical research has recently confirmed the precise role of dietary fibres in the prevention and therapy of many of these diseases, and it has furthermore identified specific treatments with particular fibre types, such as the hydrosoluble ones (guar gum, glucomannan, pectin) for dismetabolical conditions and non hydrosoluble ones for gastrointestinal pathologies (4).

In fact, the efficacy of non soluble fibres, and in particular of wheat bran, in the treatment of diverticulosis chronic constipation and irritable bowel syndrome is well proved, due to their marked bulk forming effect (5).

The aim of the present experiment is to evaluate the efficacy, the safety and the compliance of the sugar beet fibre, which contains in high proportions soluble and non soluble components.

Materials and Methods

Twenty-seven patients (18 females and 9 males) ranging in age 21 - 42 years (mean age 42 years), complaining of chronic constipation, in absence of organic pathologies potentially responsible for stool alterations, entered the study. All the patients were treated with sugar beet fibre (Fibrex; SIRC, Rome) integrated by organic salts of iron, zinc and copper, according to the FDA requirements (Table 1, 1 bis) (8).

The fibre preparation obtained by means of drying and grinding the sugar beet pulp after sugar extraction, was pressed in tablets containing 750 mg of product.

The daily dose was 9 tbs divided in three equal consumptions at the main meals. This dosage was reached after 4 days of a progressive consumption (4tbs at the 1st and 2nd day, 6 tbs at the 3th and 4th day), for the purpose of obtaining a patient's adaptation to the dietetic integration.

All the patients were invited to avoid any variation of their alimentary habits and to quit drugs or dietary products active on constipation and previously consumed.

Before starting therapy and after 15 and 30 days from full dose consumption (9tbs/d) the stool frequency and

Table 1 Pulp Composition of Sugar Beet (after Saccharose Extraction)

Alimentary fibre	80±5 g
Proteins	10±1 g
Saccharose	3±2 g
Mineral substances	1 g
Lipids	0.5 g
Water	5±1 g

Table 1 bis. Percentage Composition of Sugar Beet Fibre

Hemicellulose	41.5%
Pectin	25 %
Cellulose	31 %
Lignin	2.5%

Table 2 Stool Frequency Following Sugar Beet Fibre Treatment

	Beginning		15th day		30th day	
Stools	N°	%	N°	%	N°	%
Severe const.	11.0	42.3	2.0	7.7	1.0	3.8
Mod. const.	15.0	57.7	8.0	30.8	3.0	11.5
Normal	0.0	0.0	16.0	61.5	22.0	84.7
Total	26.0	100.0	26.0	100.0	26.0	100.0

composition were assessed according to precise criteria (severe constipation: 1 evacuation every >5 days; moderate constipation: 1 evacuation every 3-4 days; normality: 1 evacuation every <2 days). Moreover, stools were classified as hard, semi-hard, soft and liquid ones. In 7 patients, who presented serum cholesterol levels higher than 230 mg/dl, a check up was performed after 30 days of sugar beet fibre treatment.

The statistical analysis of the obtained data was assessed by means of the student's T test for paired data and chi square test.

Results

The study was completed in 26 out of 27 cases. The drop out case was a 69 years old male, who was admitted to hospital for a hip fracture after 12 days of treatment with sugar beet fibre.

The stool behaviour evaluation, reported in Table 2 and Figure 1, shows a marked decrease of severe and moderate constipation at both the 15th and the 30th day of treatment; with a significant increase in faecal frequency normalization, which reached the 61.5% after 15 days and 84.7% after 30 days ($P<0.001$ when compared to time 0, and $P<0.05$ when compared to the 15th day of treatment).

Moreover, it is observed that stool composition turned in better with the appearance of soft stools in 69.3% ($P<0.001$) and semihard stools in 26.9% of patients, at the 30th day. The cholesterol study performed in the 7 patients, at the 30th day. The cholesterol study performed in the 7 patients with more than 230 mg/dl at the starting moment, demonstrates a significant reduction after 30 days of treatment ($P<0.05$) (Table 4).

The tolerance and compliance of patients to sugar beet fibre were good. No case presented collateral effects which induced therapy suspension. In two cases the clinical efficacy was exaggerated: multiple liquid stools appeared respectively after 12 and 14 days of treatment: the problem recovered after reducing the dose from 9 to 6 tbs per day, with a complete stool normalization.

Discussion

The analysis of the obtained data proves the efficacy of sugar beet fibre in the therapy of chronic constipation. This is confirmed by the increase in evacuation frequency, with persistence of severe constipation in only 3.8% after 1 month of therapy (Table 2, Figure 1).

The pharmacological effect of sugar beet fibre is associated to the increase in faecal bulk, which subsequently stimulates colon activity, as it happens with other bulk forming products (5,7).

Another relevant parameter is the modification of faecal consistence, with the crossing over from hard and semi-hard stools to soft ones, which is observed in significant percentage in comparison to the total cases (69.3%) ($P<0.01$) (Table 3, Figure 2). Therefore, the present research confirms the possible role of sugar beet fibre in the clinical management of chronic constipation and, possibly, of all the GI diseases treatable by non hydrosoluble fibres (diverticulosis, IBS, haemorrhoids, etc.). These results show a particular clinical interest by considering their high frequency at any age and sex.

The clinical interest is also stressed by the possibility of using a dietetic integration with a vegetable product such as the sugar beet, which presents a particularly high fibre content (Table 1). In fact, the alimentary fibre percentage in dried sugar beet pulp ranges from 75 to 85% of the whole weight, with a mean protein content of 10±1% and a minimal water (5±1%) and saccharose (3±2%) content.

The fibre content in sugar beet appears much higher than other commonly used fibre sources, such as wheat bran, which presents a fibre concentration of 35-40% of total weight.

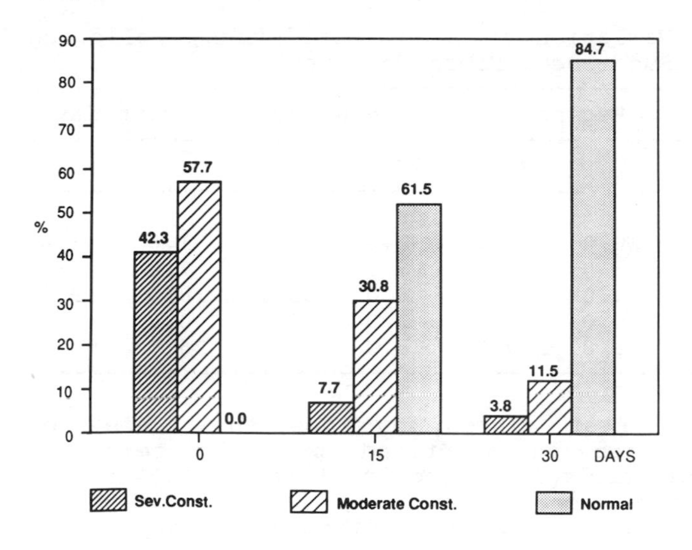

Figure 1. Stool Frequency Variation (percentual values)

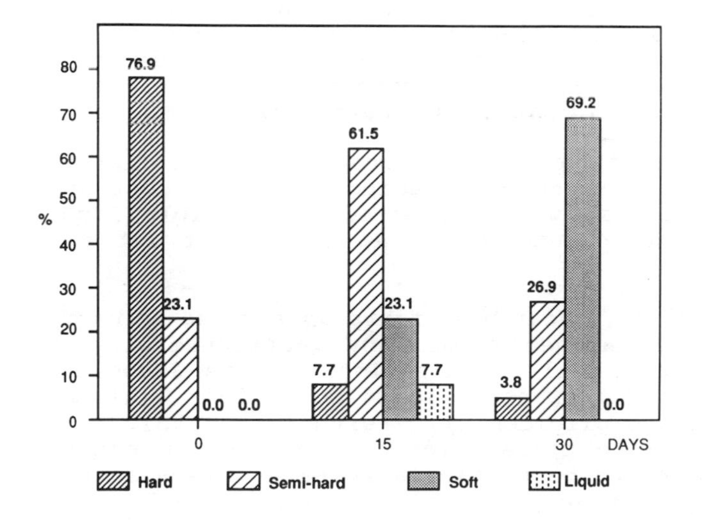

Figure 2. Stool Consistence Variation following Sugar Beet Fibre (percentual values).

Table 3 Variation in Stool Consistency Following Sugar Beet Fibre Treatment

	Beginning		15th day		30th day	
Stools	N°	%	N°	%	N°	%
Hard	20.0	76.9	2.0	7.7	1.0	3.8
Semihard	6.0	23.1	16.0	61.5	7.0	26.9
Soft	0.0	0.0	6.0	23.1	18.0	69.3
Liquid	0.0	0.0	2.0	7.7	0.0	0.0
Total	26.0	100.0	26.0	100.0	26.0	100.0

The sugar beet fibres, besides their elevated concentration, are characterized by a contemporary presence of both non soluble (hemicellulose: 41.5% of the whole fibre pool, cellulose: 31%, lignin: 2.5%) and soluble (pectin: 25% of the whole fibre pool) components.

The high content in global fibres and in the soluble component allows the clinical use of sugar beet fibre in dismetabolic conditions and particularly in glucose intolerance and dislipidemia. This is confirmed by multiple literature data and by the present experience, which proves a significant decrease in serum cholesterol in hypercholesterolic patients treated for one month (6).

Sugar beet fibre has not presented any collateral effects except the need of reducing the daily dose in 2 out of 26 patients, who complained of multiple liquid stools after the consumption of 6.75 gr/d of fibre (9 tbs).

In conclusion, sugar beet fibre appears to be useful in the clinical management of many gastroenterological diseases (first of all constipation and its associated pathologies) and dismetabolical conditions (glucose intolerance and dislipidemia). Furthermore, this fibre may be used in the preparation of many common dishes, since it appears to be particularly palatable also when it is cooked.

The ordinary use of sugar beet fibre could enter therefore in "healthy life" programs, because it finds a precise placement in prevention strategy for many diseases of our civilization.

Table 4 Serum Cholesterol Behaviour Following Sugar Beet Fibre in 7 Cases with Cholesterolemia > 230 mg/dl

Time	0	30
Cholesterol (mg/dl)	258±14	220±18

REFERENCES

1. D.P. Burkitt, A.R.P. Walker, N.S. Painter, 'Effect of dietary fibre on stools and transit-times and its role in the causation of disease', Lancet, 1972, II, 1408.
2. H. Trowell, D. Burkitt, K. Heaton, 'Dietary fibre, fibre-depleted foods and disease', Academic Press, New York, 1985.
3. A.M. Stephen, 'Should we eat more fibre?', J. Hum. Nutr., 1001, 35, 403.
4. E. Del Toma, C. Lintas, 'Fibra solubile e insolubile: indicazioni cliniche differenziate', Terapia Moderna, 1987, 1, 173.
5. A.M. Connell, 'Effect of dietary fibre on gastrointestinal motor function', Am. J. Clin. Nutr., 1978, 31 (5), 192.
6. B. Hagander, Asp N. G. Efendie S., P. Nilsson-Eble, I. Lundquist, and B. Schersten, 'Reduced glycemic response to beet fibre meal in non-insulin-dependent diabetics and its relation to plasma levels of pancreatic and gastrointestinal hormones', Diabetes Research, 1986, 3, 91.
7. A.J.M. Brodries, 'Treatment of symptomatic diverticular diseases with a high fibre diet', Lancet, 1977, I, 664.
8. J.F. Thibault, F. Guillon, F.M. Rombouts, C. Bertin, F. Michel and X. Rouau, 'Potential uses of sugar-beet pulps as a source of new pectins and fibre', Atti di: Int. Congr. Sciences des aliments, Versailles, 1989, 416.

ENRICHMENT OF ETHNIC FOODS WITH EXOGENOUS SOURCE OF DIETARY FIBER TO IMPROVE THE PLASMA GLUCOSE CONTROL IN NIDDM

Z. Madar, K. Indar-Brown, N. Feldman, Y. Berner and C. Norenberg

Department of Biochemistry and Human Nutrition
Faculty of Agriculture, The Hebrew University, Rehovot 76100, Israel
Diabetic Unit, Kupat Holim, Netania, Israel

1 INTRODUCTION

Ample evidence exists indicating that plasma glucose response to equivalent amounts of carbohydrate varies in the function of the specific carbohydrate-rich food consumed (1). This led to the ranking of individual carbohydrate foods according to their postprandial glucose excursions, termed, glycemic index (GI) (2,3). Although notable differences in the glycemic responses to carbohydrate foods solely consumed have been reported, experiments delineating the effect of regular meals composed of various carbohydrate foods with GI on postprandial glucose in NIDDM subjects have yet to be satisfactorily defined. The purpose of the present research is to examine the short-term effect of representative major Israeli ethnic foods on postprandial glucose levels. In addition, we evaluated incorporation of dietary fiber into ethnic foods with high GI on postprandial plasma glucose patients with NIDDM.

2 MATERIALS AND METHODS

Seventeen NIDDM volunteers, aged 42 - 77 years of age, BMI - 28.6 $\pm$ 0.9 were recruited for this study. Informed Consent from all subjects and approval of The Committee for the Protection of Human Subjects, "Meir" Hospital, Kfar Saba, Israel were obtained. Six subjects were treated by diet alone and eleven, by oral hypoglycemic agents. Each subject consumed at least four test meals in randomized order on separate mornings, at least one week apart, following an overnight fast. The test meals were composed of four different Jewish ethnic origins: Melawach (Yemenite) - a fried bread; Majadrah (Syrian) -lentils and rice; Kugel (Polish) - egg noddles, cheese and raisins and Couscous (Moroccan) - a stew of semolina and vegetables. The Melawach was enriched with 15 g of dietary fiber derived from carbohydrates and offered in the same manner. Aside from the four meals, each subject also consumed a Standard Meal, consisting of white bread, cheese and milk. Each meal provided 50 g carbohydrates and were compared to a glucose load of 50 g.

Blood samples were drawn before and at 30, 60, 120 and 180 min following completion of the test meal. Plasma was removed and glucose level was determined.

From area under the glucose curve, observed and predicted Glycemic Index were calculated. Results are expressed as Mean ± SE.

3 RESULTS AND DISCUSSION

The effect of the various ethnic foods on the glycemic responses in NIDDM subjects is shown in Figure 1.

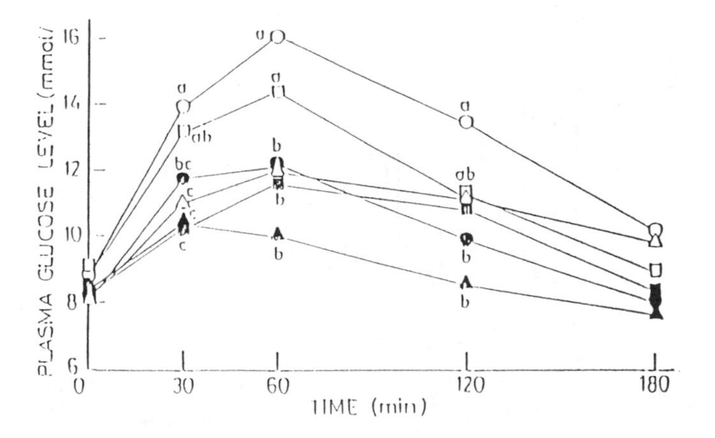

Figure 1 Glucose response to various kinds of ethnic foods compared to glucose given to NIDDM subjects. The meals ingested included: (●) Standard Meal (Δ) Melawach (▲) Majadra (▭) Kugel (▬) Couscous (○)

Glucose. Values are Mean ± SE. For each time, values not sharing the same letter were significantly different ($P < 0.05$)

No significant differences were seen in the fasting glucose levels among subjects. The peak glucose levels following the meals was reached at 60 min with the exception of Majadrah which peaked at 30 min. Ingestion of glucose and Kugel led to a sharp rise in glucose levels while the Standard Meal, Melawach, Couscous and Majadra only moderately affected glucose response. The plasma glucose levels following glucose loading was significantly higher than that of the Yemenite, Moroccan, Standard Meal and Majadra at 30, 60 and 120 min. Majadra elicited the lowest glucose response curve. These results indicate that glucose response of NIDDM subjects differed according to the various meals. These findings are consistent with those of Chew et al. (3) in which significant differences were noted in glycemic responses. The integrated area under the glucose curve in NIDDM is shown in Figure 2.

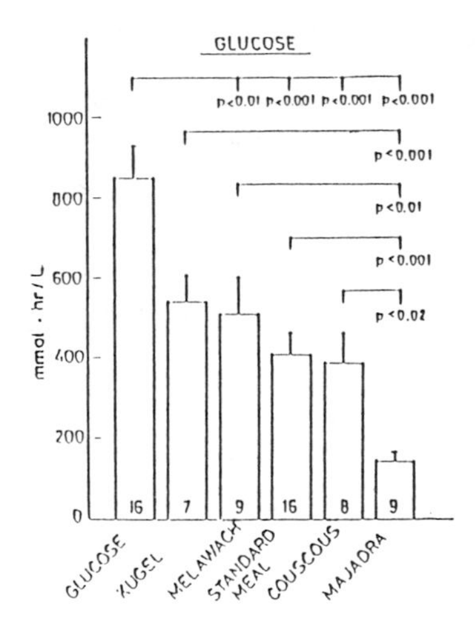

Figure 2 Area under the curve of NIDDM subjects after consumption of different meals. Bars indicate Mean ± SE. The number of subjects studied following each meal is indicated in the bars.

All the meals provided a lower incremental area under the glucose curve compared to glucose (P < 0.05). Moreover, ingestion of Majadra also led to a lower incremental area under the glucose curve compared to all the other foods (P < 0.01). The glycemic indices in NIDDM subjects are presented in Figure 3.

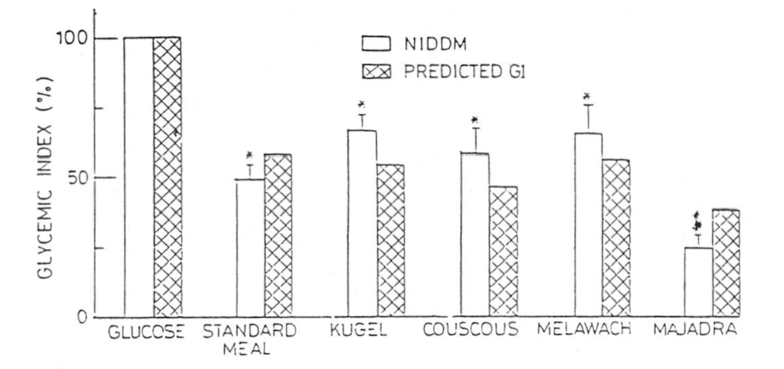

<u>Figure 3</u> Observed and predicted glycemic index of various ethnic foods given to NIDDM subjects. Bars indicated Mean ± SE. *P < 0.05 compared to glucose † P < 0.05 compared to the other meals.

Although carbohydrate quantities were equal in all ethnic foods, the glucose response curves were different. The low response following Majadra consumption indicate the nutritional components of each meal might contribute to the glucose response following the meals. Majadra is a dish rich with dietary fiber and other anti-nutrients, such as phytic acid which attributed to the reduction in glucose response (6).

The high content of lentils in Majadra (containing a relatively high amylose content) may contribute to the low glycemic response seen in NIDDM subjects (7). This finding confirmed previous reports stating that various sources of the same amount of carbohdyrate led to varied glycemic responses (4,5). Based on our results, we have suggested the main nutritional factor in the mixed meal influencing the glycemic response is the type and composition of the starch (8). The observed and predicted GI were closely correlated.

Differences in the GI may be attributed, in part, to the dietary fiber shown with Majadara. While Melawach (low in dietary fiber) exhibited high GI, Kugel, made of egg noddles, is known to be a weak provocative for glycemic response and also presented an elevated GI. We believe the other components, i.e. sugar and the interaction between the components are to blame for this discrepancy.

In conclusion: (1) The GI concept is valid and potentially useful in diet planning. (2) The GI of food enriched with legumes should serve as a carbohydrate source when diets are formulated for NIDDM patients.

Since Melawach (white flour and fat, mainly butter) demonstrated a high GI, two types of dietary fiber (15 g) were incorporated into this dish and given to NIDDM subjects to determine whether the dietary fiber, in fact, can decrease the high GI of this mixed meal. The glucose response is shown in Figure 4.

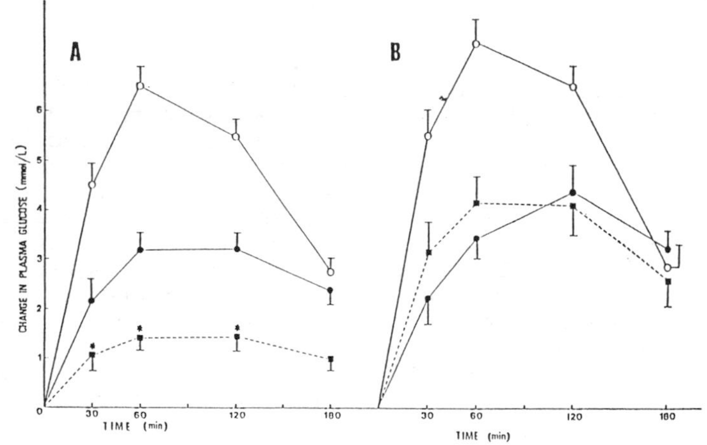

Figure 4. **A.** Plasma glucose response of diabetic patients following ingestion of Melawach with (■) or without (●) locust bean fiber. *P < 0.05. ○ - glucose. **B.** Plasma glucose response of diabetic subjects following ingestion of Melawach with (■) or without (●) corncob fiber. (○) - glucose.

Note that locust bean (carob) significantly decreases the plasma glucose at 30, 60 and 120 min following the meal test when compared to Melawach without fiber whereas corncob did not affect plasma glucose levels.

The effect of dietary fiber of GI has been evaluated. Incorporation of locust bean significantly decreased the GI from 60 to 30%. However, the GI remained unchanged when corncob was included into the Melawach. These results indicate that incorporation of soluble dietary fiber into foods with high GI may reduce the glucose response and GI index as shown previously (9) and might be beneficial as a means of controlling plasma glucose.

REFERENCES

1. P.A. Crapo, G.M. Reaven and J.M. Olefsky. Diabetes, 1976, 25, 741.
2. T.M.S. Wolever, A. Casima, D.J.A. Jenkins, G.S. Wond and R.G. Josse. J. Am. Coll. Nutr., 1989, 3, 25.
3. I. Chew, J.C. Brand, A.W. Thorburn and A.S. Truswell, Am. J. Clin. Nutr., 1988, 47, 53.
4. T.M.S. Wolever and D.J.A. Jenkins. Am. J. Clin. Nutr., 1986, 43, 167.
5. A.M. Coulston, C.B. Hollenbeck, A.L.M. Swislocki and G.M. Reaven, Diabetes Care, 1987, 10, 395.
6. M.J. Thorne, L.U. Thompson and D.J.A. Jenkins, Am. J. Clin. Nutr., 1983, 38, 481.
7. S. Sud, A. Siddhu, R.L. Bijlani and M.G. Karmarkar, Brit. J. Nutr., 1988, 59, 5.
8. K.M. Behall, D.J. Schofield and J. Canary, Am. J. Clin. Nutr., 1988, 47, 428.
9. D. Kritchevsky, Ann. Rev. Nutr., 1988, 8, 301.

EFFECTS OF SUGAR-BEET FIBRE ON BLOOD GLUCOSE, SERUM LIPIDS AND APOLIPOPROTEINS IN NON-INSULIN-DEPENDENT DIABETES MELLITUS

J.S. Travis, L.M. Morgan, J.A. Tredger and V. Marks

Department of Biochemistry
University of Surrey
Guildford
Surrey GU2 5XH

1. INTRODUCTION

Diabetes is often associated with abnormalities in lipid and lipoprotein metabolism[1]. In non-insulin-dependent diabetes mellitus (NIDDM) these can include elevated triglyceride and total-cholesterol and reduced HDL-cholesterol. In long-term diabetics, apolipoprotein AI (apo AI), and apo AI to apo B ratio are also lowered. These changes have been associated with coronary heart disease [2].

Some soluble fibres have hypoglycaemic and hypocholesterolaemic effects in normal hypercholesterolaemic and diabetic subjects[3-5]. Sugar-beet fibre (SBF) contains approximately 40% soluble fibre (soluble non-starch polysaccharide determined according to Englyst et al[6]). When incorporated into the diet of healthy subjects it lowers total- but not HDL-cholesterol[6] with reduction in apo B but not apo AI. This study assesses the effect of SBF supplementation on glycaemic control and serum lipid, lipoprotein and apolipoprotein levels in subjects with NIDDM.

2 METHODS

Seven NIDDM subjects (4 male and 3 female) participated. Mean age was 66 years (range 42-84 years) and average body mass index was 26.7kg/m^2 (SEM 1.61kg./m^2). All subjects were treated with oral hypoglycaemic agents. Mean duration of diabetes was 9 years. All subjects gave their informed consent before participating in the study. Approval for the study was given by the Ethical Committees of the University of Surrey and SW Surrey District Health Authority. The subjects maintained their usual diet for a two-week basal period. During that time two fasting blood samples were taken and the subjects' height and weight recorded. Nutritional intake was assessed by a three-day dietary record using estimated weights. The subjects were provided with bread, seed thins and lemon cookies, containing SBF (British Sugar plc, Peterborough) at approximately 2g/slice or per biscuit. By using a combination of bread and biscuits, they were asked to consume the equivalent of 18g dietary fibre from SBF per day for six weeks. Dietary advice was given to minimise changes to total energy and fat intake during the SBF period. Three fasting blood samples were taken at two-week intervals during the SBF period and each subject completed a further one or two dietary records. A final fasting blood sample was taken two weeks after the end of the supplementation period.

Plasma glucose, serum total- and HDL-cholesterol, triglyceride and apo AI and B were measured as described previously[8]. Plasma fructosamine was measured with an automated colorimetric assay[9].

Dietary records were analysed using Compeat Nutritional Analysis Program, based on food composition tables prepared by Paul and Southgate[10]. Results were compared using Student's t-test for paired values (one-tailed).

3 RESULTS

Body weights of the subjects did not change during the study. Intakes of energy, carbohydrate, fat and dietary fibre tended to be higher during the SBF period; however only energy intake differed significantly from basal values. The percentage of energy intake as protein was reduced ($p<0.01$), possibly due to a non-significant ($p>0.05$) increase in the percentage of energy intake as fat (See Table 1). Five of the seven subjects found the SBF supplemented bread and biscuits less palatable than normal products. This led to a reduction in compliance during the last two weeks of SBF supplementation. An average of measurements taken after two and four weeks of SBF supplementation was taken for comparison with basal values. No adverse abdominal symptoms were reported.

Fasting plasma glucose and fructosamine levels were similar throughout the study. During the SBF period, serum total-cholesterol, triglyceride and apolipoprotein B levels showed reductions ($p<0.05$) of 6.2% ± 2.81, 10.6% ± 5.09 and 6.0% ± 1.72 (mean ± SEM), respectively (see Table 2). Serum apo B was correlated ($p<0.001$) with both total cholesterol ($r = 0.9173$) and triglyceride ($r = 0.8666$). The change in apolipoprotein B was correlated with both the change in total cholesterol ($r = 0.8301$, $p<0.001$) and in triglyceride ($r = 0.8200$, $p<0.001$). HDL-cholesterol and serum apolipoprotein AI levels were not altered during the study ($p>0.05$). Apo AI concentrations were correlated significantly with HDL-cholesterol ($r = 0.9225$, $p<0.001$).

4 DISCUSSION

Inclusion of SBF in the diet of NIDDM subjects had no effect on glycaemic control, in agreement with other studies[7,11]. This may be because, despite the high soluble-fibre content, the fibres did not form solutions of sufficient viscosity to impede glucose absorption in the small intestine[11].

Reductions in serum cholesterol of between 3.6% and 7.8% have previously been reported[7,11,13]. In these studies, the dietary fibre intake of the participants was significantly increased compared to their normal diets. In the present study a decrease in serum cholesterol was observed when subjects attained only a modest increase in total dietary fibre intake, as they mainly substituted SBF for other mixed fibres in their diet. In accordance with previous results[7,11], SBF had no effect on HDL-cholesterol levels, implying that the predominant cholesterol fractions lowered were the atherogenic LDL and VLDL fractions.

Serum triglyceride levels were lowered by SBF as has been reported previously[11]. This may represent some reduction in the formation of triglyceride-rich particles by the liver.

In the fasted state, virtually all serum apo B will be in the LDL and VLDL fractions[14]. The tendency for SBF supplementation to lower serum apo B is, therefore, consistent with the assumption made above that the circulating levels of these fractions were reduced.

Inclusion of SBF into the diet of our NIDDM subjects was achieved largely by substitution for other mixed dietary fibres. The effects of SBF were therefore

Table 1 Composition of diet and mean daily nutrient intakes, calculated from three-day records before and during SBF supplement periods for seven NIDDM*.

Constituent	Basal Period	SBF Period
Energy, total kcal	1643 ± 110	2058 ± 184 †
MJ	6.9 ± 0.46	8.6 ± 0.77 †
Protein g	80 ± 4.2	77 ± 5.1
% E	20 ± 1.2	15 ± 1.0 ††
Carbohydrate g	168 ± 10.7	197 ± 21.9
% E	41 ± 3.5	39 ± 3.7
Total dietary fibre g	25 ± 2.7	32 ± 2.2 †
from SBF g	-	18 ± 1.6
Fat g	72 ± 10.4	105 ± 16.1
% E	38 ± 3.5	45 ± 4.0
Saturated % E	11 ± 1.8	14 ± 1.7
Monounsaturated % E	11 ± 1,3	15 ± 2.1
Polyunsaturated % E	5 ± 0.6	6 ± 2.0
P:S ratio $	0.5 ± 0.11	0.5 ± 0.09
Cholesterol mg	256 ± 64	252 ± 22

* Values are presented as means ± SEM
† Significantly different from value for basal period † (p<0.05) and †† (p<0.01)
$ Ratio of polyunsaturated to saturated fatty acids

Table 2 Effect of SBF supplementation on plasma glucose and fructosamine and serum lipids and apolipoproteins in seven NIDDM *

	Basal Period	SBF Period
Glucose mmol/l	10.8 ± 1.79	10.6 ± 1.77
Fructosamine mmol/l	2.5 ± 0.26	2.5 ± 0.23
Total Cholesterol mmol/l	6.5 ± 0.57	6.1 ± 0.46 †
HDL-Cholesterol mmol/l	1.1 ± 0.26	1.0 ± 0.21
Triglyceride mmol/l	2.3 ± 0.50	1.9 ± 0.33 †
Apolipoprotein B mg/dl	116 ± 18	107 ± 14 †
Apolipoprotein AI mg/dl	148 ± 16	144 ± 14

* Values are presented as means ± SEM
† P<0.05 for comparison with basal period

assessed in free-living subjects following realistic dietary regimens. Under these circumstances the SBF had no adverse side-effects, tended to reduce serum total cholesterol, triglyceride and apo B without altering HDL-cholesterol and apo AI levels. These changes represent an improvement in the lipid and apolipoprotein risk profile for CHD.

5 ACKNOWLEDGEMENTS

The authors are grateful to British Sugar plc for supplying the sugar-beet fibre supplemented bread and biscuits used, and for their financial support.

6 REFERENCES

1 D.J.Betteridge, Diabetic Med., 1989, 6, 195.
2 V.J.Schauer, D.Pissarek and G.Panzram, Acta diabetol.lat., 26, 35.
3 T.A.Miettinen, Am.J.Clin.Nutr., 1987, 45, 1237.
4 R.W.Kirby, J.W.Anderson, B.Sieling, E.D.Rees, W.L.Chen, R.E.Miller and R.M.Kay, Am.J.Clin.Nutr., 1981, 34, 824.
5 B.Hagander, N-G.Asp, S.Efendic, P.Nilsson-Ehle and B.Scherstén, Am.J.Clin.Nutr., 1988, 47, 852.
6 H.Englyst, H.F.Wiggins and J.H.Cummings, Analyst, 1982, 107, 307.
7 L.M. Morgan, J.A.Tredger, C.A.Williams and V. Marks, Proc.Nutr.Soc., 1988, 47, 185A.
8 L.M.Morgan, J.A.Tredger, J.Wright and V.Marks, Br.J.Nutr., 1990, in press.
9 J.R.Baker, P.A.Metcalf, R.N. Johnson, D.Newham and P.Rietz, Clin.Chem., 1985, 31, 1550.
10 A.A.Paul and D.A.T.Southgate, McCance and Widdowson's The Composition of Foods, 1978.London:HMSO.
11 B.Hagander, N-G.Asp, R.Ekman, P.Nilsson-Ehle and B.Schersten, Euro.J.Clin.Nutr., 1989, 43, 35.
12 D.J.A.Jenkins, A.R.Leeds, M.Houston, L.Hinks, K.G.M.M. Alberti and J.H.Cummings, Proc.Nutr.Soc., 1977, 36, 60A.
13 B.Israelsson, G.Jänblad and K.Persson, Dietetics in the 90s. Role of the dietitian/nutritionist. Ed. M.F.Moyal, 1988. John Libbey Eurotext Ltd. pp. 167-170.
14 J.J.Albers, J.D.Brunzell and R.H.Knopp, Clin.Lab.Med., 1989, 9, 137.

Subject Index